Essential Graph Theory

Concepts and Algorithms

T. Asir

M. Evangeline Prathibha

B. Surendranath Reddy

CAMBRIDGE
UNIVERSITY PRESS

Shaftesbury Road, Cambridge CB2 8EA, United Kingdom

One Liberty Plaza, 20th Floor, New York, NY 10006, USA

477 Williamstown Road, Port Melbourne, VIC 3207, Australia

314–321, 3rd Floor, Plot No.3, Splendor Forum, Jasola District Centre, New Delhi – 110025, India

103 Penang Road, #05–06/07, Visioncrest Commercial, Singapore 238467

Cambridge University Press is part of Cambridge University Press & Assessment, a department of the University of Cambridge.

We share the University's mission to contribute to society through the pursuit of education, learning and research at the highest international levels of excellence.

www.cambridge.org
Information on this title: www.cambridge.org/9781009559379

First published 2025

Printed in India by Nutech Print Services, New Delhi 110020

A catalogue record for this publication is available from the British Library

ISBN 978-1-009-55937-9 Paperback

Essential Graph Theory

Contents

Figures

Tables

Preface

The world of graph theory is expanding at a pace at which it is hard to keep track of the various disciplines of study that have somehow been irrevocably affected by the techniques owned and created by this versatile subject. However, the beauty of graph theory and its grandeur can also be intimidating for a beginner who wishes to explore its realms. The primary aim of the book is therefore to help a student understand and master the tools and techniques that are inherent to the subject and to serve as a handbook for anyone who wishes to explore the amazing world of graph theory.

We started writing this book with the goal of creating an undergraduate graph theory textbook that would be used by a broad audience, including students and teachers of Mathematics, Computer Science, Economics and perhaps Business Administration. However, mathematical training in proofs and logic is a prerequisite for understanding and following this book.

We have included topics in a structured manner, giving definitions, examples and sufficient grounding in the basics before moving on to the next topic. Learning objectives and an introduction have been included at the beginning of every chapter of the book, so that the reader knows what to expect from it. The exercises at the end of each chapter will help the reader to test their knowledge and solve problems.

We are now living in an era where interdisciplinary and multidisciplinary approaches are celebrated and where the lines between different disciplines of study have started to blur. It is for this reason that we included sections on *algorithms* in this book, so that a reader can get a sense of their application in Computer Science while remaining firmly grounded in theory. In a world increasingly driven by algorithms, this parallel study might prove beneficial to a student to explore the graph theoretical perspective on algorithms. A novel effort on our part was to include the pseudocodes of all the algorithms covered in this book. We have included the classic algorithms of Dijkstra, Kruskal and Prim, in addition to many others. A Mathematics student may not be familiar with algorithms and pseudocodes, but we hope it might provide readers with an introduction to coding. A brief introduction to algorithms and their complexities has been included in Chapter 2 for the sole purpose of initiating an algorithm novice into the fundamental principles of problem-solving in Computer Science. We have not included the theorems that prove the correctness of the algorithms, since it is beyond the scope of our book. But we have included references to most of those theorems.

This book can serve as a textbook on graph theory alone, if one were to omit the sections devoted to algorithms. It is a gentle introduction to graph theory for beginners, a practical resource for undergraduate courses, and a foundational guide for Computer Science students exploring applied graph theory.

We have taken care not to intimidate the anxious reader exploring the lush lands of graph theory for the first time. In this book, we aim to achieve a balance that allows an undergraduate student to follow the topics discussed without being overwhelmed by additional applications and extensions, while a postgraduate student finds it a trusted guide to reinforce their understanding, thereby gathering courage to embark upon journeys to newer and advanced topics in graph theory and graph algorithms.

A note to readers

While we have taken great care in preparing this book, we recognize that occasional errors may still be present. If you come across any inaccuracies, unclear explanations or have suggestions for improvement, we would greatly appreciate your feedback. Your insights help us enhance the quality of this work for future readers. Please feel free to write to us at: asirjacob75@gmail.com.

November 2024

T. Asir
M. Evangeline Prathibha
B. Surendranath Reddy

Acknowledgements

Writing this book has been a journey, and we are deeply grateful to all those who have supported us along the way. Many people have been instrumental in the creation of this book, and we take this moment to thank them from the bottom of our hearts.

We are profoundly grateful to *Dr J. Chithra*, Associate Professor (Retired), Department of Mathematics, Lady Doak College, Madurai, for serving as the reviewer of the early drafts of the book. Her insightful comments and feedback have been truly invaluable.

We are immensely grateful to *Dr Subbarao Venkatesh Guggilam*, Postdoctoral Researcher, Department of Mathematics and Statistics, UiT The Arctic University of Norway, Norway, for his help in editing the pseudocodes and algorithms of the book. His astute comments and logical review have greatly helped to ensure the accuracy of each algorithm and pseudocode.

We are thankful for the thoughtful suggestions of *Dr Rupali Jain*, Associate Professor, and the research scholars *Shaik Wajid* and *Sumedha Mehtre* from the Department of Mathematics, SRTM University, Nanded, for helping with the rough drafts and proofreading of the manuscript. We are also thankful to *Prof. V. Suresh*, Professor, Emory University, Atlanta and *Prof. S. Kumaresan*, Professor (Retired), University of Hyderabad, Hyderabad.

A special thanks to the research scholars *Cheri Paul, Akhil, Sanusha, Esha* and *Ankitha*, Department of Mathematics, Pondicherry University, Puducherry, for their extensive help with research, mathematical validation and feedback on the early drafts. Their keen eyes and valuable suggestions greatly improved the quality of the manuscript, and their dedication has been an inspiration. In particular, we thank *Cheri Paul* and *Esha* for going above and beyond.

We were influenced by many outstanding textbooks on graph theory, known for their in-depth knowledge, well-crafted proofs, explorations of related topics and comprehensive coverage of essential concepts. Notably, the works of Gary Chartrand and Ping Zhang, Gary Chartrand and Ortrud R. Oellermann, and John Clark and Derek Allan Holton have profoundly influenced us, both as students and now as authors of this book. We gratefully acknowledge the impact of their contributions on our thinking and approach.

We wish to express our heartfelt gratitude to *Mr Ankush Kumar* of Cambridge University Press for his unwavering support and diligent follow-up, which was invaluable throughout this journey.

We owe an incredible debt of gratitude to our families for their patience, love and support. Their understanding and encouragement sustained us through long hours of writing, rewriting and proofreading. We could not have reached the finish line without them, cheering us on.

T. Asir
M. Evangeline Prathibha
B. Surendranath Reddy

Notation

Throughout, unless otherwise stated, our notation will be as given below:

$\mathbb{N}$	Set of positive integers
$\mathbb{R}$	Set of real numbers
$\mathbb{Q}$	Set of rational numbers
$\mathbb{C}$	Set of complex numbers
$\mathbb{Z}$ ($\mathbb{Z}^+$)	Set of integers (nonnegative integers)
$A \setminus B$	Removing the elements of the set B from the set A
$|A|$	the cardinality of the set A
$\binom{n}{r}$	$\frac{n!}{r!(n-r)!}$
$p \mid q$	p divides q
$p \nmid q$	p does not divide q
$\lceil k \rceil$	smallest integer less than or equal to k
$\lfloor k \rfloor$	largest integer less than or equal to k
$G = (V, E)$	Graph with vertex set $V(G)$ and edge set $E(G)$
$deg_G(v)$ or $deg(v)$	Degree of the vertex $v \in V(G)$
$\Delta(G)$	Maximum degree of G
$\delta(G)$	Minimum degree of G
K_n	Complete graph on n vertices
$K_{m,n}$	Complete bipartite graph with partitions having m and n vertices
P_n	Path on n vertices
C_n	Cycle on n vertices
W_n	Wheel on n vertices
Q_n	n-cube
kG	k copies of the graph G
$G[S]$ or $\langle S \rangle$	Subgraph induced by $S \subseteq V(G)$
$G[X]$ or $\langle X \rangle$	Subgraph induced by $X \subseteq E(G)$
$\overline{G}$	Complement graph of G
$N_G(v)$ or $N(v)$	Open neighborhood of $v \in V(G)$
$N_G(S)$ or $N(S)$	Open neighborhood of $S \subseteq V(G)$

$N[v] = N(v) \cup \{v\}$	Closed neighborhood of $v \in V(G)$
$gr(G)$	Girth of G
$c(G)$	Number of connected components in G
$d(u,v)$	The distance between the vertices u and
$e(v)$	Eccentricity of $v \in V(G)$
$rad(G)$	Radius of G
$diam(G)$	Diameter of G
$cir(G)$	Circumference of G
$C(G)$	Center of a graph G
$Per(G)$	Periphery of a graph G
$G \cong H$	Group G isomorphic to a group H
$Aut(G)$	Automorphism group of G
$\chi(G)$	Vertex chromatic number of G
$\chi'(G)$	Edge chromatic number of G
$f(G)$	Number of faces of G
$d(\varphi)$	Number of edges on the boundary of a face φ
$G * e$	Edge contraction graph contracted by the edge e from G
G^*	Dual of G
G^{**}	Double dual of G
$\omega(G)$	Clique number of G
$\alpha(G)$	Vertex independence number of G
$\alpha'(G)$	Edge independence number or matching number of G
$\beta(G)$	Vertex covering number of G
$\beta'(G)$	Edge covering number of G
$\kappa(G)$	Vertex connectivity number of G
$\lambda(G)$	Edge connectivity number of G
$G - \{v\}$ or $G - v$	Removal of a vertex v from G
$G - \{e\}$ or $G - e$	Removal of an edge e from G
$G - \{S\}$ or $G - S$	Removal of a vertex subset S from G
$G - \{X\}$ or $G - X$	Removal of an edge subset X from G
$G \cup H$	Union of the graphs G and H
$G + H$	Join of the graphs G and H
$G \square H$	Cartesian product of the graphs G and H
$\tau(G)$	Number of spanning trees of the graph G
$O(f(n))$	Big oh of a function f
$w(e)$	Weight of an edge e in a weighted graph
$D = (V, E)$	Directed graph
$\vec{e}$	Directed edge of D
$id(v)$	indegree of a vertex $v \in V(D)$
$od(v)$	outdegree of a vertex $v \in V(D)$

1

Introduction to Graphs

- Develop a foundational understanding of graphs.
- Identify different types of graphs.
- Understand the different modes of traversing a graph.
- Build subgraphs using various processes.
- Distinguish between connected and disconnected graphs.
- Identify common classes of graphs by their structures.

The world of graph theory owes its birth to **Leonhard Euler (1707–1782)** who employed a new strategy to settle a then-unsolved problem called the Königsberg Bridge problem. There were two islands in the middle of the Pregel river, which were connected to each other and also to the mainland by means of seven bridges. The structure of Königsberg and the bridges are described in Figure 1.1.

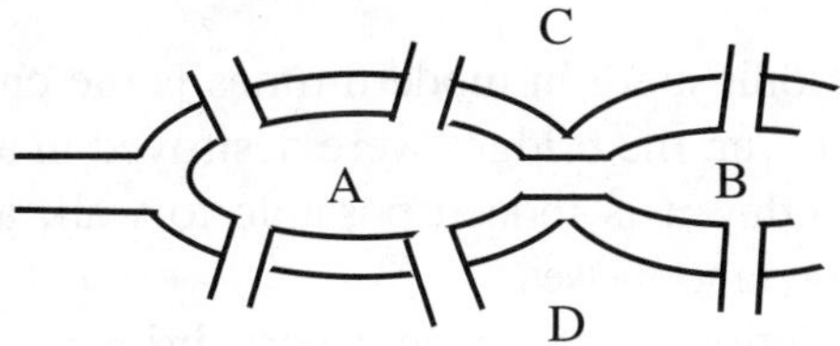

Figure 1.1   Königsberg bridges diagram

The question was, "Can a person start at any one of the land masses, walk across each bridge exactly once, touch all land masses and return to the land mass where the person started?" In 1735, Euler correctly identified that there were 4 landmasses and each land mass was connected to the other landmass by means of seven bridges. He intuitively decided that he would model the land masses as "vertices" or "dots" and the seven bridges as "edges" or "lines" connecting the vertices.

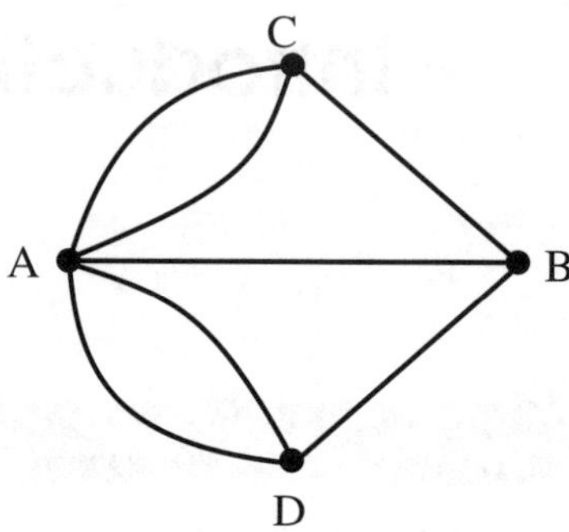

Figure 1.2   Mathematical modeling of the Königsberg bridge

Figure 1.2 is a rudimentary mathematical modeling of the Königsberg bridge problem. Using this model, Euler was able to conclusively say that it is not possible to traverse along all the seven bridges exactly once while touching all land masses too. As we progress into the world of graph theory, we will learn that Figure 1.2 is what we call a multigraph, and in order to traverse all "lines" while touching all "dots", each "dot" should have an even number of "lines" arising out of it, and that there should not be more than 2 "dots" having an odd number of "lines" arising from them. Essentially, Euler proved the first theorem of graph theory. In modern graph theoretical terminology, the "dot" became the "vertex" and the "line" became the "edge". The number of edges that arise from a vertex is called the *degree* of the vertex. Euler's theorem about the Königsberg bridges can be restated as: "If there is a path that traverses each edge once, then there exists at most two vertices of odd degree. If the path starts and ends at the same vertex, then no vertices will have an odd degree".

On an interesting side note, Königsberg in modern times is the city of Kaliningrad in Russia. During the Second World War, the bridges were destroyed in a bombing, and only five out of seven were rebuilt. Today it is indeed possible to walk across these bridges exactly once while touching all the landmasses.

Although historically, the solution to the Königsberg bridge problem marked the beginning of graph theory as a subject, it would be another hundred and fifty years before another mathematician by the name of Hamilton would do something to take the subject forward. We will now dive into the preliminary definitions of graphs and other allied concepts that will help to lay the foundation for a better understanding of graph theory.

We will keep it simple in Chapter 1 by delivering only the content that will help to develop an intuitive understanding of graph theory.

DEFINITION 1.0.1 A *simple graph G* consists of a vertex set $V(G)$, an edge set $E(G)$ and a relation φ that associates each edge $e \in E(G)$ with two distinct vertices $u, v \in V(G)$ such that $e = uv$.

EXAMPLE 1.0.2 Let $V(G) = \{v_1, v_2, v_3, v_4\}$ and $E(G) = \{e_1, e_2, e_3, e_4, e_5\}$. Let φ be the relation on $(V(G) \times V(G), E(G))$ such that $\varphi(v_1, v_2) = e_1$, $\varphi(v_1, v_3) = e_2$, $\varphi(v_2, v_3) = e_3$, $\varphi(v_1, v_4) = e_4$ and $\varphi(v_3, v_4) = e_5$. The resulting graph is in Figure 1.3.

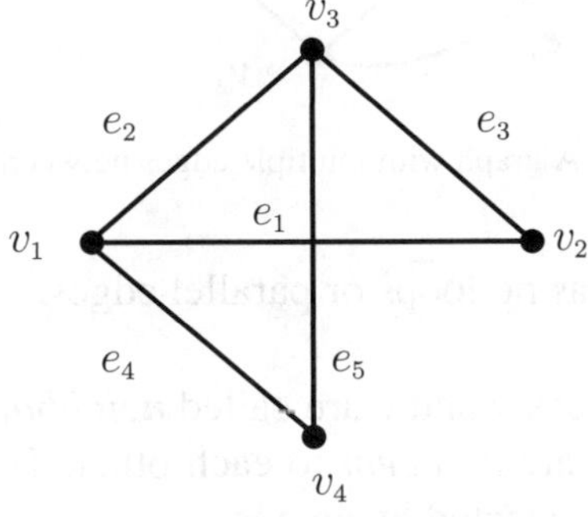

Figure 1.3   A simple graph

DEFINITION 1.0.3 A graph is said to contain a *loop* if there exists an edge $e \in E(G)$ such that both the vertices or endpoints of this edge are the same. In this case, $\varphi(v, v) = e$.

EXAMPLE 1.0.4 In Figure 1.4, the edge $e_5 = \varphi(v_2, v_2)$ is a loop.

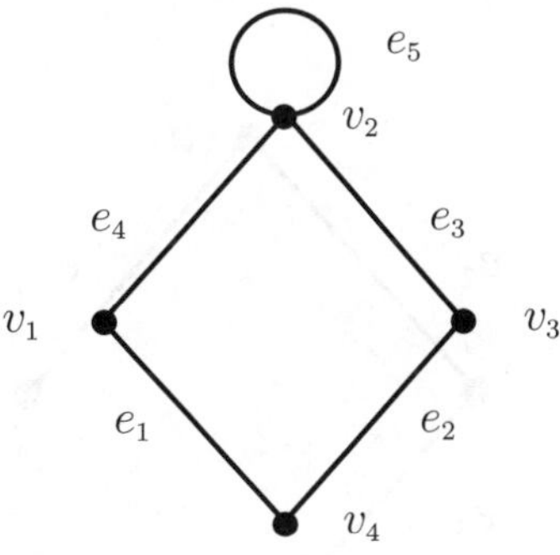

Figure 1.4   A graph with a loop

DEFINITION 1.0.5 A graph is said to contain a *parallel edge* or *multiple edge* if there is more than one edge between two vertices.

EXAMPLE 1.0.6 In Figure 1.5, $\varphi(v_3, v_4) = \{e_2, e_3, e_4\}$ and $\varphi(v_1, v_2) = \{e_6, e_7\}$.

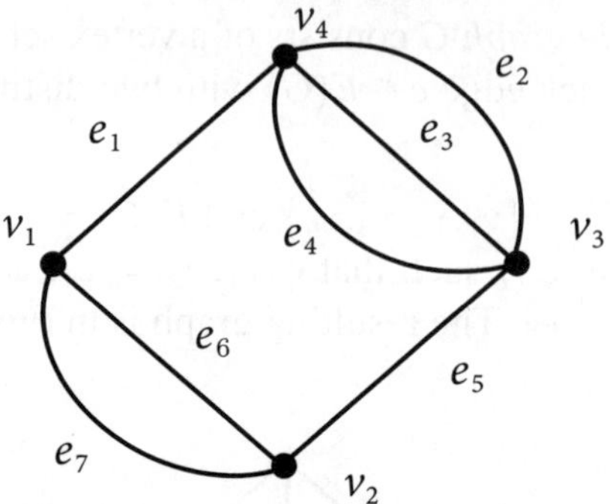

Figure 1.5   A graph with multiple edges between two vertices

Note that a simple graph has no loops or parallel edges.

DEFINITION 1.0.7 Two vertices u and v are called *neighbors* if they are connected by an edge. In other words, u and v are *adjacent* to each other. Two vertices u and v are called *non-adjacent* if they are not connected by an edge.

DEFINITION 1.0.8 The cardinality of the vertex set of a graph G is the *order* of a graph. The cardinality of the edge set of a graph G is the *size* of a graph.

EXAMPLE 1.0.9 For the graph in Figure 1.6, u_1 and u_2 are adjacent, whereas u_2 and u_3 are not-adjacent. Also the order and size of the graph given in Figure 1.6 are 4 and 3 respectively.

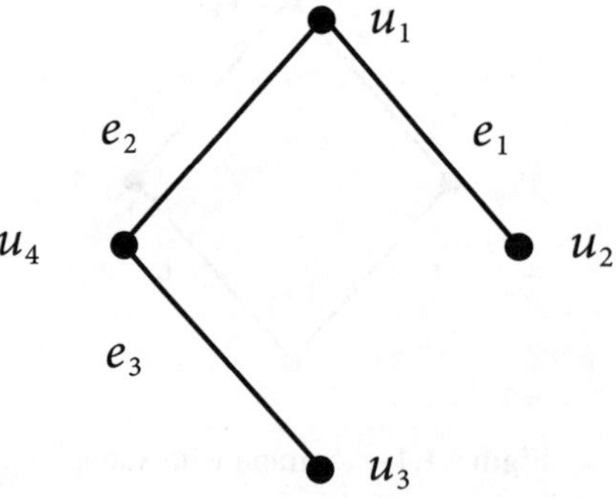

Figure 1.6   Graph with adjacent and non-adjacent vertices

1.1 Types of graphs

We have already defined a simple graph. In this section, we shall define some other types of graphs as well.

DEFINITION 1.1.1 A graph is said to be a *finite graph*, if both its vertex set and its edge set are finite.

REMARK 1.1.2 The definition of a graph as finite, provokes the question, "Is there an infinite graph?". The answer to that is "Yes! We do have infinite graphs". There are a large number of graphs defined on infinite sets of vertices and/or edges. Such graphs are known as *infinite graphs*. An infinite graph is countable if both its vertex set and edge set are countable. Some well known countable graphs are the square lattice, the triangular lattice and the hexagonal lattice as seen in Figure 1.7. We also have uncountably infinite graphs.

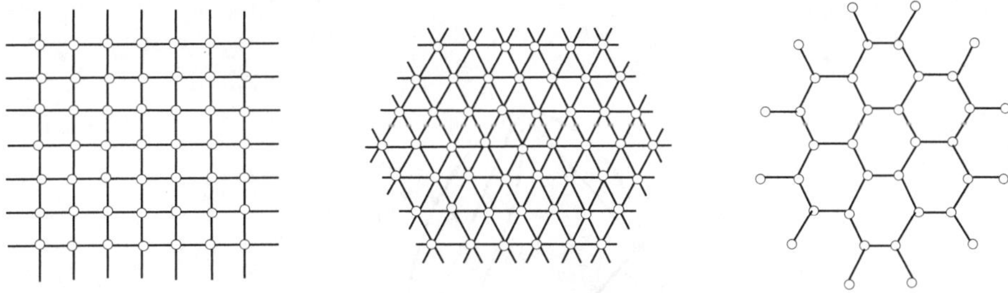

Figure 1.7   Countably infinite graphs

A more relatable example of an infinite graph can be constructed this way. In this age of social media, if we imagine ourselves as a vertex, then all our Facebook friends or Instagram followers can also be represented as vertices and connected to the primary vertex, which is oneself. Now each of these "friends" or "followers" as vertices is also connected to several other friends or followers. It is not hard to imagine an infinite graph with the 7 billion people on earth as vertices being connected to their friends by means of edges on social media. In fact, there is a name for this graph, representing friendships or connections between people and it is the *social graph*.

For the sake of simplicity, the graphs that will be discussed in this book here on, will be assumed to be simple and finite unless mentioned otherwise.

DEFINITION 1.1.3 A *null graph* is a graph, where vertex set and edge set are empty. A *trivial graph* is a graph containing only one vertex and naturally, no edges.

DEFINITION 1.1.4 A *multigraph* is a graph with multiple edges or parallel edges between any two vertices.

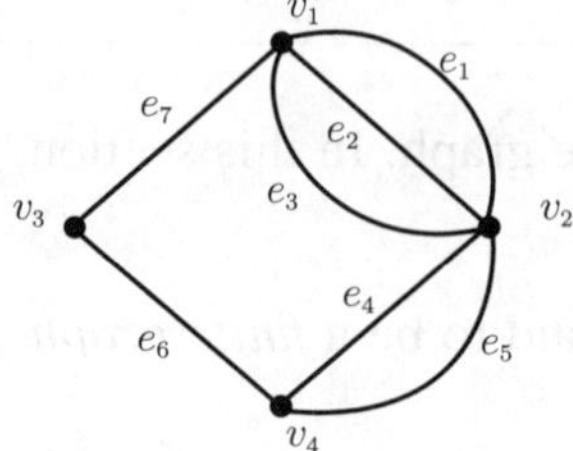

Figure 1.8   A multigraph

An example of a multigraph is given in Figure 1.8.

DEFINITION 1.1.5 A *pseudo graph* is a graph that contains both loops and parallel edges. For example, refer Figure 1.9.

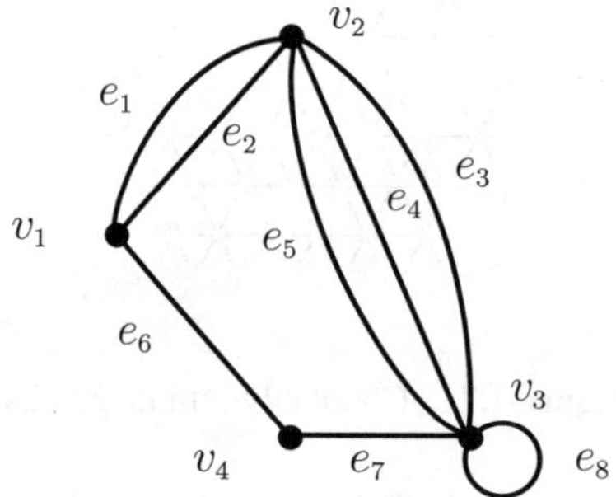

Figure 1.9   A pseudo graph

Some mathematicians like to distinguish between pseudo graph and multigraph, whereas some don't. We will allow for that distinction and define a pseudo graph as a multigraph with loops.

DEFINITION 1.1.6 A *non-directed graph* is a graph, with edges that are non-directed. That is, the edges do not have a direction or orientation.

All the graphs that we have seen so far from Figure 1.3 to Figure 1.9 are non-directed graphs.

DEFINITION 1.1.7 A graph in which all the edges are directed or have some direction is called a *directed graph*. For instance, refer the graph in Figure 1.10.

DEFINITION 1.1.8 A pseudo graph which contains directed edges is called a *quiver*.

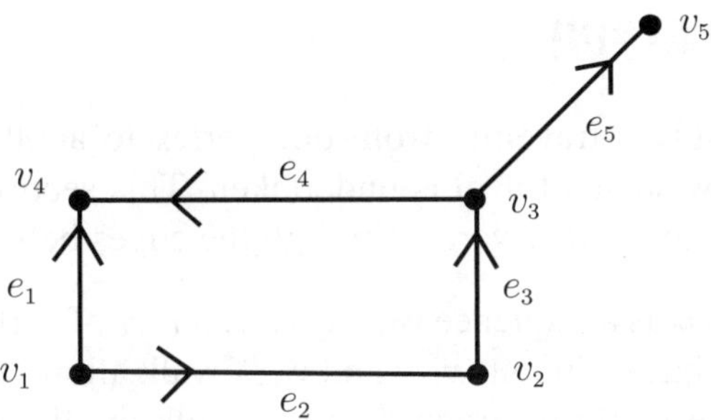

Figure 1.10   A directed graph

One usually wonders if there is any use in defining pseudo graphs and multi graphs, since we have rarely come across any use for them. But there are uses for pseudo graphs in studying network flow, modeling real-world traffic, or in GPS (Global Positioning System), where there is more than one edge between two vertices. A pseudo graph becomes very handy in the mathematical modeling of such systems.

DEFINITION 1.1.9 A *weighted graph* is a graph with a non-negative numerical value assigned to its edges, see Figure 1.11.

A weighted graph is most often used when one has to consider some "cost" of going from one vertex to another. This weight could be distance, time, or cost of travel between two vertices. For example, in Figure 1.11, the "cost" of traveling from v_1 to v_5 is the sum of the weights $5 + 6 + 2 = 13$. In the following chapters, we will be studying some algorithms that can maximize or minimize the sum of these non-negative edge weights.

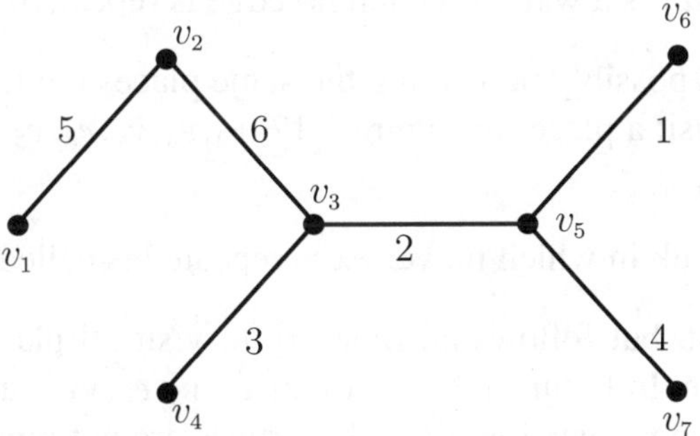

Figure 1.11   A weighted graph

1.2   Traversing a graph

Traversal of a graph involves traveling from one vertex to another. We will have to set certain guidelines on how such a travel is undertaken. This section deals with all manner of travel from one vertex to another vertex through the edges between them.

DEFINITION 1.2.1 A *walk* is a sequence $v_0, e_1, v_1, \ldots, e_k, v_k$ of vertices and edges such that for $1 \leq i \leq k$, the edge e_i has endpoints v_{i-1} and v_i. A walk allows for the repetition of both vertices and edges. In the world of graph theory, a walk implies aimlessly traversing the vertices and edges of a graph like a lost tourist. A walk is *closed* if the starting and ending points of a walk are the same (that is $v_0 = v_k$), or *open* otherwise.

EXAMPLE 1.2.2 In Figure 1.12, $v_1, e_1, v_5, e_8, v_4, e_4, v_2, e_3, v_1$ is a closed walk and $v_1, e_3, v_2, e_4, v_4, e_2, v_1, e_1, v_5$ is an open walk.

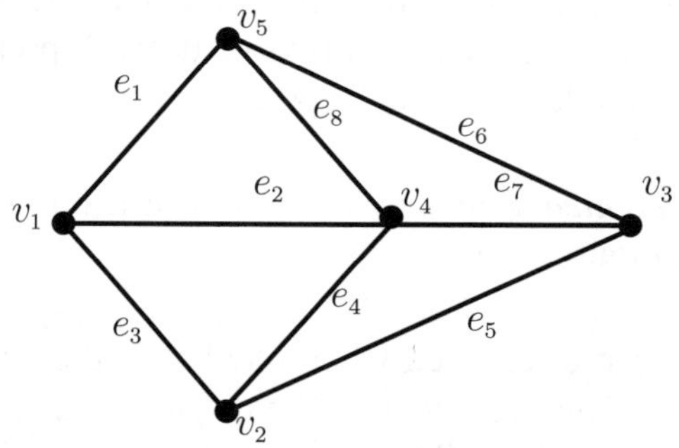

Figure 1.12   A graph G with labeled vertices and edges

DEFINITION 1.2.3 A *trail* is a walk in which no edge is repeated.

Our lost tourist could possibly be visiting the same places but takes care not to take the same route or roads to visit a place. In Figure 1.12, $v_2, e_4, v_4, e_8, v_5, e_6, v_3, e_7, v_4, e_2, v_1$ is an example of an open trail.

DEFINITION 1.2.4 A walk in which no vertex is repeated is called a *path*.

Our tourist is not lost, but follows an itinerary to visit all places, making sure not to visit the same place twice. In Figure 1.12, $v_2, e_4, v_4, e_8, v_5, e_6, v_3$ is an example of a path.

Note that a path has no repeated vertices. If vertices are not repeated in a walk, then of course no edge is repeated in the walk. So every path is basically a trail (which allows for repetition of vertices but not edges). But not every trail is a path in a graph. For instance, in Figure 1.12, the walk $v_5, e_8, v_4, e_7, v_3, e_5, v_2, e_4, v_4, e_2, v_1$ is a trail but not a path.

In a $u - v$ path, the vertices u and v are called the *end vertices* of the $u - v$ path, while all the other vertices are called the *internal vertices* of the $u - v$ path.

The *length of a path* is measured in terms of the number of edges lying between its vertices. In Figure 1.12, the path $P : v_1, e_1, v_5, e_6, v_3$ is a path of length 2, since there are two edges in the path. Mathematicians like to use the word "$u - v$ path" or "$u - v$ walk" in order to express the path or walk or trail entirely in terms of vertices, while omitting the edges. This is because, mentioning the edges, as we did so far, is somewhat redundant. We also use the same terminology hereafter, which leads us to our first theorem.

THEOREM 1.2.5 *If a graph G contains a $u - v$ walk of length k, then G contains a $u - v$ path of length at most k.*

Proof. Assume that G contains a $u - v$ walk of length k. Let us remember that a $u - v$ walk of a graph G can contain repetitions of a vertex or multiple vertices. Choose P, a $u - v$ walk of smallest length say m. Let $P : u = u_0, u_1, u_2, \ldots, u_m = v$. Therefore $m \leq k$. We shall prove that P is the $u - v$ path of length atmost k that we are looking for. Suppose not, then P contains at least one vertex which is repeated. Let this vertex be $u_i = u_j$ for some i and j such that $0 \leq i < j \leq k$. Then we delete the vertices $u_{i+1}, u_{i+2}, \ldots, u_j$ from P, then the walk P becomes the path P' which contains the vertices $P' : u = u_0, u_1, u_2, \ldots, u_i = u_j, u_{j+1}, \ldots, u_m = v$ whose length is now less than m, contradicting our assumption that P is the walk of smallest length m. Hence P is a $u - v$ path of length $m \leq k$. $\square$

DEFINITION 1.2.6 A *circuit* starts and ends at the same vertex, with no repetitions of edges. In other words, a circuit is a closed trail of three or more vertices.

DEFINITION 1.2.7 A circuit in which no vertices are repeated, except the first and last vertex is called a *cycle*. In other words, a cycle is a closed path of three or more vertices.

For example, in Figure 1.12, v_1, v_2, v_3, v_4, v_1 is a cycle. Also, $v_1, v_5, v_4, v_3, v_2, v_4, v_1$ is a circuit but not a cycle.

Furthermore a *k-cycle*, $k \geq 3$, is a cycle of length k. In other words, a k-cycle has k vertices and k edges and hence the traversal of those k edges count towards the length of the cycle. A cycle of odd length is an odd cycle and a cycle of even length is an even cycle. A 3-cycle is usually a triangle, whereas v_1, v_2, v_3, v_5, v_1 in Figure 1.12 is a quadrilateral.

DEFINITION 1.2.8 The length of a shortest cycle is known as the *girth* of a graph. It is generally accepted that graphs with no cycle have infinite girth. The girth of a graph G is denoted by $gr(G)$. The length of any longest cycle in a graph G is its *circumference* and is denoted by $cir(G)$.

The graph in Figure 1.12, v_1, v_5, v_4, v_1 is a shortest cycle and $v_1, v_2, v_3, v_4, v_5, v_1$ is a longest cycle. So the girth and circumference of the graph in Figure 1.12 are 3 and 5 respectively.

1.3 Subgraphs

In this section, let us familiarize with the concept of subgraph of a graph.

DEFINITION 1.3.1 A graph H is called a *subgraph* of a graph G if $V(H) \subseteq V(G)$ and $E(H) \subseteq E(G)$. If so, denoted by $H \subseteq G$, is to be read as "subgraph" and not "subset".

For example, the graphs H and H_1 in Figure 1.13 are subgraphs of the graph G. We have different kinds of subgraphs. Let us define one by one.

DEFINITION 1.3.2 A subgraph H in which $V(H)$ is a proper subset of $V(G)$ and $E(H)$ also follows suit, is said to be a *proper subgraph* of G.

The graph H in Figure 1.13 is an example of a proper subgraph of G.

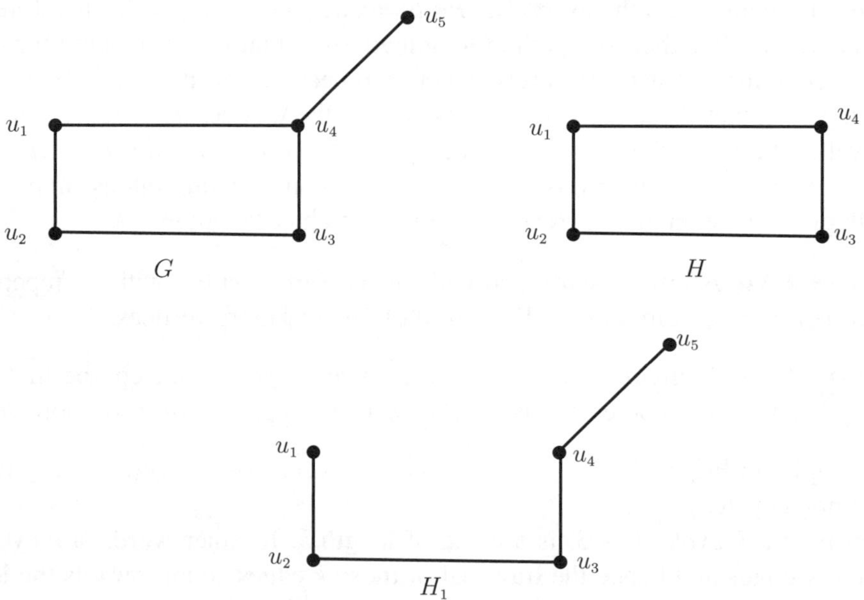

Figure 1.13   Subgraphs

DEFINITION 1.3.3 A subgraph H of G is said to be a *spanning subgraph* of G, if $V(H) = V(G)$ but $E(H) \subseteq E(G)$.

The graph H_1 in Figure 1.13 is an example of a spanning subgraph of G.

DEFINITION 1.3.4 A subgraph H of G is an *induced subgraph* of G, if whenever u and v are vertices of H, the edges connecting these vertices in G are present in H as well.

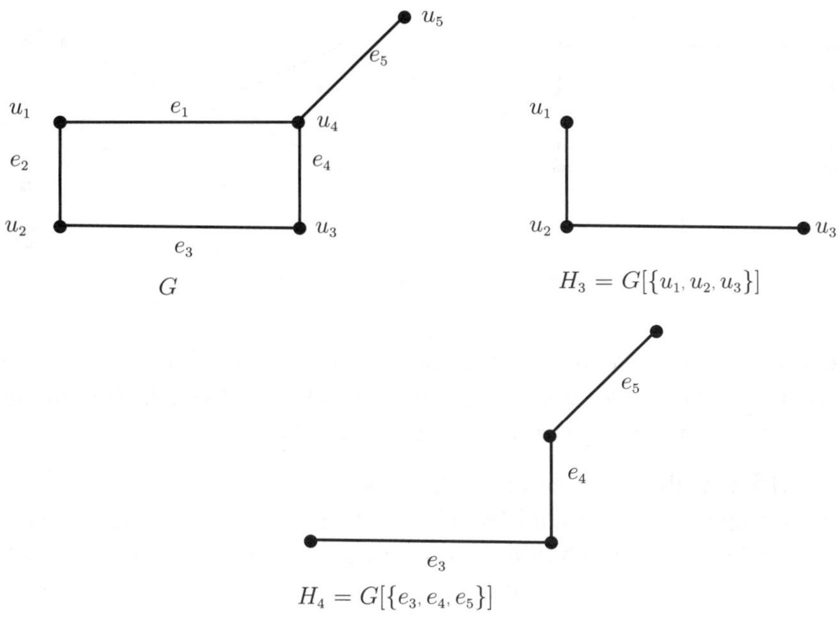

Figure 1.14   Induced subgraphs

In Figure 1.13, H is an example of an induced subgraph of G, whereas H_1 is not.

There is one more way of defining an induced subgraph and that is a subgraph induced by a specific set of vertices taken from G.

DEFINITION 1.3.5 If S is a non-empty set of vertices of a graph G, then the *subgraph induced by S in G* is the induced subgraph with the vertex set S. The induced subgraph is denoted by $G[S]$. It is also called *vertex induced subgraph* of a graph.

For the graph G in Figure 1.14, consider $S = \{u_1, u_2, u_3\}$. Then the subgraph induced by the non-empty set S of vertices of G is the graph with the vertex set S. The graph H_3 in Figure 1.14 is the resulting subgraph.

DEFINITION 1.3.6 If X is a non-empty set of edges of a graph G, then the *subgraph induced by X in G* consists of all the vertices which are incident with some edge in X and all the edges of X. It is also called *edge induced subgraph* of a graph and it is denoted by $G[X]$.

The subgraph H_4 in Figure 1.14 is an edge induced subgraph of the set of edges $X = \{e_3, e_4, e_5\}$. It is useful to discuss certain other graph theoretic operations on graphs that helps one to deconstruct a graph or reconstruct a new one.

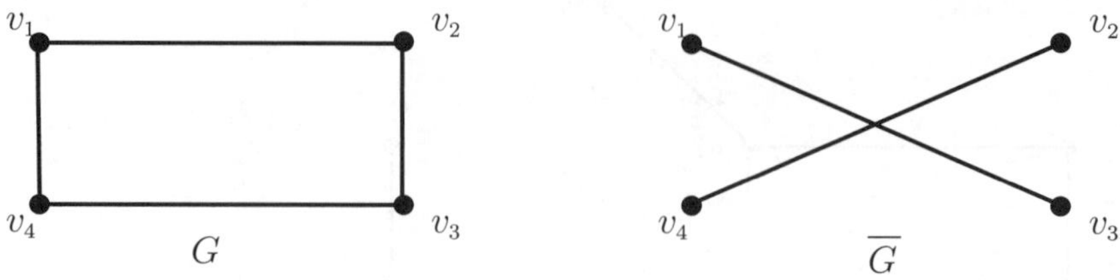

Figure 1.15   Complement graph

DEFINITION 1.3.7 A *complement* of a graph $\overline{G}$ of G is defined to be the graph with the same vertex set as G, but the vertices in G which are adjacent to each other are not so in $\overline{G}$ and those vertices that are not adjacent in G are adjacent in $\overline{G}$.

In Figure 1.15, $\overline{G}$ is the complement of G.

Note that a graph can be created by adding (or removing) a vertex or an edge. For instance, $H \cup \{v\}$ leads to a graph with a new vertex addition. Similarly $H \cup \{e\}$ creates a graph with a new edge addition. These graphs fall under the classification of supergraphs.

DEFINITION 1.3.8 Let G be a graph with $u \in V(G)$ and $e \in E(G)$. A *vertex removal* subgraph $G - \{u\}$ of G is the subgraph of G created by removing the vertex u as well as all its incident edges. An *edge removal* subgraph $G - \{e\}$ of G is the subgraph of G created by just removing the edge e.

A diagrammatic explanation is given in Figure 1.16.

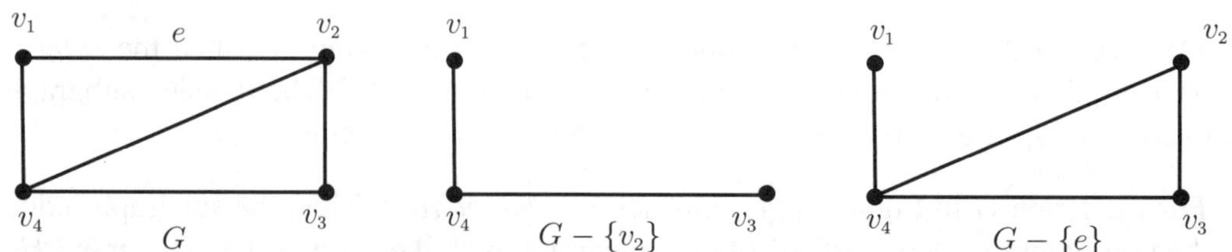

Figure 1.16   Vertex and edge removal subgraphs

We also have several ways to produce new graphs from a given pair of graphs. Let us explore a few of them.

DEFINITION 1.3.9 If G and H are two different vertex disjoint and edge disjoint graphs, the *union* of these graphs, $G \cup H$ is merely a display of the two graphs side by side without any interaction between them.

In the above definition, vertex disjoint means $V(G) \cap V(H) = \emptyset$ and edge disjoint means $E(G) \cap E(H) = \emptyset$.

DEFINITION 1.3.10 If G and H are two vertex disjoint and edge disjoint graphs, then the *join* of these graphs, $G + H$ consists of the vertex set $V(G) \cup V(H)$ and the edge set $E(G) \cup E(H)$ together with all edges that join each vertex of G with each vertex of H.

DEFINITION 1.3.11 If G and H are two vertex disjoint and edge disjoint graphs, then the *Cartesian product* of these graphs, $G \Box H$ consists of the vertex set $V(G) \times V(H) = \{(x,y) : x \in V(G) \text{ and } y \in V(H)\}$ and two distinct vertices (x,y) and (x',y') are adjacent if $x = x'$ and $yy' \in E(H)$, or $xx' \in E(G)$ and $y = y'$.

Example for join of two graphs and Cartesian product of two graphs is given in Figure 1.17.

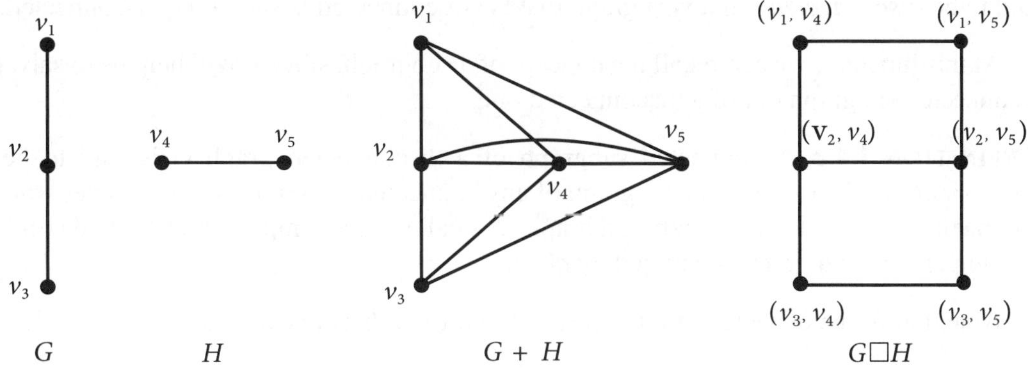

Figure 1.17   Join and Cartesian product of two graphs

When discussing the concept of subgraphs, it useful to know the maximality (or minimality) condition.

DEFINITION 1.3.12 A subgraph H of G is *maximal* if a property of a graph is satisfied for H but not for any other graph $H \cup \{v\}$ where $v \in V(G)$.

The minimality condition has to do with the removal of a vertex. A subgraph is *minimal* if a loss of a vertex results in a loss of the property of a graph.

DEFINITION 1.3.13 A subgraph H of G is *edge-maximal*, if a property of a graph is satisfied for H but not for any other graph $H \cup \{e\}$ where $e \in E(G)$.

The edge minimality condition can be similarly arrived at.

1.4   Connected graphs

While traversing a graph, we have automatically assumed connectivity so far, that is it is possible to start from one vertex and traverse the graph and reach another vertex, which leads us to the question, "Is it possible to do so for all graphs?" The answer is no. We have a class of graphs called the connected graphs, in which such traversal is easy and another class of graphs called the disconnected graphs where such a traversal is akin to time travel, which we know to be impossible(at least for now!).

DEFINITION 1.4.1 A graph is said to be *connected* if there exists a path between every two vertices of a graph.

All graphs that we have seen so far in the figures are connected graphs. The negation of the above definition leads us to the definition of a disconnected graph.

DEFINITION 1.4.2 A graph that does not have a path for at least one pair of vertices is said to be *disconnected*. That is, a graph that is not connected is said to be disconnected.

At this juncture, we can recall the concept of a subgraph, since it will help us to salvage a connected subgraph out of a disconnected one.

DEFINITION 1.4.3 A connected subgraph of a disconnected graph G is said to be a *component*, if it is not a proper subgraph of any connected subgraphs of G. In other words, the maximal connected subgraph of a graph G is called as a component of G. The number of components of a graph is denoted by $c(G)$.

Note that for a connected graph, $c(G) = 1$ and for a disconnected graph, $c(G) > 1$.

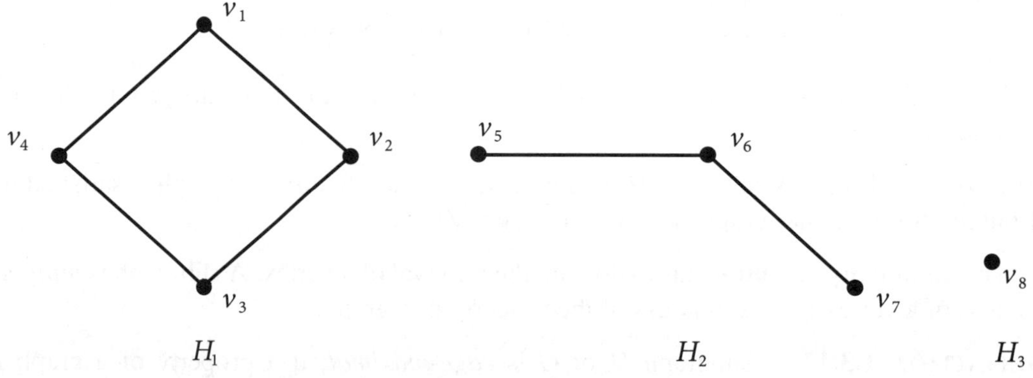

Figure 1.18   The graph $H = H_1 \cup H_2 \cup H_3$

In Figure 1.18, there is no $v_2 - v_5$ path or a $v_7 - v_8$ path in H. Note that the graph H has three components or maximal connected subgraphs. Since H_1, H_2 and H_3 are the three components of the graph H, we get $c(H) = 3$.

REMARK 1.4.4 We may define a relation on the vertex set of a graph G, by connectivity. To elaborate, we confirm the relation between two vertices u, v only if they are connected by a $u - v$ path. In this case we obtain an equivalence relation. In Figure 1.18, there are three components of a graph H, namely H_1, H_2 and H_3. These three components form three equivalence classes under the relation of connectivity, since the vertex set of the three components are pairwise disjoint. We will discuss more about connectivity and connectedness in Chapter 4.

1.5 Common classes of graphs

In continuation of subgraphs, it is better for the reader to be introduced to certain classes of graphs that we will repeatedly visit in the sections to come. In a world where various classes of graphs abound, some graphs are "celebrities" of graph theory in the sense that all graph theorists will know about them. Any graph theorist will reach for these graphs to first test out a property, and hence their popularity. We will discover and learn about more such "celebrities" in the upcoming chapters. But first we will discuss about path graphs, cycle graphs, complete graphs, bipartite graphs and complete bipartite graphs.

Path, cycle and complete graphs

We are familiar with the definition of path and cycle. Now, let us formally define the path graph and cycle graph.

DEFINITION 1.5.1 A *path graph* is a graph with vertex set $\{v_1, v_2, \ldots, v_n\}$ such that the edges are $v_i v_{i+1}$ for all $1 \leq i \leq n$. Thus, a graph that can be drawn with all of its edges and vertices falling on a single straight line is called a path graph. The path graph on n vertices is denoted by P_n.

The path graphs of order up to 5 are given in Figure 1.19. The size of P_n is $n - 1$.

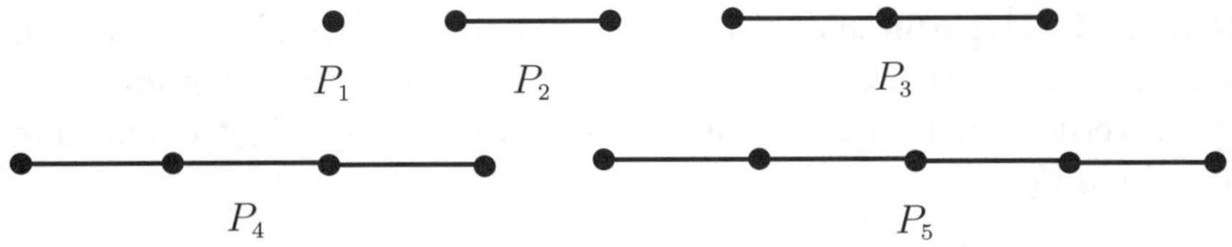

Figure 1.19   Path graphs

DEFINITION 1.5.2 A graph with just one cycle is called a *cycle graph*. The cycle graph on n vertices, where $n \geq 3$, is denoted by C_n.

The cycle graph of order up to 5 is given in Figure 1.20. The size of C_n is n.

DEFINITION 1.5.3 A *complete graph* is a graph in which every pair of vertices is adjacent. It is commonly denoted by K_n, where n denotes the number of vertices.

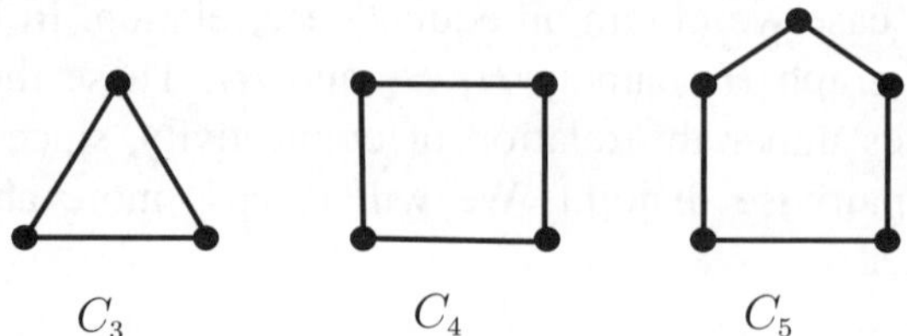

Figure 1.20   Cycle graphs

Complete graph of order up to 5 is given in Figure 1.21. Note that the size of K_n is $(n-1)+(n-2)+\cdots+2+1 = \frac{n(n-1)}{2} = \binom{n}{2}$. On the other hand, the complement of complete graph K_n is a graph with n vertices and no edges. Such a graph is called a *totally disconnected graph* and is denoted by $\overline{K_n}$. So the size of $\overline{K_n}$ is 0. Therefore the size of any simple graph of order n is at least 0 and at most $\binom{n}{2}$. That is $0 \leq |E(G)| \leq \binom{n}{2}$.

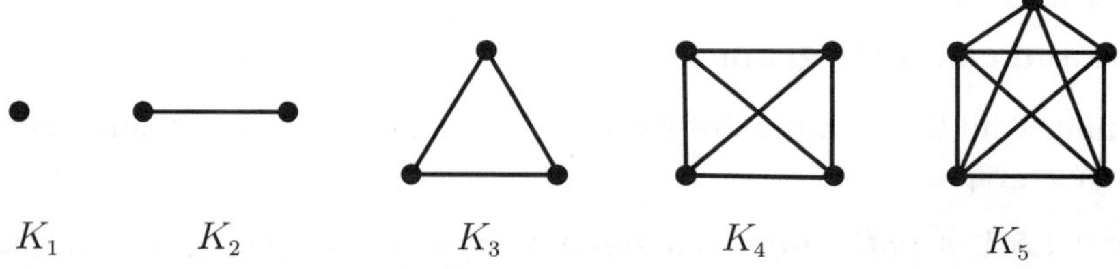

Figure 1.21   Complete graphs

Bipartite graphs

DEFINITION 1.5.4 A bipartition of G is the specification of two disjoint independent sets of vertices of V_1 and V_2 such that $V_1 \cup V_2 = V(G)$. A *bipartite graph* consists of edges connecting the vertices of V_1 and V_2 but does not contain any edges connecting vertices within V_1 or within V_2.

Refer Figure 1.22 for a diagrammatic representation of a bipartite graph.

The concept of bipartite graphs sprang from job assignments. If there are m jobs and n people of which only a few have the requisite skills, we need to assign the right people to the jobs. An example of a task accomplished would be like Figure 1.23.

An interesting aspect of bipartite graphs is that we can characterize them using cycles. The following statement is a useful tool to prove it.

Note that the existence of a cycle in a closed walk W is always guaranteed. For, if we consider an edge $e = xy$ in W, then $W - e$ results in an $x - y$ walk that avoids e. Since every $x - y$ walk contains an $x - y$ path, the $x - y$ path together with the edge e results in a cycle.

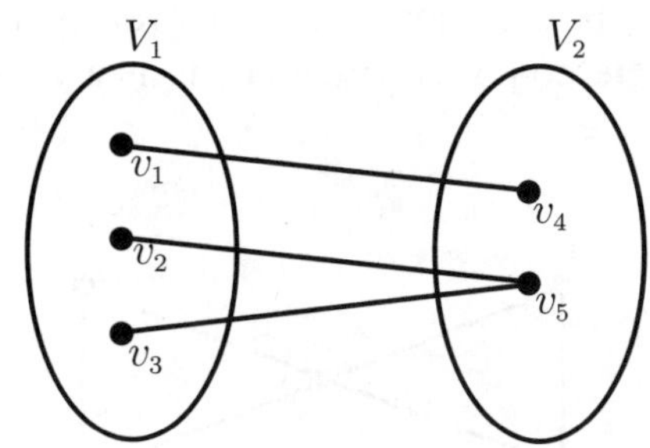

Figure 1.22 A bipartite graph

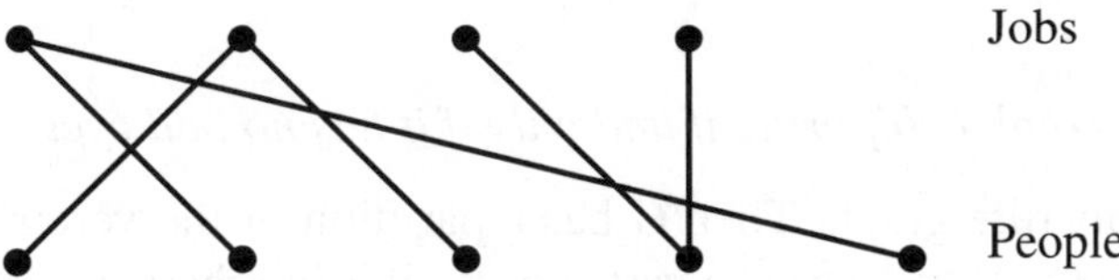

Figure 1.23 Job assignments

LEMMA 1.5.5 *Every closed odd walk contains an odd cycle.*

Proof. Consider a closed odd walk of length ℓ. Let us prove the result by induction on n. The case $\ell = 1$ is trivial. Let $\ell > 1$. Assume that every closed odd walk contains an odd cycle when $\ell \leq n$ for some odd integer n. Let us suppose $\ell > n$. If W has no repeated vertex other than the first and the last, then W itself forms a cycle of odd length. Suppose W has a repeated vertex v. Then W can be broken into the two $v - v$ walks, say X and Y, refer Figure 1.24. Since W is of odd length, without loss of generality we can assume that X is of odd length and Y is of even length. Since the length of the odd walk X is less than W through induction, we can state that X contains an odd cycle, which also appears in W.

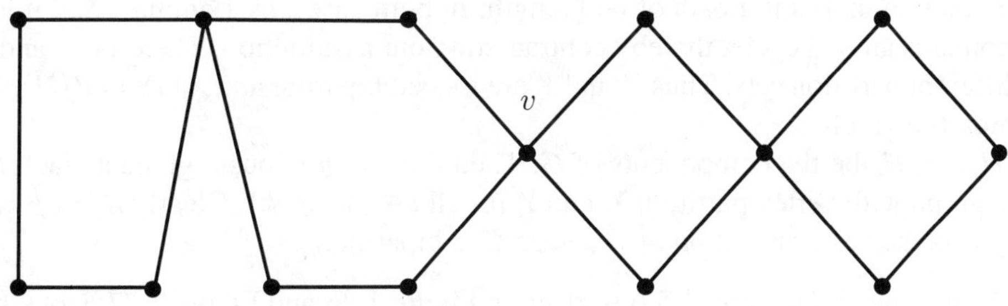

Figure 1.24 Diagrammatic representation for Lemma 1.5.5

$\square$

Note that a closed even walk need not contain an even cycle. For instance, consider the graph given in Figure 1.25. Here $v_1, v_2, v_3, v_4, v_2, v_5, v_1$ is a walk of length 6. But this walk does not contain any even cycle.

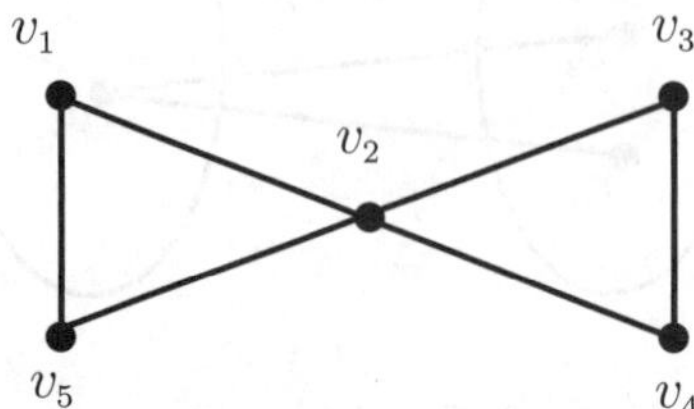

Figure 1.25   Even closed walk but not having even cycle

THEOREM 1.5.6 *A graph is bipartite if and only if it has no odd cycle.*

Proof. Let G be a bipartite graph. Then G has a partition of the vertex set into V_1 and V_2 such that $V_1 \cup V_2 = V(G)$. If we start at a vertex of V_1, it is immediately followed by a vertex in V_2. Therefore the walk alternates between the two partite sets. To construct a cycle, we have to return to V_1 at the end and so it can be done only after an even number of vertices. Hence there is no odd cycle.

Conversely, suppose that G is a graph with no odd cycles. We shall prove that G is bipartite, by constructing a bipartition of each non-trivial component. Let H be a non-trivial component and let $u \in H$. Now for every $v \in H$, define $f(v)$ to be the length of the shortest $u - v$ path P.

Let $X = \{v \in V(H) : f(v) \text{ is even}\}$ and $Y = \{v \in V(H) : f(v) \text{ is odd}\}$.

Let v and v' be two adjacent vertices in H. Then the $v - u$ path P followed by the $u - v'$ path P_1, along with the edge vv' forms a closed walk. Suppose $v, v' \in X$. Since the lengths of P and P_1 are even, the combined length is also even. This coupled with edge vv' makes it a walk of odd length. If $v, v' \in Y$, then the path P is of odd length and the path P_1' is also odd and so the combination of which is a path of even length. The edge vv' when added to this combination makes it a path of odd length. In both cases, by Lemma 1.5.5, it is clear that H contains an odd cycle, thereby contradicting our assumption. Therefore v and v' are in the different partition sets. Thus X and Y are indeed bipartite and $X \cup Y = V(H)$. Hence H is a bipartite graph.

Let $H_1, \ldots, H_k$ be the components of G. Then, by the previous argument, each H_i is a bipartite graph with vertex partition X_i and Y_i for all $i \in \{1, \ldots, k\}$. Clearly $X = \cup_{i=1}^{k} X_i$ and $Y = \cup_{i=1}^{k} Y_i$ is the vertex partition of G and so G is bipartite. $\qquad\square$

An illustration of Theorem 1.5.6 is given in Figure 1.26 and Figure 1.27. Note that the graph G in Figure 1.26 is not bipartite since G has a cycle of length 5; whereas the graph G' in Figure 1.27 is bipartite since G' has no odd cycle.

Next, let us see what is complete bipartite graph.

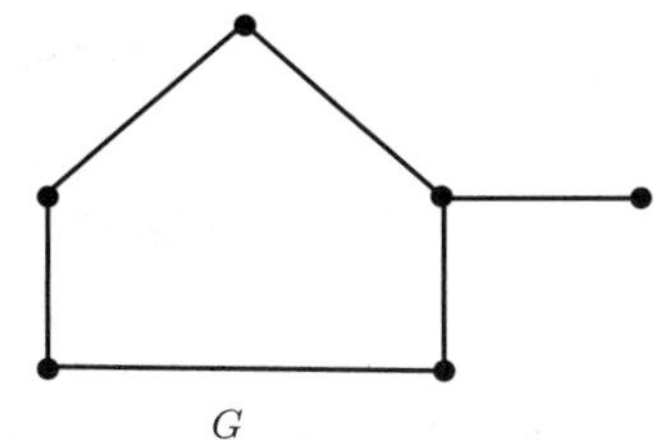

Figure 1.26   A non-bipartite graph

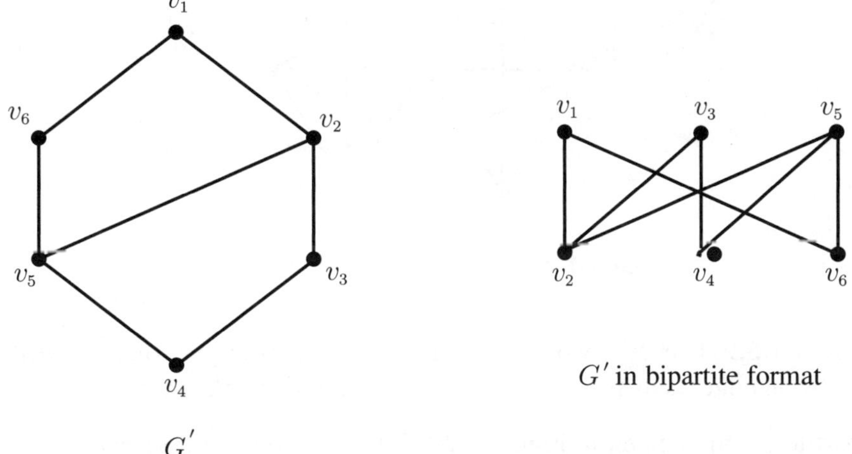

Figure 1.27   A bipartite graph

DEFINITION 1.5.7 If G is graph with two partite sets V_1 and V_2 and if all the vertices of V_1 are adjacent to all vertices of V_2, then G is said to be a *complete bipartite graph*. If V_1 has m vertices and V_2 has n vertices, the complete bipartite graph is denoted by $K_{m,n}$.

In Figure 1.28, we can see an example of some complete bipartite graphs. The complete bipartite graph $K_{1,n}$ is called as *star graph* because of its structure consisting of a single vertex in one partite set while connected to all the n vertices in the other partite set. Note that the number of vertices in $K_{m,n}$ is $m+n$ and the number of edges in $K_{m,n}$ is mn.

k-**partite graphs**

In general the bipartite graphs belong to the more general class of k-partite graphs.

DEFINITION 1.5.8 A graph G is a k-*partite graph* if $V(G)$ can be partitioned into k subsets $V_1, V_2, \ldots, V_k$, called k-partite sets of G, such that uv is an edge of G only if u and v belong to different partite sets of G.

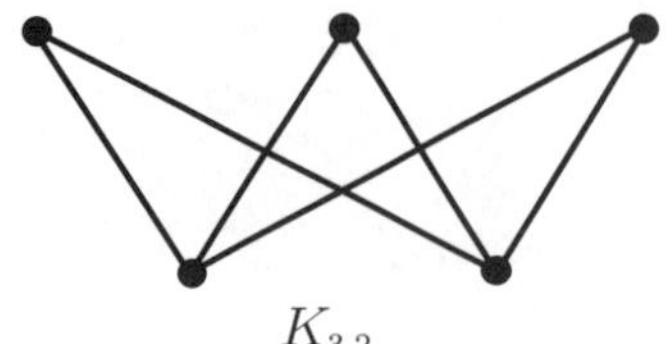
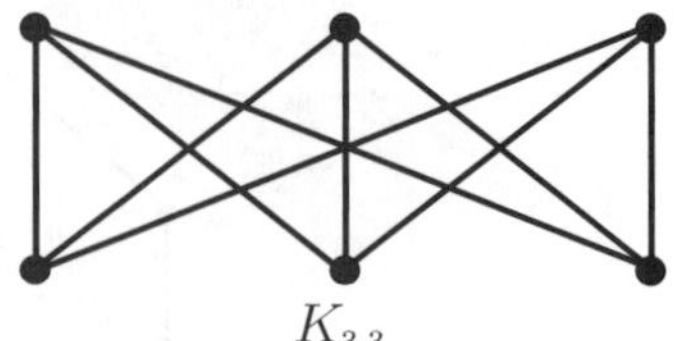

Figure 1.28   Complete bipartite graphs

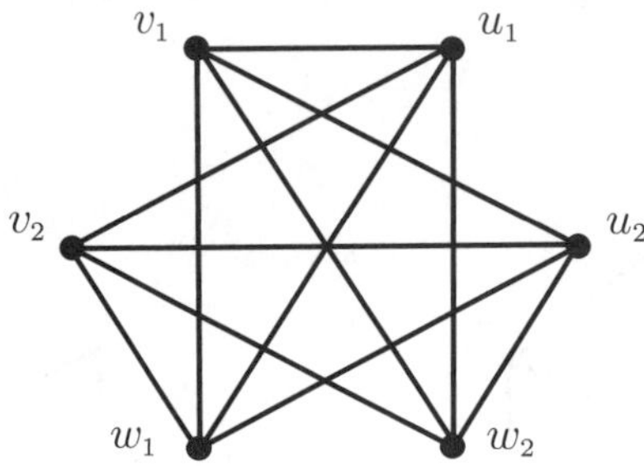

Figure 1.29   Complete tri-partite graph $K_{2,2,2}$

DEFINITION 1.5.9 If every two vertices in distinct partite sets of a *k*-partite graph is connected by an edge, then such a graph is a *complete k-partite graph.*

A 3-partite graph is given in Figure 1.29. Here $\{u_1, u_2\}$, $\{v_1, v_2\}$ and $\{w_1, w_2\}$ are the 3-partite sets.

Summary

In this chapter, we introduce the reader to the concept of graphs, while providing historical background on the origins of graph theory. Various types of graphs, their characteristics have been discussed with suitable examples. Graph traversals such as walk, path and trail are clearly differentiated and explained with examples. In addition, theorems related to graph traversals are discussed.

The definition of a subgraph and the different techniques to construct various subgraphs are discussed. The concept of connectedness is touched upon and the distinguishing characteristics of connected graphs are explained. Common classes of graphs such as path graphs, cycle graphs, complete graphs, bipartite graphs and the general class of *k*-partite graphs are defined with examples.

1.6 Exercises

Section 1.1 Types of graphs

1. Find the join of G_1 and G_2, as provided in Figure 1.30.

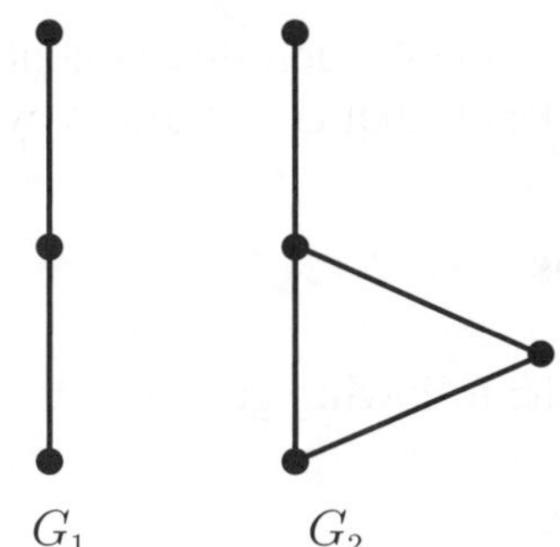

Figure 1.30   Find the join of G_1 and G_2

2. Determine the number of edges of $G_1 + G_2$ if n_1 and n_2 are the number of edges of graphs G_1 and G_2 respectively.

Section 1.2 Traversing a graph

3. The "costs" of going from v_3 to v_4 is provided in the graph G given in Figure 1.31. Can you identify all the paths that lead from v_3 to v_4 and determine which one has the least "cost"?

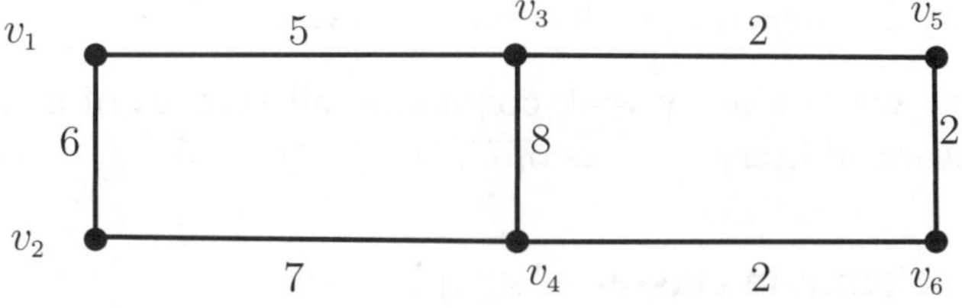

Figure 1.31   Graph G

4. Examine whether K_4 contains the following (give an example or a proof of non-existence).

   (i). A closed trail that is not a cycle.

(ii). A trail that is not closed and is not a path.

(iii). A walk that is not a trail.

5. Prove or disprove: the union of two distinct walks joining two distinct vertices of a simple graph G contains a cycle.

6. Find the circumference and girth of $K_{3,3}$ and C_n.

7. Let G be a graph having a cycle C such that G contains a path of length at least k between two vertices of C. Prove that G contains a cycle of length at least $\sqrt{k}$.

Section 1.3   Subgraphs

8. Draw the complements of the following graphs:

   (i) C_5

   (ii) $K_{1,3}$

   (iii) P_4.

9. Draw a graph having K_4 as an induced subgraph.

10. Prove/disprove: the complement of a simple disconnected graph should be connected.

11. Draw the graph $2K_4 \cup C_5 \cup 3P_3$.

Section 1.4   Connected graphs

12. Prove that any two longest paths in a connected graph G with at least three vertices have a vertex in common.

13. Prove that every 2-connected graph contains a cycle.

14. Prove that there exists a $u - v$ walk containing all vertices of a connected graph G where, u and v are arbitrary vertices of G.

Section 1.5   Common classes of graphs

15. Consider the graph in Figure 1.31 by neglecting the weights and viewing it as a simple graph.

   (i). Construct a walk from v_1 to v_6.

   (ii). Construct a path from v_1 to v_6.

   (iii). Determine whether the graph contains any cycles or circuits.

16. Show that the union of two distinct paths joining two distinct vertices in a simple graph G contains a cycle.

17. Determine the number of edges by drawing the following graphs:

    (i). K_7

    (ii). $K_{3,2}$

    (iii). $K_{1,5}$.

18. Let G be a graph of order at least 5. Show that at most one of G and $\overline{G}$ is bipartite.

19. Given a bipartite graph with 10 vertices, what is the maximum number of edges it can have?

20. For what values of $n \geq 3$, is the graph C_n bipartite?

21. Prove that there exist three vertices u, v, w of G such that uv and vw are edges of G, but uw is not an edge of G, where G is a simple connected graph that is not complete.

22. Explain whether a graph can be bipartite and have K_3 as a subgraph.

2

Basic Properties of Graphs

LEARNING OBJECTIVES

- Calculate the degree of a vertex in a graph.
- Perform worst case complexity analysis for a given algorithm.
- Differentiate between efficient algorithms and inefficient algorithms.
- Use Havel–Hakimi algorithm to realize a degree sequence into a graph.
- Create adjacency and incidence matrices for a given graph.
- Recognize key characteristics of isomorphic graphs.

We have discussed the traversal of graphs and now it is time to learn about yet another parameter, that will help to visualize and represent a graph even better. We are familiar with the order and size of a graph which are basic parameters of a graph, and we are aware about the connectedness of a graph. Closely related to the concept of connectedness, is the question of how "connected" a graph is. The answer to that question comes from the *degree* of a graph which gives an idea as to how densely or sparsely connected a graph can be. We will introduce the reader to algorithms and complexity analysis. The question of whether a degree sequence can be realized into a graph, is settled by the Havel–Hakimi algorithm, which will be covered in detail. The reader is also introduced to the process of identifying whether two graphs are alike, through the concept of isomorphism. An alternate method of representing graphs, apart from drawing them, using the adjacency matrix will be explained in this chapter.

2.1 Degree

A graph is said to be *dense* or *sparse*, depending on the relation between the number of vertices and edges. For a simple graph with n vertices, the maximum number of edges possible is $\frac{n(n-1)}{2}$. A dense graph is one in which, the total number of edges is $\frac{n(n-1)}{2}$ or close to this number. For example, a complete graph is dense since every vertex is connected to every other vertex through an edge. A sparse graph is one in which the number of edges is much lesser than the maximum possible number of edges. We shall see in this section how the *degree* of the vertices of a graph can also be reliable indicator of the nature of the graph.

DEFINITION 2.1.1 The *degree* of a vertex v in a graph G is the number of edges incident with the vertex v and is denoted by $deg(v)$.

The degree of a vertex v is also the number of vertices adjacent to v.

DEFINITION 2.1.2 Two adjacent vertices are referred to as *neighbors*. The set of all neighbors of a vertex is called a *neighborhood* and it is denoted by $N(v)$.

The connection between degree and neighborhood is that $deg(v) = |N(v)|$. For an isolated vertex, $deg(v) = 0$. If $deg(v) = 1$, the vertex v is called an *end vertex*. Also the *closed neighborhood* of v is $N[v] := N(v) \cup \{v\}$.

It is possible to determine the degree of every vertex in a graph and it leads to the following definition.

DEFINITION 2.1.3 The least value of $deg(v) = |N(v)|$, among all the vertices of a graph G is consigned as the *minimum degree* of a graph G and is denoted by $\delta(G)$. Similarly, the largest value of $deg(v)$, among all the vertices of a graph G is consigned as the *maximum degree* of a graph G and is denoted by $\Delta(G)$.

That is

$$\delta(G) = \min_{v \in V(G)} deg(v) \text{ and } \Delta(G) = \max_{v \in V(G)} deg(v).$$

It follows naturally that for a simple graph of order n,

$$0 \leq \delta(G) \leq deg(v) \leq \Delta(G) \leq n - 1.$$

The following theorem earns the title of "First theorem of graph theory" although we have dealt with theorems in the last chapter. It is called so, because it is a theorem due to Leonhard Euler published in 1736, in relation to the Königsberg bridge problem. The fact that Euler intuitively identifies that the sum of the degrees of a graph is even, is of special interest because the degree of a vertex is the first graph theory concept to be defined and Euler created this theorem to prove that Königsberg bridge problem can only be negatively resolved. The following theorem is also known as the *Handshake Lemma*.

THEOREM 2.1.4 (Fundamental theorem of graph theory) *Let G be a simple graph of size m. The sum of the degrees of every vertex in a graph is even, being twice the number of edges. That is*

$$\sum_{v \in V(G)} deg(v) = 2|E(G)| = 2m.$$

Proof. Since each edge has a starting vertex and end vertex, each edge contributes a count of two to the sum of the degrees of a graph. Hence the theorem follows.   □

THEOREM 2.1.5 *In any graph, the number of vertices with an odd degree is an even number.*

Proof. Let G be a graph and let $deg(v)$ be the degree of any vertex, $v \in V$. Let m be the size of the graph. Then by Theorem 2.1.4, $\sum_{v \in V(G)} deg(v) = 2m$. Clearly

$$\sum_{v \text{ has odd degree}} deg(v) + \sum_{v \text{ has even degree}} deg(v) = \sum_{v \in V(G)} deg(v) = 2m.$$

It follows that $\sum_{v \text{ has odd degree}} deg(v) + \sum_{v \text{ has even degree}} deg(v)$ is also an even number. It can be summarized that the sum of the degrees which are even is once again even. Hence it easily follows that the sum of the odd degrees of the vertices is also even, which implies that the number of vertices with odd degree are even.   □

Note that in any graph G, the average vertex degree is $\frac{2|E(G)|}{|V(G)|}$ and hence $\delta(G) \leq \frac{2|E(G)|}{|V(G)|} \leq \Delta(G)$.

As we proceed into the study of the degree of a graph, we come across a certain class of graphs called regular graphs for which the maximum degree $\Delta(G)$ is the same as $\delta(G)$. Regular graphs have some attractive properties, which are not usually found in other graphs. The concept of regularity also appeals to a sense of symmetry that the humankind finds aesthetically pleasing.

DEFINITION 2.1.6 A *regular graph* is one in which all the vertices have the same degree. In other words $\Delta(G) = \delta(G) = deg(v)$ for all $v \in V(G)$.

An extension of regularity is k-regularity. A graph is *k-regular* if the common degree is k. For example, refer Figure 2.1.

Note that the cycle graph C_n is a 2-regular graph and the complete graph K_n is an $n-1$-regular graph. A 3-regular graph is also called a *cubic graph*. A famous cubic graph is the *Petersen graph*, which is given in Figure 2.2.

Figure 2.1 Regular graphs

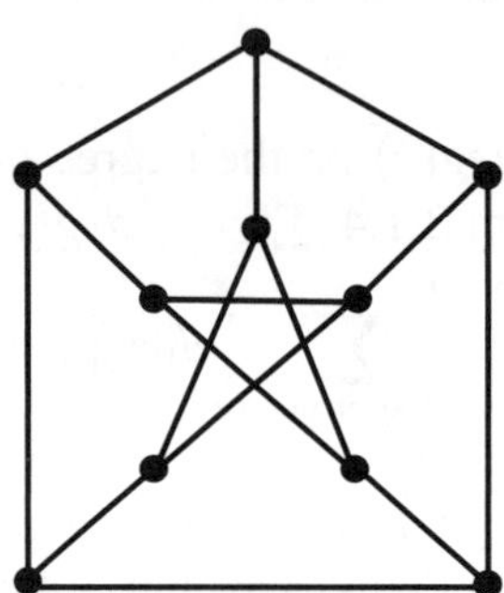

Figure 2.2 Petersen graph

THEOREM 2.1.7 *A k-regular graph with order n has $\frac{nk}{2}$ edges.*

Proof. By Theorem 2.1.4, we have $\sum_{v \in V(G)} deg(v) = 2m$ where m is the size of the graph. For the k-regular graph, $deg(v) = k$ for all $v \in G$. Therefore $\underbrace{k + k + \ldots + k}_{n \text{ times}} = 2m$. Thus $\frac{nk}{2} = m$. $\qquad\square$

An interesting thing to note is that, there are no r-regular graphs of order n, if r and n are both odd, which leads us to the next theorem. Note that the following theorem presents a proof by construction instead of a proof which merely claims the existence of the desired r-regular graph.

THEOREM 2.1.8 *Let r and n be integers with $0 \leq r \leq n - 1$. Then there exists an r-regular graph of order n if and only if at least one of r and n is even.*

Proof. By Theorem 2.1.5, the number of vertices with odd degree is an even number and so it follows that there is no r-regular graph of order n, if r and n are both odd. Conversely, let us choose r and n to be integers such that at least one of r and n is even and $0 \leq r \leq n - 1$. We will now begin the construction of $H_{r,n}$ which is a r-regular graph of order n. Let $V(H_{r,n}) = \{u_1, u_2, \ldots u_n\}$.

Case 1: Suppose r is even. Then $r = 2k \leq n-1$ for some non-negative integer $k \leq \frac{n-1}{2}$. For each i, where $1 \leq i \leq n$, we join u_i to $u_{i+1}, u_{i+2}, \ldots, u_{i+k}$. If we arrange the vertices $u_1, u_2, \ldots, u_n$ cyclically, then each vertex u_i is adjacent to k vertices that precede u_i and k vertices that follow u_i. Thus $H_{r,n}$ is r-regular.

Case 2: Suppose r is odd. In this case, n must be even. Then $r = 2k+1 \leq n-1$ and $n = 2\ell$ for some $k, \ell \in \mathbb{N}$ with $k \leq \frac{n-2}{2}$. We join u_i to the $2k$ vertices described in Case 1 as well as to u_{i+l}. Arranging the vertices $u_1, u_2, \ldots, u_n$ cyclically and joining each vertex u_i to the k vertices immediately, following it, the k vertices immediately preceding it and the unique vertex opposite to u_i. Thus $H_{r,n}$ is r-regular. $\qquad \square$

The graphs $H_{r,n}$ constructed in Theorem 2.1.8 are called *Harary graphs* and Figure 2.3 gives an example of both even and odd Harary graphs.

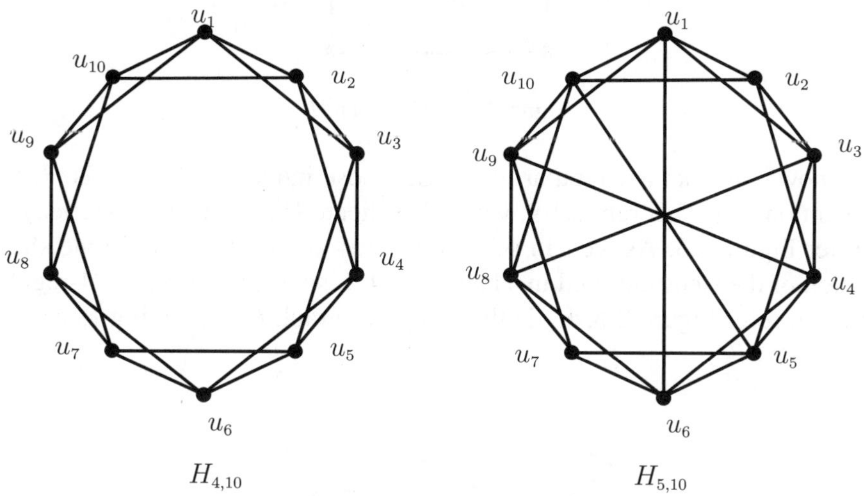

Figure 2.3   Harary graphs

If G is an r-regular graph, then we cannot construct a subgraph H of G if $\Delta(H) > r$. But the next result says that H is an induced subgraph of some r-regular graph G when $\Delta(H) \leq r$ for some $r \in \mathbb{N}$.

THEOREM 2.1.9 *For every graph H and every integer $r \geq \Delta(H)$, there exists an r-regular graph G containing H as an induced subgraph.*

Proof. If we tackle the trivial case of H also being r-regular, then we let $H = G$. Suppose H is not r-regular graph with order n and $V(H) = \{v_1, v_2, \ldots, v_n\}$. Let H' be a replication of H with n vertices such that $V(H') = \{v'_1, v'_2, \ldots, v'_n\}$. We use the word "replication" to mean an exact copy of H and not in any isomorphic sense. Each vertex v_i in H corresponds to v'_i in H' for $1 \leq i \leq n$. We now construct a graph H_1 from H and H' by connecting v_i with vertices v_i' for $1 \le i \le n$ whenever $deg_H(v_i) < r$. Then H is the induced subgraph of H_1 with $\delta(H_1) = \delta(H) + 1$ and $\Delta(H_1) = \Delta(H)$. If H_1 is r-regular, then we let $G = H_1$. Otherwise we continue this process until we arrive at an r-regular graph H_k where $k = r - \delta(H)$. The graph H_k is the desired graph G. $\qquad\square$

Demonstration: Consider a graph H with $\Delta(H) = 4$ and $\delta(H) = 2$, see Figure 2.4.

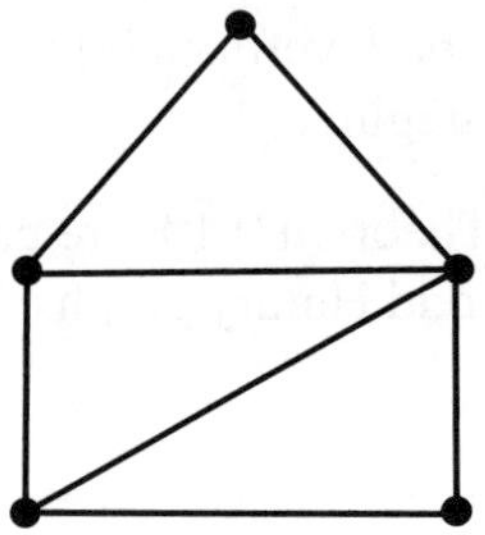

Figure 2.4   The graph H

We will now construct a 4-regular graph G, whose induced subgraph is H. Following the steps mentioned in Theorem 2.1.9, we will replicate H', connect the vertices and name it as H_1, refer Figure 2.5. As we can see, $\Delta(H_1) = 4$ and $\delta(H_1) = 3$. So we shall do one more iteration of the steps outlined in Theorem 2.1.9, so as to construct a 4-regular graph. As we can see, refer Figure 2.6, H_2 is the 4-regular graph G containing H as an induced subgraph.

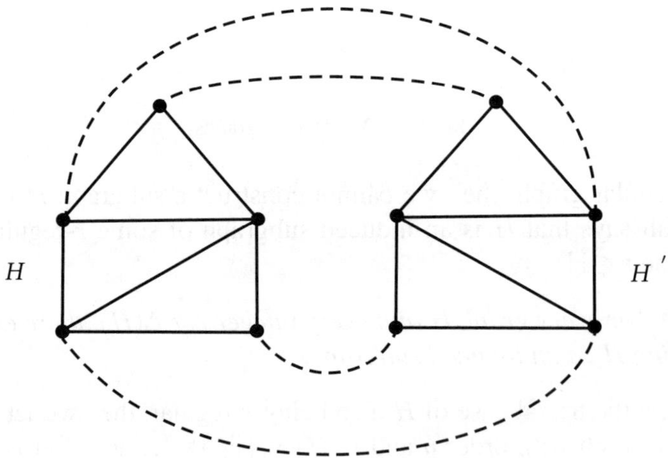

Figure 2.5   The graph H_1

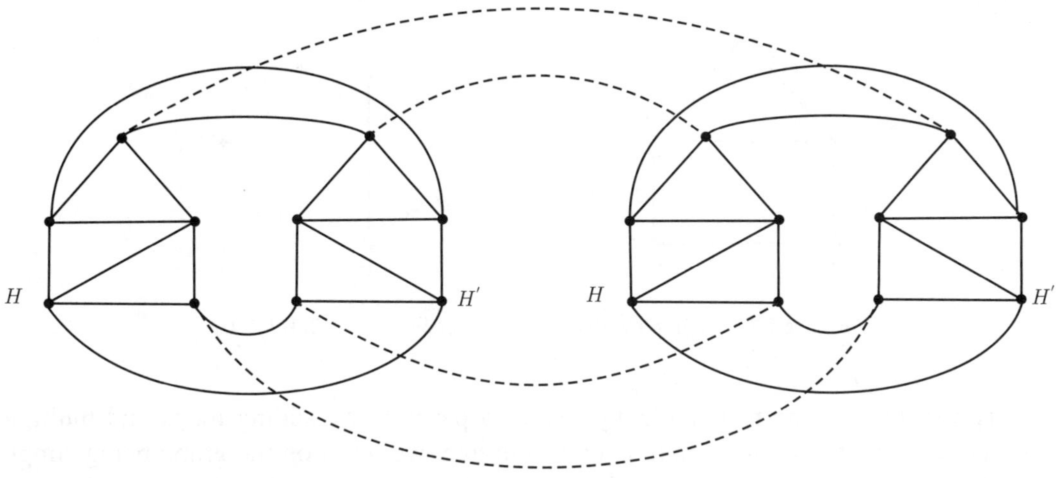

Figure 2.6   The graph $H_2 = G$

2.2 Degree sequences

If we were to represent a graph merely by means of the degree of their vertices, it would look something like this: 3, 4, 2, 2, 2, 1, 1, 1, 0. From this sequence, we can perceive that there are 9 vertices in this graph, written in random order, with their respective degrees. It is sometimes advantageous to perceive graphs in this manner because we break down a graph in terms of its neighbors.

DEFINITION 2.2.1 A *partition* of a non-negative integer n is a finite sequence of non-negative integers whose sum is n.

We know from Theorem 2.1.4 that the sum of the degrees of a graph is an even integer. So we arrive at the question, "Can we fashion a graph out of every partition of an even integer? Can we prove that it is impossible to do so, if one is not given a right mix of numbers"?

DEFINITION 2.2.2 The *degree sequence* of a graph of order n is the list of vertex degrees $(d_1, d_2, \ldots, d_n)$, usually written in non-increasing order as $d_1 \geq d_2 \geq \ldots \geq d_n$.

Let us consider a degree sequence $s : (4, 3, 2, 2, 2, 1, 1, 1, 0)$. This is the same as the one mentioned earlier, except arranged in a non-increasing order. If we were to try and mold a graph out of these numbers, we would arrive at something like the graph in Figure 2.7.

DEFINITION 2.2.3 Let $s = (d_1, d_2, \ldots, d_n)$ where $d_1 \geq d_2 \geq \ldots \geq d_n$. Then s is a *graphical sequence* if there exists a simple graph G whose degree sequence is s. The graph G is the *graphical realization* of the partition s.

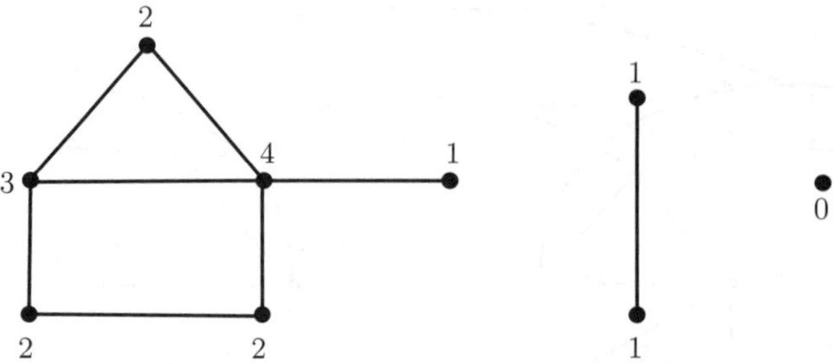

Figure 2.7   A graph with degree sequence $(4,3,2,2,2,1,1,1,0)$

It is possible for one to graphically realize a partition by adding loops and multiple edges between vertices. Since there is no fun in that, we insist on the graph being simple for a degree sequence to be graphical. Another thing to note is that the realization of a sequence or partition, need not be unique. The reader is encouraged to try out a sequence of random non-negative integers to see if it is graphical or not.

The following theorems will help us to efficiently determine if a degree sequence is graphical or not. A graphical degree sequence is basically a partition of an even integer.

THEOREM 2.2.4 *If the degree sequence* $s : (d_1, d_2, \ldots, d_n)$ *is graphical, then* $\sum_{i=1}^{n} d_i$ *is even.*

Proof. Suppose the degree sequence $s : (d_1, d_2, \ldots, d_n)$ is graphical. Then $d_1, d_2, \ldots, d_n$ are the degrees of the vertices of some graph G. Then by Theorem 2.1.4, $\sum_{i=1}^{n} d_i = 2|E(G)|$, which clearly indicates that the sum of the degrees is even. $\qquad\qquad\square$

EXAMPLE 2.2.5 Consider the sequence $s : (3,3,3,1)$. If s is graphical, it would be a graph G with 4 vertices such that $\Delta(G) = 3$ and $\delta(G) = 1$. There are 4 vertices with odd degrees and hence satisfies Theorem 2.2.4. Let $V(G) = \{v_1, v_2, v_3, v_4\}$ with $deg(v_1) = deg(v_2) = deg(v_3) = 3$ and $deg(v_4) = 1$. This implies that v_1, v_2 and v_3 are adjacent to all other vertices of G so that their degrees are 3, but v_4, however is connected to only one vertex. It is impossible to construct a graph according to this degree sequence. Thus s is not graphical and hence Theorem 2.2.4 is only a necessary condition to see if a sequence is graphical.

The following theorem is a necessary and sufficient condition to prove whether a degree sequence is graphical or not. This theorem is also known as the *Havel–Hakimi theorem*, based on which the Havel–Hakimi algorithm was designed.

THEOREM 2.2.6 *A non-increasing sequence* $s: (d_1, d_2, \ldots, d_n)$ *of non-negative integers, where* $n \geq 2$ *and* $d_1 \geq 1$, *is graphical if and only if the sequence* $s_1 : (d_2 - 1, \ldots, d_{d_1+1} - 1, d_{d_1+2}, \ldots, d_n)$ *is graphical.*

Proof. First assume that s_1 is graphical. Then there exists a graph G with $V(G) = \{v_2, v_3, \ldots, v_n\}$ such that

$$deg(v_i) = \begin{cases} d_i - 1 & \text{if } 2 \le i \le d_1 + 1, \\ d_i & \text{if } d_1 + 2 \le i \le n. \end{cases}$$

We construct a graph G' from G by adding a new vertex v_1 and the d_1 edges $v_1 v_i$ for $2 \le i \le d_1 + 1$, refer Figure 2.8. Since $deg_{G'}(v_i) = d_i$ for $1 \le i \le n$, it follows that s is a degree sequence of G' and so s is graphical.

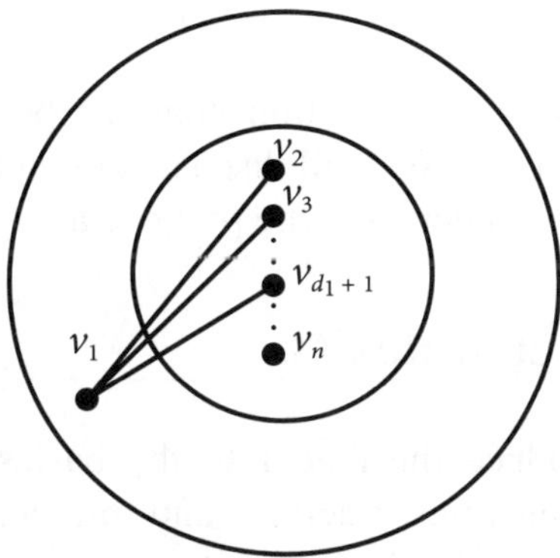

Figure 2.8 Construction of the graph G' from G

To prove the converse, let us assume that $s : (d_1, d_2, \ldots, d_n)$ is graphical. Let G be a graph with $V(G) = \{v_1, v_2, \ldots, v_n\}$ such that $deg(v_i) = d_i$ for $1 \le i \le n$. If v_1 is adjacent to the vertices $v_2, \ldots, v_{d_1+1}$, then s_1 is the degree sequence of $G - v_1$ and the proof of the converse is complete. Suppose not, let v_j be the first vertex which is not adjacent to v_1 for some $j \in \{2, \ldots, d_1 + 1\}$. Since $deg(v_1) = d_1$, there exists v_k, where $d_1 + 2 \le k \le n$, with v_k is adjacent to v_1. Since $deg(v_j) \ge deg(v_k)$ together with the fact that $v_1 \notin N(v_j)$ and $v_1 \in N(v_k)$, we have a vertex v_ℓ with v_ℓ adjacent to v_j but v_ℓ not adjacent to v_k. Now, construct a graph G_1 from G by removing the edges $v_1 v_k$ and $v_j v_\ell$ and instead adding the edge $v_1 v_j$ and $v_k v_\ell$, refer Figure 2.9. It is obvious that our graphs G_1 and the graph G have the same vertex set and the same degree sequence. However v_1 is adjacent to v_j in G_1. Now repeat the same process for all the vertices v_j, $2 \le j \le d_1 + 1$, which are not adjacent to v_1. Finally, we get a graph G_k whose degree sequence is same as the degree sequence of G and v_1 is adjacent to the vertices $v_2, \ldots, v_{d_1+1}$. Now s_1 is the degree sequence of the graph $G_k - v_1$. Hence s_1 is graphical.

$\square$

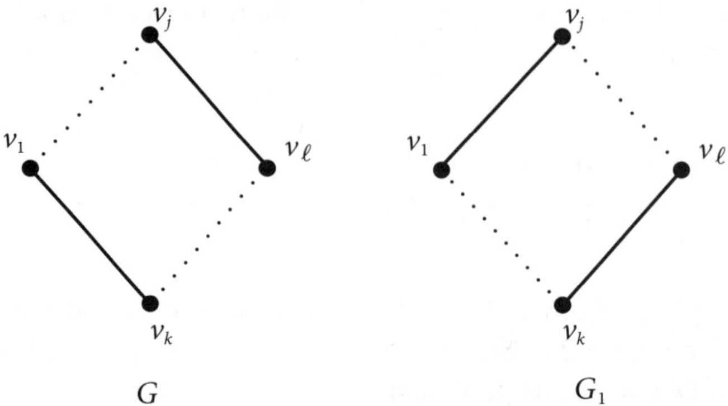

Figure 2.9   Graph G_1 obtained from G

Theorem 2.2.6 lends itself as an algorithm, that can be used to determine whether a degree sequence is graphical or not. We will discuss it in an upcoming section. But before we delve into algorithms, let us understand the purpose and parameters of algorithms.

2.3 Algorithms and complexity

In this section, we will introduce the reader to the basics of algorithms such as data structures, pseudocodes, complexities, tractable and intractable problems. Many of these concepts may already be known to the computer-literate reader, but for the beginner, the ideas and definitions in this section, will prove to be a gentle introduction into the extraordinary world of algorithms.

DEFINITION 2.3.1 An *algorithm* is a sequence of precise instructions with the objective of accomplishing a specific task.

Algorithms are a key component of computer programming and are used in many systems and applications such as navigation systems, search engines, social media and music streaming services. Algorithms can be executed manually by people or automatically by a machine. Algorithms are an integral part of this book, since it reaps multiple benefits for one to study discrete structures such as a graph, while also learning to navigate, analyze and build such structures. In a world increasingly driven by algorithms, it is now time for an aspiring graph theorist to dip his/her toes into the fascinating world of graph algorithms. Some well-known graph algorithms will be covered in various chapters of this book. In this book, we will also show an algorithm as written in *pseudocode* (pronounced as soo-doh kode). Usually algorithms are written in *pseudocode* to make it easy to understand.

DEFINITION 2.3.2 A *pseudocode* is an informal way of describing an algorithm using simple, readable language and structured steps. A pseudocode does not follow the strict syntax rules of a programming language.

A pseudocode serves as a bridge between the concept of an algorithm and its implementation in code, making it easy to understand the logical flow without focusing on specific programming details. It is usually a mixture of usual language and programming constructs to yield a more crisp and succinct conceptualization of an algorithm. There are different styles in writing a pseudocode but anyone with a basic understanding of programming languages, can follow them without much difficulty.

Once an algorithm is designed, we need a yardstick to measure the merit and efficiency of an algorithm. This is where *complexity* of an algorithm, comes into play.

DEFINITION 2.3.3 The *complexity* of an algorithm refers to the quantitative measure of the resources, primarily time and space, that an algorithm requires to execute as a function of the input size.

It indicates how efficiently an algorithm performs particularly in terms of

Time complexity: The amount of time or the number of steps an algorithm takes to complete as the input size grows.

Space complexity: The amount of memory or storage space an algorithm uses relative to the input size.

Complexity analysis: To predict the behavior of an algorithm in terms of scalability and efficiency, providing insight into how it will handle larger inputs, we need complexity analysis. In the real world, there are usually more than one algorithm to solve a problem. In this case, complexity analysis allows for comparisons between algorithms to determine the most efficient solution for a given problem. The sorting algorithm is the most common algorithm used to sort an unarranged list either alphabetically or numerically. There are plenty of algorithms such as bubble sort, sequential sort, merge sort, insertion sort etc. to perform the same operation.

Let us suppose algorithm X will sort a list of n items in $0.5n^3$ units of time, while another algorithm Y will sort a list of n items in $2.5n^2$ units of time. Consider a list with 4 unsorted items. Algorithm X will sort the list in $0.5 \times (4^3) = 32$ units of time while algorithm Y will sort it in $2.5 \times (4^2) = 40$ units of time. But if we are given an unsorted list of 8 items algorithm X will sort the list in $0.5 \times (8^3) = 256$ units of time while algorithm Y will sort the list in $2.5 \times (8^2) = 16$ unit of time. From this comparison, we can predict that algorithm Y will do a better job of sorting bigger lists than algorithm X.

Additionally, one must remember that even the complexity $0.5n^3$ or $2.5n^2$ is actually the *worst-case complexity* where we measure the time taken for an algorithm to work in the worst case scenario. Even in an unsorted list, there can be order in some entries, which makes the process work a little faster than the worst case. Although worst-case complexity is the most common measure of complexity, we can describe other measures of complexity such as the *average-case complexity*, where the average running time of an algorithm is measured over all inputs.

An algorithm is *efficient* if the complexity of the algorithm is a polynomial function of the input size (say) n or if it is bounded by a polynomial in n. Algorithms whose complexities are $n^4, n^2, logn, nlogn, \sqrt{n}$ are efficient while those with complexities such as $2^n, n!, n^n$ are not efficient.

Is it possible, that one can design an efficient algorithm for all problems?.The answer to that question is sadly in the negative. We have an entire class of problems for which an efficient algorithm does not exist. A computational problem is called *tractable* if there exists an efficient algorithm for solving the problem. A computational problem is *intractable* if it can be established that there are no efficient algorithms to solve the problem. Fortunately, we will be dealing with only tractable problems and their algorithms throughout this book, helping the reader to identify and revel in the beauty of a well-designed algorithm for a tractable problem. Once the reader has mastered the algorithms for tractable problems, he/she can confidently take a plunge into the vast world of intractable problems, armed with this knowledge.

For a tractable problem, when there is more than one algorithm to a computational problem, we have choices before us and we must choose an algorithm to solve the problem depending on our needs. In order to compare the complexities of algorithms, we describe the *order* of a function.

DEFINITION 2.3.4 Let f and g be two functions defined on the set of positive integers. The *order* of f is said to be lower than or equal to the order of g if there exists a positive real constant C and a non-negative integer n_0 such that

$$f(n) \leq Cg(n) \text{ for all } n > n_0.$$

If the order of f is lower than or equal to the order of g we write $f(n) = O(g(n))$ or we say $f(n)$ is $O(g(n))$(read as "big oh" of $g(n)$). It means, f is bounded by g or f does not grow bigger than g. It can also be interpreted as: f may grow slowly than g or at the same rate as g. Two functions are of the *same order* if $f(n) = O(g(n))$ and $g(n) = O(f(n))$.

Suppose $f(n) = 8n + 15$, and $g(n) = n^2$. Then $f(n) = O(g(n))$. For the constant $C = 3$,

$$8n + 15 \leq 3n^2 \text{ for all } n > 2. \text{ Here } n_0 = 2.$$

This inequality can be verified by mathematical induction. We can also use other methods to verify our intuitive choices. Also it is not hard to see that other choices of C and n_0 can also work.

Going back to our sorting algorithms X and Y, whose complexities are $f(n) = 0.5n^3$ and $g(n) = 2.5n^2$, it easy to prove that $g(n) = O(f(n))$; in fact the complexity of algorithm Y is $O(n^2)$. It follows that the complexity of algorithm X is $O(n^3)$, however $f(n) \neq O(g(n))$. We can conclude that Algorithm Y is more efficient that Algorithm X, because of this reason. To sum up, if two algorithms A and B have complexity function $f(n)$ and $g(n)$ respectively, we say algorithm A is *more efficient* than Algorithm B if $f(n) = O(g(n))$ but $g(n) \neq O(f(n))$. There is a heuristic rule, based on a hierarchy of increasing orders, to help one compare the complexities of algorithms.

$$O(1) \leq O(logn) \leq O(n) \leq O(nlogn) \leq O(n^2) \leq O(n^3) \leq O(2^n) \leq O(n!)$$

Since $logn < n$ for all $n \geq 1$, $logn$ is also $O(n)$. Also 2^n is $O(n!)$ for $n \geq 4$. However 2^n is not of polynomial order since for every positive integer k and every positive real constant c, the inequality $2^n > Cn^k$ holds for sufficiently large n. The general rule for efficient algorithms are that they have polynomial time complexities while inefficient algorithms have exponential time complexities. So a problem is tractable if there exists an algorithm to solve it in polynomial time. It is important to note that intractable problems, do have algorithms to solve them. But the complexity of such algorithms is exponential. *In other words, the existence of an algorithm does not guarantee its feasibility as a solution.*

Calculating the complexity: To calculate the complexity of any algorithm, we must identify the key operations, within the algorithm that contribute significantly to its run time, such as arithmetic operations, memory access and function calls. We must also calculate the frequency of the operations, or the number of times it runs, based on the input size n. For example, if a loop runs n times and each iteration takes a constant time $O(1)$, the loop contributes $O(n)$ to the total time complexity. It is important to consider nested loops, recursive calls and conditional branches. For example consider the nested loop: The outer

```
for i=1 to n do
    for j=1 to n do
        some constant time-work;
    end
end
```

loop runs n times and for each iteration of the outer loop, the inner loop also runs n times. Each operation within the inner loop is $O(1)$, hence the total time complexity is $O(n^2)$. For recursive algorithms, we use recurrence relations to analyze complexity. For example if an algorithm splits a problem into two smaller sub problems each time (as in the merge sort algorithm), it might have a recurrence relation like

$$T(n) = 2T(n/2) + O(n)$$

which shows that the complexity of the merge sort algorithm is $O(nlogn)$.

It is important to add the complexity of each part of the algorithm. For example, consider the following algorithm. The outer For-loop runs from $i = 1$ to $i = n$ and hence it

```
for i =1 to n do
    if some condition is true then
        | do something;
    else
        for j=1 to n do
            for k=1 to n do
                | Do something else;
            end
        end
    end
end
```

iterates n times. Inside the outer loop, there is an "if" condition. These are the two possible branches.

- If the condition is true: The code executes a single operation (do something) with $O(1)$ complexity.

- If the condition is false: The code enters two nested loops on j and k, each running from 1 to n.

In the "Else" branch, there are nested loops on j and k. If the "If" condition is false, we enter the nested loops:

- The first "for" loop on j runs from 1 to n, so it iterates n times.

- The second "for" loop on k also runs from 1 to n, within, each iteration of j so it iterates n times, for each value of j.

Thus the nested loops on j and k together execute $n \times n = n^2$ times, each time, performing the operation "Do something else", which takes $O(1)$ time.

Overall complexity calculation:

Case 1: In the case where the "if" condition is true for each i from 1 to n, the outer loop does $O(1)$ work per iteration, resulting in $O(n) \times O(1) = O(n)$.

Case 2: If the "if" condition is false, for each i, then for each iteration of i, we enter the nested loops which take $O(n^2)$ time, per iteration of i. So the complexity is $O(n) \times O(n^2) = O(n^3)$.

Mixed case: If the "if" condition is true for some values of i and false for others, the worst case complexity still arises from the times, the nested loop runs, resulting in a complexity of $O(n^3)$.

Hence we can conclude that the complexity of this algorithm is $O(n^3)$.

Data structures We must become familiar with methods of input and data structures because they enable efficient data organization and management. A poorly thought out data structure can lead to inefficient data access and operations, excessive memory usage, poor handling of large datasets and increased complexity of the algorithm. A brief description of the various data structures commonly used, follows. The reader can access books [9] or other sources for in-depth information on data structures and algorithms.

Arrays: Fixed size, ordered collection of elements of the same type, with direct access by index

Linked lists: Sequences of nodes, where each node points to the next, allowing dynamic resizing where memory is managed by

> i **Reallocation:** to expand or shrink the structure or
>
> ii **Doubling or Halving:** where the lists double in size, when the structure is full and halves it, when usage drops below a certain threshold.

Stacks: Last-in-first-out (LIFO) structures

Queues: First-in, First-out (FIFO) structures

Trees: Hierarchical structures with nodes connected in parent-child relationships, ideal for representing sorted data(binary trees)

Graphs: set of nodes(vertices), connected by edges, representing networks, paths and connections

Hashtables: key-value storage allowing efficient data retrieval by mapping keys to values using hash function.

We would be using a few of these data structures, such as arrays, lists, stacks and queues in the pseudocode of some of our algorithms, in this book.

2.4 Havel–Hakimi algorithm

Now that we have a basic understanding of what algorithms are, let us start with a simple one that helps to check and realize the graphical nature of a degree sequence. If it is graphical, the degree sequence is used to realize the graph. The algorithm is known as

Havel–Hakimi algorithm and it was independently discovered by two mathematicians *V. Havel* and *H. L. Hakimi* in 1955 and 1962 respectively. We will provide the description and steps of the Havel–Hakimi algorithm that helps one to check if a degree sequence is graphical or not and in the process, provide an output of vertices and edges that help to graphically realize the degree sequence.

The algorithm works by following these steps:

Initialization: We initialize an empty graph G as well as the degree sequence array $D[i]$.(If it is arranged as a non-increasing sequence, it is easier). Each number in $D[i]$ is assumed to be the degree $d(v)$ of any vertex in the graph realization G.

Algorithm 1: Havel–Hakimi Algorithm

Data: A degree sequence in an array:$D[i]$
Result: $V(G), E(G)$ of the graphical realization of $D[i]$
$V(G) \leftarrow \emptyset$;
$t \leftarrow 0$;
$x \leftarrow 0$;
$v_i \leftarrow v_0$;   /* The vertex label is initialized as v_0 */
$E(G) \leftarrow \emptyset$;
while $D[i] \neq 0$ *for all i* **do**
    $D[i]$ is searched for the largest value
    $x \leftarrow$ largest value of $D[i]$
    $t \leftarrow$ position of x
    $D[t] \leftarrow 0$
    $v_i \leftarrow v_t$
    ;   /* The vertex at position t with degree x is assigned as v_t */
    if v_t *is not in* $V(G)$ **then**
      | $V(G) \leftarrow V(G) \cup \{v_t\}$
    end
    Decrease the largest x degrees in $D[i]$ by 1
    if $D[i]$ *contains all numbers* ≥ 0 **then**
      | $V(G) \leftarrow V(G) \cup \{$all vertices whose degrees were decremented by 1 $\}$
      | $E(G) \leftarrow E(G) \cup \{$all edges between v_t and all vertices whose degrees were decremented by 1 $\}$
    else
      | $D[i]$ is NOT GRAPHICAL
    end
end

We repeat the following steps until we arrive at an array $D[i]$ where every entry is zero.

Step 1: We set t to the position of the largest degree x in $D[i]$.

Step 2: We construct a graph G with vertex set $\{v_t\}$, in such a way that we first set $d(v_t)$ to zero in $D[i]$ and then add the corresponding vertex v_t with degree x to G.

Step 3: Decrease the next x largest degrees in $D[i]$ by 1. (We decrease the degrees of the next x largest degrees in the degree sequence by 1.) If one of the decremented degrees is negative, we terminate the algorithm and state that the degree sequence is not graphical. Otherwise, the vertices corresponding to the next x largest degrees are added to G and connected to v_t.

EXAMPLE 2.4.1 Determine by use of the Havel-Hakimi algorithm whether the sequence $(6,6,5,4,3,3,1)$ is graphical or not.

Let $D[i] : (6,6,5,4,3,3,1)$ be the degree sequence. Let us now initialize the partition $D[i]$ as well as the integers t and x. A graph G whose vertex set is empty is also initialized.

Initialization: $D[i] : (6,6,5,4,3,3,1)$, $t = 0$, $x = 0$, $V(G) = \emptyset$ and $E(G) = \emptyset$. Since $D[i]$ is not all zero, we are free to determine the largest number in $D[i]$, which is assigned to x.

Iteration 1: Since 6 is the largest integer, x is assigned 6 and since its position is 1 in $D[i]$, 1 is assigned to t. $D[1]$ becomes 0 and because we do not have v_1 in $V(G)$, the vertex v_1 is added to G. At this point, we have removed a vertex with the maximum degree 6 which results in a decrease of 1 in the degrees of the next largest 6 vertices in $D[i]$ and leads us to a new partition $D[i] = (0, 5, 4, 3, 2, 2, 0)$. Since we do not have any number < 0 in this sequence, we now add all the vertices whose degrees were decremented by 1 to $V(G)$. This results in the addition of vertices v_2, v_3, v_4, v_5, v_6 and v_7 to $V(G)$, and also edges $\{v_1v_2, v_1v_3, v_1v_4, v_1v_5, v_1v_6v_1v_7\}$. Since the algorithm continues until all the terms of $D[i]$ are zero, we move to the next iteration.

Iteration 2: Now the maximum degree is 5 and it is in the second position, which leads to x being assigned the value 5 and $D[2]$ being marked 0. Since v_2 is already in G, we move to the next step of reducing the degree of the next five vertices by 1. But when we decrease the next 5 largest terms in $D[i]$ by 1, it results in $D[i] = (0,0,3,2,1,1,-1)$. Now $D[i]$ contains a number < 0 and therefore the algorithm returns "NOT GRAPHICAL" and terminates.

We shall see one more demonstration of the algorithm through an example of a graphical degree sequence.

EXAMPLE 2.4.2 Determine by using Havel–Hakimi algorithm whether $(4,4,4,2,2,2)$ is graphical.

Let $D[i]$: $(4,4,4,2,2,2)$. As per the usual process, we initialize t, x and an empty graph G. Although the Havel–Hakimi algorithm does not explicitly call for a degree sequence which has been arranged in non-increasing order, it makes it significantly easier if it is so.

Initialization $D[i]$: $(4,4,4,2,2,2)$, $t=0$, $x=0$, $V(G) = \emptyset$ and $E(G) = \emptyset$.

Iteration 1: $D[i]$: $(0,3,3,1,1,2)$, $t=1$ and $x=4$. Also v_1 is added to $V(G)$. Since v_2, v_3, v_4 and v_5 are decremented by 1, $V(G) = \{v_1\} \cup \{v_2, v_3, v_4, v_5\}$. Edges are drawn from v_1 to v_2, v_3, v_4, v_5, refer Figure 2.10.

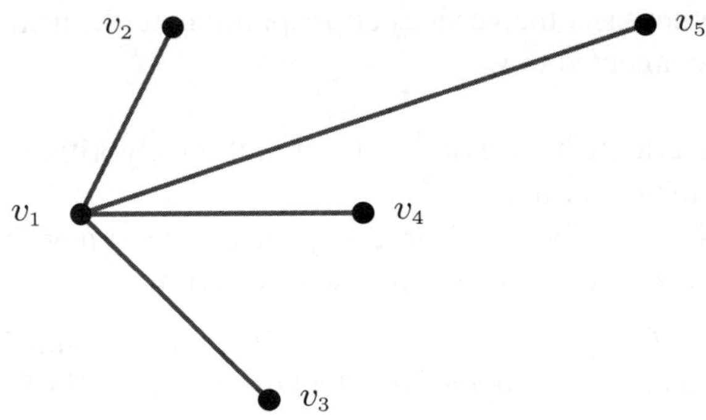

Figure 2.10 After 1st iteration of degree sequence $(4, 4, 4, 2, 2, 2)$

Iteration 2: $D[i]$: $(0,0,2,0,1,1)$, $t=2$ and $x=3$. Also $V(G) = \{v_1, v_2, v_3, v_4, v_5\} \cup \{v_6\}$. Note that v_2 is already in $V(G)$. Edges are drawn from v_2 to v_3, v_4 and v_6 because these are the vertices whose degrees are decremented by 1, refer Figure 2.11.

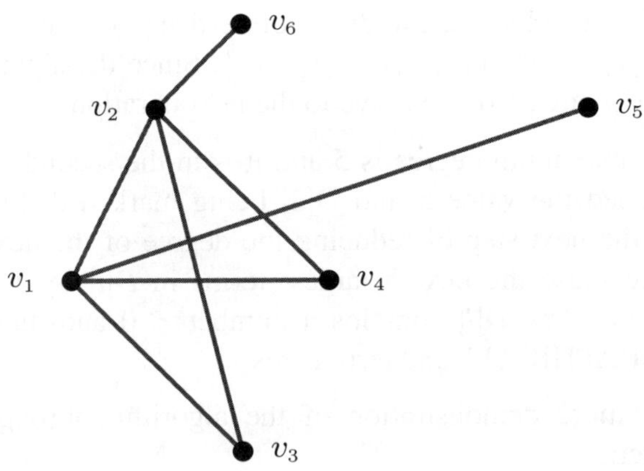

Figure 2.11 After second iteration of degree sequence $(4, 4, 4, 2, 2, 2)$

Iteration 3: $D[i] : (0,0,0,0,0,0)$, $t = 3$ and $x = 2$. Also $V(G) = \{v_1, v_2, v_3, v_4, v_5, v_6\}$. Since v_3 is already in $V(G)$, edges are drawn from v_3 to v_5 and v_6. Now $D[i]$ is all zero and the algorithm terminates. It is clear that $D[i]$ is graphical and the graph given in Figure 2.12 is proof that the degree sequence is graphical.

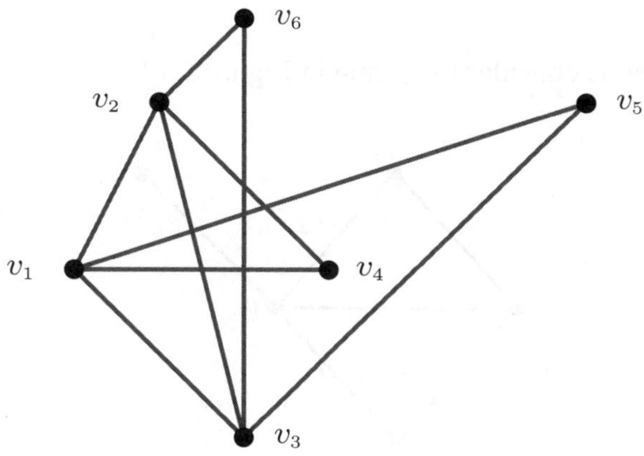

Figure 2.12   After third iteration of degree sequence $(4,4,4,2,2,2)$

The complexity of the Havel–Hakimi algorithm is $O(n^2)$. This is due to the fact that the 'While' loop has a complexity of $O(n)$ and since it can run up to n times, the overall complexity of the algorithm becomes $O(n) \times O(n) = O(n^2)$.

2.5 Matrix representation of a graph

The matrix representation of a graph helps to lend itself to an algebraic structure. We have a distinct specialization of graph theory called the Algebraic graph theory, where algebraic problems are explored and solved using graphs, and vice-versa. Hence by learning to represent a graph in terms of a vertex's adjacency to another vertex or the number of edges that emanate from a vertex is a useful tool to characterize a graph.

DEFINITION 2.5.1 Let G be a graph of order n and let $V(G) = \{v_1, v_2, \ldots, v_n\}$. The *adjacency matrix* is a $n \times n$ matrix, that assigns a value of 1 to a_{ij} if and only if v_i and v_j are adjacent and, 0 otherwise. That is the adjacency matrix $A = [a_{ij}]$ where

$$a_{ij} = \begin{cases} 1 & \text{if } v_i v_j \in E(G), \\ 0 & \text{otherwise.} \end{cases}$$

DEFINITION 2.5.2 Let G be a graph with $V(G) = \{v_1, v_2, \ldots, v_n\}$ and $E(G) = \{e_1, e_2, \ldots, e_m\}$. Then the *incidence matrix* of G is a $n \times m$ matrix $B = [b_{ij}]$, that assigns a value of 1 to b_{ij} if and only if v_i is incident on e_j and, 0 otherwise. That is

$$b_{ij} = \begin{cases} 1 & \text{if } e_j \text{ emanates from } v_i, \\ 0 & \text{otherwise.} \end{cases}$$

EXAMPLE 2.5.3 Let us consider the graph in Figure 2.13.

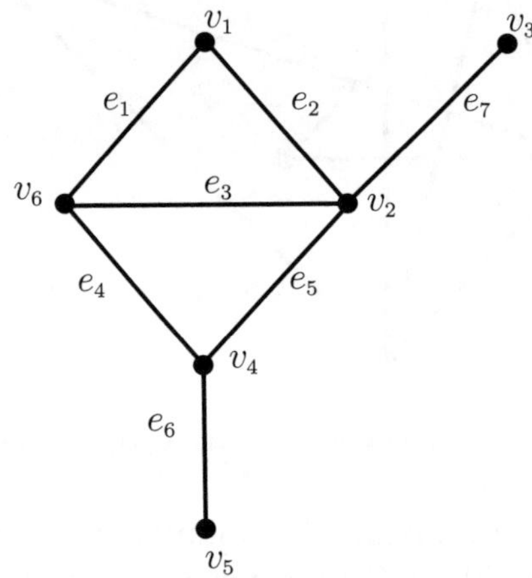

Figure 2.13   Graph G

The adjacency matrix of the graph G is $A = [a_{ij}]$ which is given by

$$
\mathbf{A} = \begin{array}{c} \\ v_1 \\ v_2 \\ v_3 \\ v_4 \\ v_5 \\ v_6 \end{array}
\begin{array}{c} \begin{array}{cccccc} v_1 & v_2 & v_3 & v_4 & v_5 & v_6 \end{array} \\
\left(\begin{array}{cccccc}
0 & 1 & 0 & 0 & 0 & 1 \\
1 & 0 & 1 & 1 & 0 & 1 \\
0 & 1 & 0 & 0 & 0 & 0 \\
0 & 1 & 0 & 0 & 1 & 1 \\
0 & 0 & 0 & 1 & 0 & 0 \\
1 & 1 & 0 & 1 & 0 & 0
\end{array} \right) \end{array}
$$

The incidence matrix of graph G, $B = [b_{ij}]$ is

$$
B = \begin{array}{c}
\\
v_1 \\ v_2 \\ v_3 \\ v_4 \\ v_5 \\ v_6
\end{array}
\begin{array}{cccccccc}
e_1 & e_2 & e_3 & e_4 & e_5 & e_6 & e_7 \\
\begin{pmatrix}
1 & 1 & 0 & 0 & 0 & 0 & 0 \\
0 & 1 & 1 & 0 & 1 & 0 & 1 \\
0 & 0 & 0 & 0 & 0 & 0 & 1 \\
0 & 0 & 0 & 1 & 1 & 1 & 0 \\
0 & 0 & 0 & 0 & 0 & 1 & 0 \\
1 & 0 & 1 & 1 & 0 & 0 & 0
\end{pmatrix}
\end{array}
$$

The matrices are dependent on how the vertices and the edges of a graph G are labeled. One can note that the adjacency matrix of a simple matrix always has every entry on its main diagonal as 0.

2.6 Isomorphism

We have heard of the term isomorphism in Algebra, where it means that two algebraic structures are similar to each other. An identical concept can be found in graphs too.

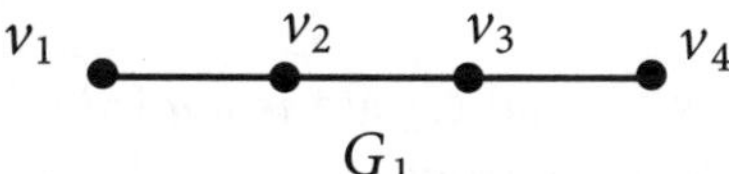

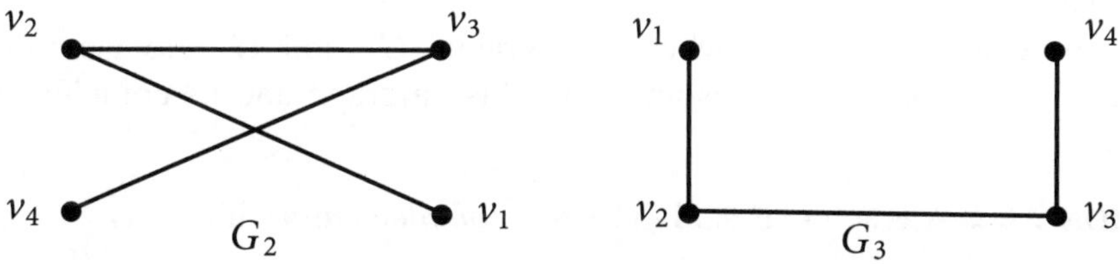

Figure 2.14   Similar graphs of order 4 and size 3

The graphs G_1, G_2 and G_3 in Figure 2.14 have order 4 and size 3. They may all look different but surprisingly they are the same. The adjacency matrix is useful to establish that all these graphs are the same. All of the graphs G_1, G_2, G_3 have the adjacency matrix

$$
\begin{pmatrix}
0 & 1 & 0 & 0 \\
1 & 0 & 1 & 0 \\
0 & 1 & 0 & 1 \\
0 & 0 & 1 & 0
\end{pmatrix}.
$$

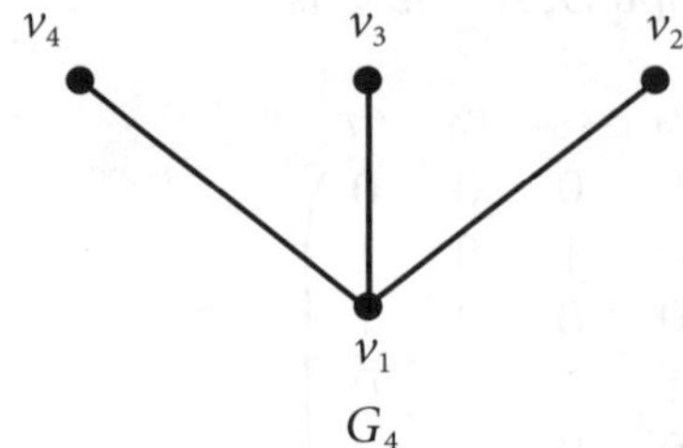

Figure 2.15   A graph of order 4 and size 3

It is quite possible for a graph of order 4 and size 3 to turn out differently. For instance, consider the graph G_4 in Figure 2.15. The adjacency matrix of G_4 is $\begin{pmatrix} 0 & 1 & 1 & 1 \\ 1 & 0 & 0 & 0 \\ 1 & 0 & 0 & 0 \\ 1 & 0 & 0 & 0 \end{pmatrix}$.

Clearly G_4 has a different matrix. So G_4 is non-isomorphic to G_1 or G_2 or G_3. Thus it is obvious that merely having the same order and size is not enough to result in isomorphic graphs. It is easy to extrapolate that isomorphic graphs have one-to-one correspondence in their vertex sets.

DEFINITION 2.6.1 Two graphs G_1 and G_2 are *isomorphic*, denoted by $G_1 \cong G_2$, if there exists a one-to-one correspondence φ from $V(G_1)$ to $V(G_2)$ such that $uv \in E(G_1)$ if and only if $\varphi(u)\varphi(v) \in E(G_2)$. The map φ is said to be an *isomorphism* from G_1 to G_2.

The graphs in Figure 2.16 show all non-isomorphic simple connected graphs on four vertices or less.

It is important to note that when two graphs G_1 and G_2 are isomorphic, their complements $\overline{G_1}$ and $\overline{G_2}$ are also isomorphic. Also a graph and its complement may be isomorphic.

DEFINITION 2.6.2 A graph G is said to be *self-complementary* if $G \cong \overline{G}$.

Example for a self-complementary graph is given in Figure 2.17.

REMARK 2.6.3 Let G be a self-complementary graph. We know that for any graph $G \cup \overline{G} = K_n$. Since G and $\overline{G}$ have the same size for the self-complementary graph G, we have $2|E(G)| = \frac{n(n-1)}{2}$. Therefore $|E(G)| = \frac{n(n-1)}{4}$. For $\frac{n(n-1)}{4}$ to be an integer, $4|n$ or $4|(n-1)$. So $n \equiv 0(mod\ 4)$ or $n \equiv 1(mod\ 4)$.

In addition to order and size, we shall see in the following theorem, that two isomorphic graphs have the same degree.

THEOREM 2.6.4 *Two isomorphic graphs G and H have the same degree sequence.*

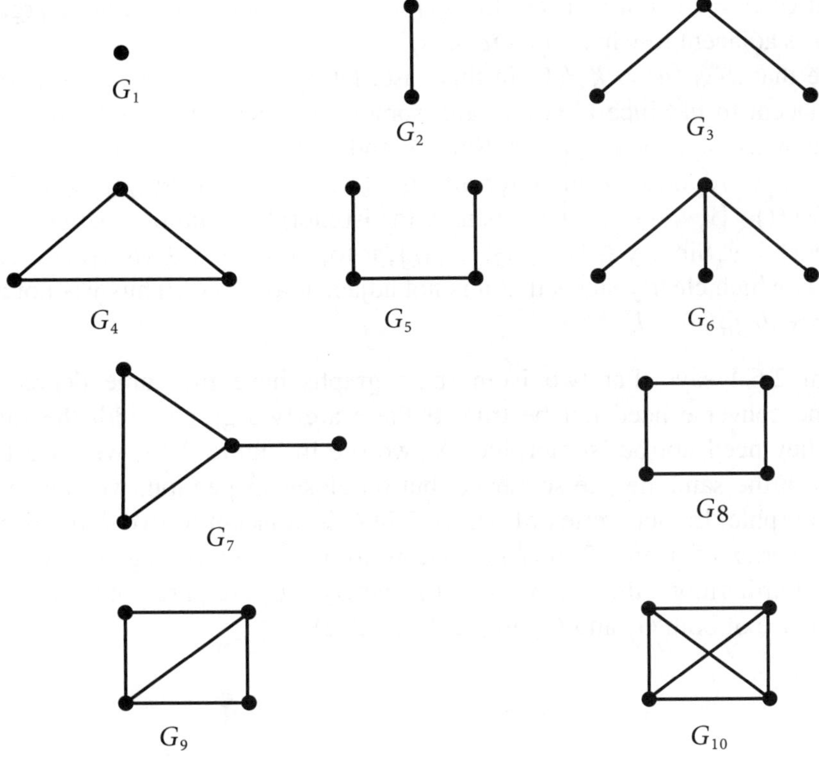

Figure 2.16   All simple connected graphs of order less than 5

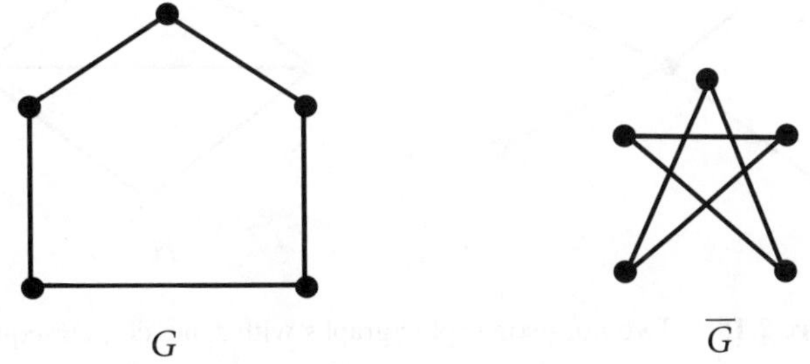

Figure 2.17   A self-complementary graph

Proof. Let G and H are isomorphic graphs. Then there exists a one-one relation $\varphi : V(G) \to V(H)$. Let $u \in G$. By nature of isomorphism, there exists a vertex $v \in V(H)$ such that $\varphi(u) = v$. We will prove that $deg_G(u) = deg_H(v)$.

Suppose $deg_G(u) = 0$. Let $v \neq y \in V(H)$. By one-one relation φ, there exists a vertex x in G such that $\varphi(x) = y$. Since u has degree 0 in G, u is not adjacent to x in G, which implies that $\varphi(u) = v$ is not adjacent to $\varphi(x) = y$ in H. Therefore it can be argued that no vertex of H is adjacent to v in H and $deg_H(v) = 0$.

Suppose that $deg_G(u) = k \neq 0$. In this case, let $N_G(u) = \{x_1, x_2, \ldots, x_k\}$, the set of k vertices adjacent to u. Since G and H are isomorphic, there exists a set of vertices y_i in H such that $\varphi(x_i) = y_i$ for $1 \leq i \leq k$. Since u and x_i are adjacent in G for $1 \leq i \leq k$, the vertices v and y_i are adjacent in H for all $i \in \{1, \ldots, k\}$. Thus $deg_H(v) \geq k$. Now pick a vertex $y \in V(H) \setminus \{v, y_1, y_2, \ldots, y_k\}$. Then by the isomorphic relation, there exists an $x \in G$ such that $\varphi(x) = y$. Since $y \notin \{v, y_1, y_2, \ldots, y_k\}$, it follows that $x \notin \{u, x_1, x_2, \ldots, x_k\}$. In G, $deg_G(u) = k$, which clearly states that u is not adjacent to x in G. Thus y is not adjacent to v in H. Hence $deg_H(v) = k$. $\qquad\square$

Theorem 2.6.4 says that two isomorphic graphs have the same degree sequence. However the converse need not be true. If there are two graphs with the same degree sequence, they need not be isomorphic. As we see in Figure 2.18, we have two graphs G and H with the same degree sequence, but on closer inspection, we can see that they are not isomorphic. In fact, vertex of degree 2 in G is adjacent to two vertices of degree 3 whereas the vertex of degree 2 in H is adjacent to one vertex of degree 3 and one vertex of degree 4. Furthermore, the graphs $G_1 = C_6$ and $G_2 = C_3 \cup C_3$ are not isomorphic but the degree sequence of both G_1 and G_2 are $(2,2,2,2,2,2)$.

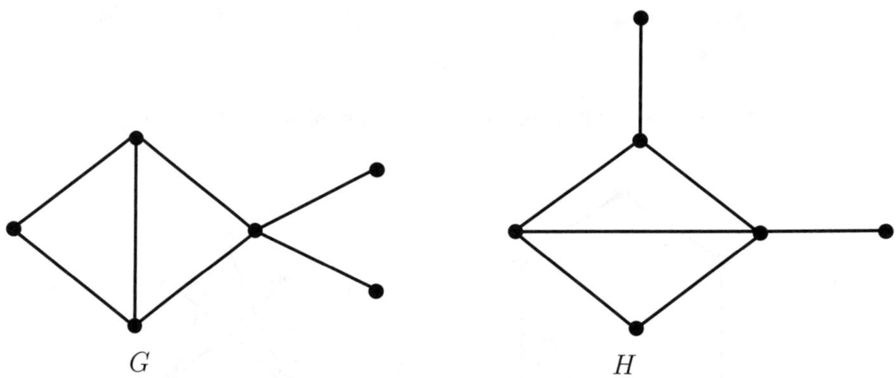

Figure 2.18   Two non-isomorphic graphs with same degree sequence

REMARK 2.6.5 If G and H are isomorphic graphs, then

(i). G is bipartite if and only if H is bipartite.

(ii). G is connected if and only if H is connected.

Summary

In this chapter, we discuss the answers to the questions of how connected a graph is, how we can represent a graph in alternate ways and whether it is possible to create a graph out of a suitable mix of numbers. Additionally, we have provided a condensed tutorial on algorithms and complexity, so that the reader can better understand the mechanism behind the algorithms. The technique of representing a graph as a matrix is demonstrated through examples. A property of significance, the isomorphism between graphs is discussed and the methods to identify isomorphism are thoroughly explained. The first algorithm in this book, the Havel- Hakimi algorithm has been discussed with a detailed illustration. Some of the basics of graph theory upon which we shall build other concepts, have been explored and explained in this chapter.

2.7 Exercises

Section 2.1: Degree

1. Show that a graph with $n \geq 2$ possesses two vertices of the same degree.

2. Consider the vertex v of a graph G. Prove that if $G - v$ is 3 regular, then G has odd order.

3. Draw the 4-regular graph of order 7. What is its size?

4. Suppose a graph G has order 14 and size 27. Let the degree of each vertex of G be 3, 4, or 5. If there are six vertices of degree 4, how many vertices of G have degree 3 and how many have degree 5?

5. Show that if a graph of order $3n$ $(n \geq 1)$ has n vertices with each of the degrees $n-1, n$, and $n+1$, then n is even.

6. Prove that if G is a disconnected graph containing exactly two odd vertices, then these odd vertices should lie in the same component of G.

7. Provide an example of a graph G of order 6 and size 10 with $\delta(G) = 3$ and $\Delta(G) = 4$.

8. Draw the following simple graphs, or provide an explanation if they do not exist.

    (i). A bipartite graph of order 5 and size 7.

    (ii). A bipartite graph of order 8 and size 10.

(iii). A graph with an isolated vertex and a vertex which is adjacent to all the other vertices.

(iv). A graph with degree of each vertex is 3 and order 5.

Section 2.2: Degree sequences

9. Show that a graph (not necessarily simple) with degree sequence $d_1, d_2, \ldots, d_n$ exists if $\sum_{i=1}^{n} d_i$ is even.

10. Let $d_1 \geq d_2 \geq \ldots \geq d_n$ and $\sum_{i=1}^{n} d_i$ be even. Show that there exists a multigraph (no loops) with degree sequence $d_1, d_2 \ldots d_n$ if and only if $d_1 \leq \sum_{i=2}^{n} d_i$.

11. What number of (non-isomorphic) graphs have the degree sequence s : $(6, 6, 6, 6, 6, 6, 6, 6, 6)$?.

Section 2.3: Algorithms and complexity

12. Write an algorithm to check if a number is prime.

13. Write the pseudocode for listing first n natural numbers.

14. Calculate the time complexity of the following algorithm:

```
for i=1 to n do
    for j=1 to n do
        for k=1 to n do
            print (i, j, k);
        end
    end
end
```

15. Calculate the time complexity of the following algorithm

```
i ← n
while i > 0 do
    print i
    i ← i − 2
end
```

Section 2.4: Havel–Hakimi Algorithm

16. Show that $(0,1,2,3,4)$ is not graphical.

17. Is $(4,4,3,2,2,1,1)$ graphical? If not, give reason. If yes, draw a simple graph with this degree sequence.

18. Is $(4,4,4,2,2)$ graphical? If not, explain why. If yes, find a multigraph (no loops) with this degree sequence.

19. Show that for every integer x with $0 \le x \le 5$, the sequence $(x,1,2,3,5,5)$ is not graphical.

20. Let the sequence $(x,7,7,5,5,4,3,2)$ be graphical. Then list out the possible values of x (with $0 \le x \le 7$).

Section 2.5: Matrix representation of a graph

21. Create an adjacency matrix and incidence matrix of the following:

    - Petersen graph
    - C_5
    - the complete graph K_5.
    - the path graph P_7.

22. Label the figures in Figure 2.19 and determine their adjacency and incidence matrix.[Sidenote: Observe if their adjacency matrices can help to prove or disprove their isomorphism]

Section 2.6: Isomorphism

23. Suppose G is a self-complementary graph with order $n = 4k$, where $k \ge 1$. Let $U = \{v \in V(G) : deg(v) \le \frac{n}{2}\}$ and $W = \{v \in V(G) : deg(v) \ge \frac{n}{2}\}$. Show that if $|U| = |W|$, then G contains no vertex v with $deg(v) = \frac{n}{2}$.

24. Construct the adjacency matrix and incidence matrix for the graph in Figure 1.12.

25. Demonstrate that graphs with the same order and size don't always have to be isomorphic.

26. Show that isomorphism is an equivalence relation on the set of all graphs.

27. Examine whether the graphs G_1 and G_2 given in Figure 2.19 are isomorphic. Justify your answer.

28. Draw all simple graphs having 4 vertices that are non-isomorphic. (Hint: There are 11).

29. Provide two non-isomorphic self complementary graphs of order 5.

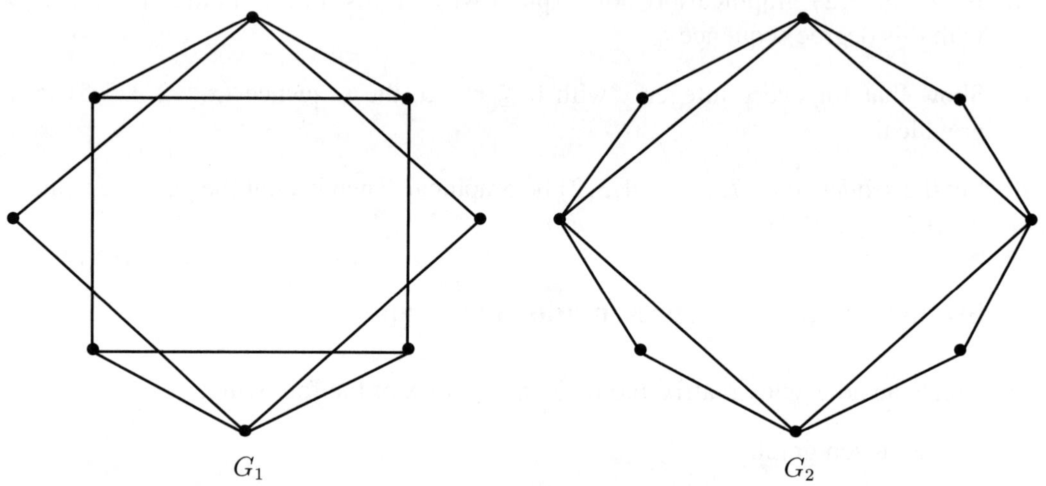

Figure 2.19   Graphs G_1 and G_2

3

Trees

Trees form an important class of graphs, in graph theory. Trees hold a special place in graph theory, due to their simplicity, versatility and widespread applications. From representing hierarchical data structures like family trees and organizational charts, their unique properties, such as having a single path between any two vertices, make them indispensable in solving complex problems with efficiency and clarity. In this chapter, we will explore different types of trees such a spanning trees, rooted trees and binary trees. We will also study different algorithms to construct minimum spanning trees.

DEFINITION 3.0.1 A graph without cycles is said to be an *acyclic graph*. Acyclic graphs are also called as *forests*. A *tree* is an acyclic connected graph.

We usually denote the tree by T instead of G.

In Figure 3.1, we may observe all trees of order less than 6. The isolated vertex is the smallest tree, while K_2 is a tree on two vertices. Based on the definition, every component of a forest is a tree. It is sometimes possible for a forest to have a single tree. In Figure 3.2, we see different types of trees of order 6. Clearly every path of length greater than 2 is a tree, as is evident from the Figure 3.2. The star graph $K_{1,n}$ is a tree.

Since a tree with no cycles has no odd cycles, it follows from Theorem 1.5.6 that all trees and forests are bipartite.

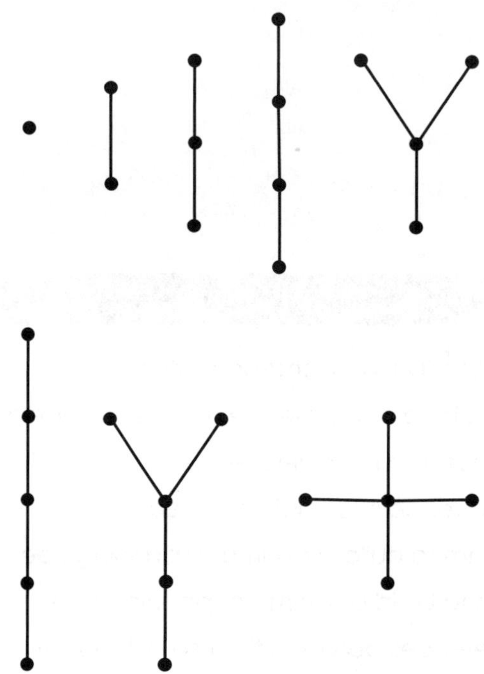

Figure 3.1   Trees of order less than 6

It is sometimes convenient to draw a tree in the form of a hierarchical structure.

DEFINITION 3.0.2 A vertex u in a tree is called as the *root* of a tree if it can be drawn as an apex vertex at the top while all the other vertices are drawn at different levels depending on the length of the path between u and the vertex. Such a tree is called a *rooted tree*.

When we have a rooted tree, we usually draw the root r at the apex, signifying a level 0. Vertices which are adjacent to r are placed one level below level 0 and are said to belong to level 1. We can continue this process, with vertices adjacent to those in level 1 and proceed until we cannot proceed further. In Figure 3.3, r is at level 0, a,b are at level 1, c,d at level 2, e at level 3 while f,g are at level 4. In general, every vertex at level $i > 0$ is adjacent to exactly one vertex in level $i-1$. We shall define some terminologies related to rooted trees.

Also a vertex x in a rooted tree with root r is at level i if and only if the $r-x$ path has length i. The largest integer h for which there is a vertex at level h in a rooted tree is called

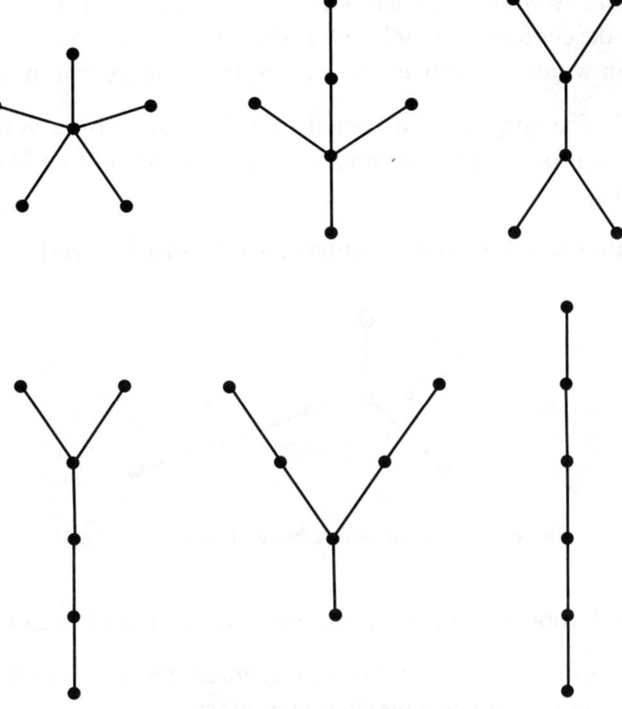

Figure 3.2   Trees of order 6

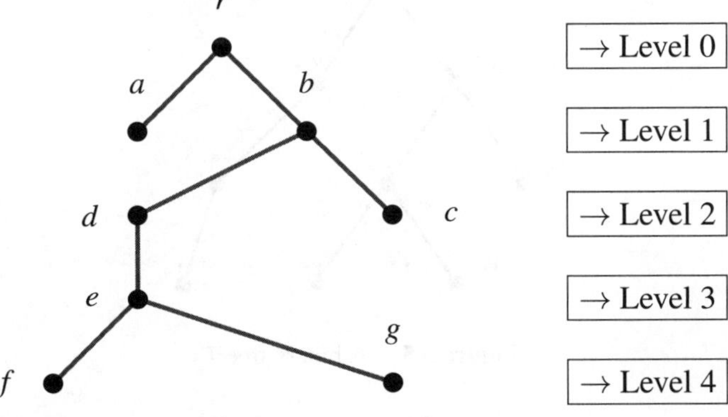

Figure 3.3   A rooted tree T

the *height of the rooted tree*. The height of the rooted tree in Figure 3.3 is 4. A vertex v of the rooted tree T, which is adjacent to a vertex x of T and lying one level below x is said to be the *child* of the vertex x. In this case, x is the *parent* of v. A vertex v is a *descendant* of x, while x is an *ancestor* of v, if there exists a $x - v$ path in T below x. For example in Figure 3.3, f is the descendant of b, while b is the ancestor of f and g. It can be said that a rooted tree is one, in which all vertices have a single parent except the root.

DEFINITION 3.0.3 The subtree of a rooted tree T induced by a vertex v and all of its descendants is also a rooted tree with root v. This induced subtree is called the *maximal subtree* of T rooted at v.

We may see in Figure 3.4 a maximal subtree with its root as vertex d.

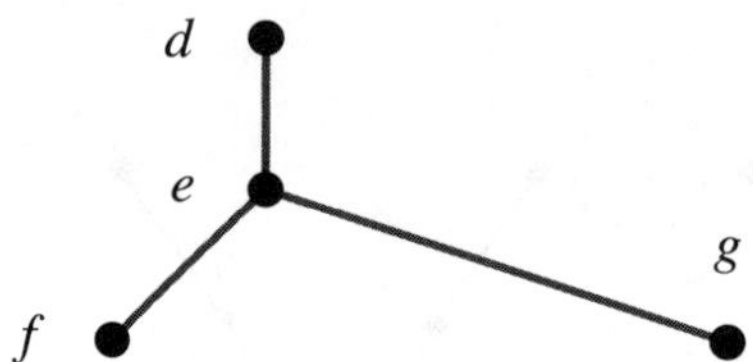

Figure 3.4 A maximal subtree of tree T rooted at d

DEFINITION 3.0.4 A rooted tree is *m-ary* if every vertex has at most m children.

DEFINITION 3.0.5 A *binary tree* is defined as a rooted tree in which the root has degree 2 and each of the remaining vertices have degree at most 2.

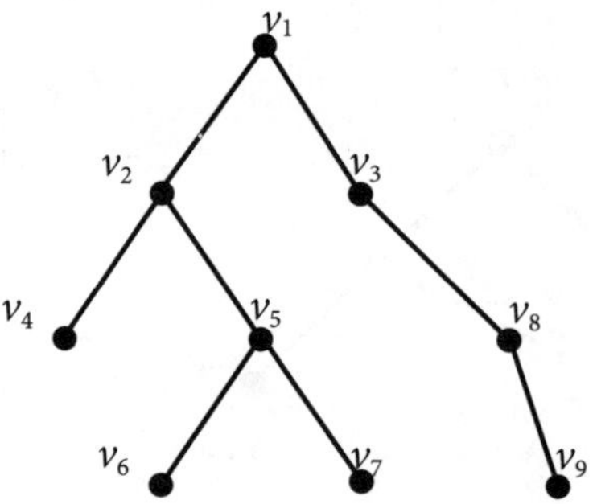

Figure 3.5 A binary tree T

Figure 3.5 shows an example of a binary tree T. Binary trees are primarily used in searching and sorting algorithms and also in decision analysis.

A vertex of degree 1 in a tree is called a *pendant vertex* or *leaf* or *end vertex*.

Now, we will discuss some basic properties of trees.

THEOREM 3.0.6 *A graph is a tree if and only if every two vertices of G are connected by a unique path.*

Proof. Let G be a tree. Then by definition, G is a connected graph. Since G is connected, every two vertices of G are connected by a path. We claim that such a path is unique. Suppose to the contrary that G has a pair of vertices $u, v \in V(G)$ which are connected by two distinct paths. Then a cycle is produced by these two distinct paths, thereby contradicting the assumption that G is a tree. Hence every two vertices of G are connected by distinct paths.

Conversely, let us assume that every two distinct vertices of G are connected by a unique path. It follows that G is connected. It remains to be proved that G is acyclic. Assume to the contrary, that G contains a cycle C. In a cycle, any two vertices are connected by two distinct paths, thereby producing a contradiction to our assumption. Thus G is acyclic. By definition, G is a tree. $\qquad\square$

The following theorem helps to create a smaller tree from any tree T, by deletion of leaves or end vertices.

THEOREM 3.0.7 *Every tree with at least two vertices has at least two end vertices.*

Proof. Let T be a tree of order $n \geq 2$. Let P be a path of greatest length in T. Suppose P is a $u - v$ path $u = v_0, v_1, v_2, \ldots, v_k = v$ where $k \geq 1$. We claim that u and v are end vertices of T. Suppose either u or v is adjacent to some $x \in V(G)$ where $x \neq v_i$ for all $i \in \{0, 1, \ldots, k\}$, say u is adjacent to x. Then the length of the $x - v$ path in G is greater than the length of P, a contradiction to the fact that P is a longest path in G. If u or v is adjacent to v_j for some $j \in \{0, 1, \ldots, k\}$, then the sector $u - v_j$ in P together with the edge $v_j u$ form a cycle in G, a contradiction to G is tree. Therefore $deg(u) = deg(v) = 1$, which proves the theorem. $\qquad\square$

Note that, by Theorem 3.0.7, $\delta(T) = 1$ for any tree T. In other words, if G is a graph with $\delta(G) \geq 2$, then G contains a cycle. Another major consequence of Theorem 3.0.7 is that if T is a tree of order $k \geq 2$, then for each end vertex v of T, the removal of the vertex results in another tree T' of order $k - 1$.

3.1   Characterization of trees and spanning trees

It follows from the definitions of tree and forest that a subgraph of a forest is a tree and a connected subgraph of a tree T is a *subtree* of T.

DEFINITION 3.1.1 A tree is said to be a *spanning tree* of a connected graph G, if T is a subgraph of G and T contains all vertices of G.

The spanning tree obtained from the graph G can be seen in Figure 3.6.

THEOREM 3.1.2 *Let T be a tree of order k. If G is a graph with $\delta(G) \geq k - 1$, then T is isomorphic to some subgraph of G.*

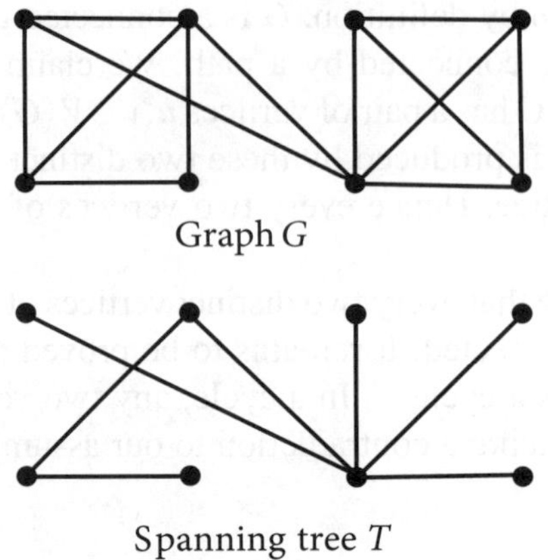

Graph G

Spanning tree T

Figure 3.6   Spanning tree T obtained from G

Proof. We will apply induction to arrive at the result. If $k = 1$, then the result is trivially true, since every graph G contains a vertex. If $k = 2$, then T is merely an edge. In this case, since $\delta(G) \geq 1$, G is a graph without isolated vertices and so contains an edge. Let us assume that every tree of order $k - 1$ is isomorphic to some subgraph of G when $\delta(G) \geq k - 2$ for $k \geq 3$. Let T be a tree of order k and let G be a graph with $\delta(G) \geq k - 1$. We will prove that T is isomorphic to a subgraph of G. Choose an end vertex x in T; existence is granted by Theorem 3.0.7. Let x be adjacent to y in T. Now the removal of x results in a tree of order $k - 1$. Since $\delta(G) \geq k - 1 \geq k - 2$, the induction hypothesis allows for $T - x$ to be isomorphic to a subgraph T' of G. Due to the isomorphism, there exists a vertex $y' \in T'$ corresponding to $y \in T$. It is obvious that $deg_G(y') \geq k - 1$ and the number of vertices in T' can only be $k - 1$. Hence the vertex y' is adjacent to a vertex x' of G that is not in T'. Since all other vertices of $T - x$ are accounted for, T is isomorphic to the subgraph $T' \cup \{x'y'\}$ of G. $\square$

The advantage of the above theorem lies in the construction of spanning trees from any given graph.

THEOREM 3.1.3 *Every connected graph contains a spanning tree.*

Proof. Let G be a connected graph. Let $\mathcal{T}$ be the collection of all connected spanning subgraphs of G. Since $G \in \mathcal{T}$, we have $\mathcal{T} \neq \emptyset$. Choose $T \in \mathcal{T}$ have the fewest number of edges among the members of $\mathcal{T}$. Suppose T is not a tree. Then T would contain a cycle C of G. The deletion of any edge e of C would give a subgraph of G. Since $V(T) = V(T - e)$, we have $T - e \in \mathcal{T}$, a contradiction to T having the fewest number of edges in $\mathcal{T}$. Thus T is a spanning tree of G. $\square$

The following result gives us a necessary and sufficient condition for a graph to be a tree.

THEOREM 3.1.4 Let G be a graph of order n. Then G is a tree if and only if the size of G is $n-1$.

Proof. The first part of the proof is due to induction on n. Let us consider the graph K_1, which has one vertex and no edges. Hence the result is true for $n=1$. Assume that the result is true for trees of order k. Now let T be a tree of order $k+1$, where $k \geq 1$. Then by Theorem 3.0.7, T has at least two end vertices. The removal of one of the end vertices in T, still results in a tree T' with k vertices. By induction hypothesis, tree T' has $k-1$ edges. Clearly $|E(T)| = |E(T')| + 1$. This leads us to the conclusion that T has $(k-1)+1$ edges, as desired.

Conversely, assume that the size of G is $n-1$. By Theorem 3.1.3, there exists a spanning tree T of G. Therefore the size of T is $n-1$. Hence $G \cong T$ and so G is a tree. $\square$

COROLLARY 3.1.5 *Every forest of order n with k components has size $n-k$.*

Proof. Let G be a forest of order n and size m. Let $G_1, G_2, \ldots, G_k$ be the k components of the forest. Let the order and size of each G_i be n_i and m_i respectively, for $i = 1, 2, \ldots, k$. Hence we can assign the values of n and m as $n = \sum_{i=1}^{k} n_i$ and $m = \sum_{i=1}^{k} m_i$. Since each component of the forest is a tree, by Theorem 3.1.4, we have $m_i = n_i - 1$ for all $1 \leq i \leq k$. It follows that $m = \sum_{i=1}^{k} m_i = \sum_{i=1}^{k} n_i - 1 = n - k$. $\square$

Note that if we remove an edge from a graph which is a tree, then the graph will become disconnected. Therefore a minimal connected subgraph of a graph is a tree. Also a minimal connected spanning subgraph of a graph is a spanning tree.

3.2 Cayley's formula

Cayley's formula helps to determine the number of spanning trees that can be constructed using n vertices. In other words, it helps to calculate $\tau(K_n)$, where K_n is the complete graph on n vertices and $\tau(G)$ gives the number of spanning trees in a graph G. Cayley's formula states that $\tau(K_n) = n^{n-2}$. If we are discussing K_3, then according to Cayley's formula, the number of spanning trees are 3 in number. As we see in Figure 3.7, the three graphs are isomorphic. But if we notice the labels, we can see that the three spanning trees are different. The number of spanning trees that we determine using Cayley's formula are actually labeled trees. So when $N = \{1, 2, \ldots, n\}$, the number of labeled spanning trees using symbols in N are n^{n-2}. In other words, Cayley's formula does not count the non-isomorphic spanning trees of K_n. However it does count the number of distinct labeled spanning trees of K_n.

There are many proofs for Cayley's formula. Some notable ones are Kirchhoff's matrix tree theorem, which uses the determinant of a matrix; the bijective proof by André Joyal; and a double counting proof technique due to Jim Pitman. But we will concentrate on the

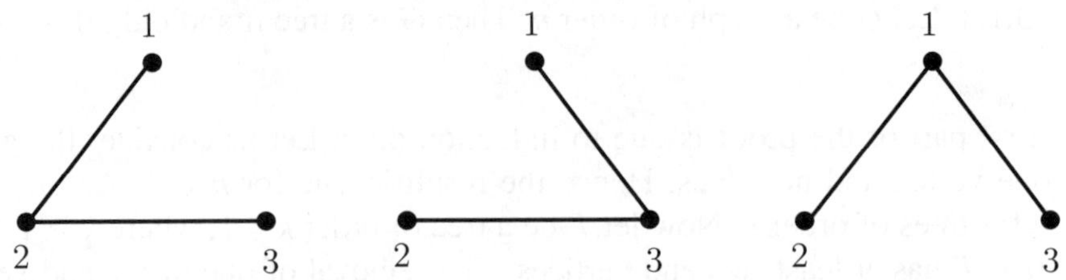

Figure 3.7   Spanning trees of K_3

bijective proof given by *Heinz Prüfer*, which involves determining a 1-1 correspondence between the labeled spanning trees and a *Prüfer sequence*.

DEFINITION 3.2.1 A *Prüfer sequence* or *Prüfer code* is a sequence of $n - 2$ numbers on n symbols or n numbers which encodes a labeled spanning tree.

For example $S : (5, 1, 1, 2, 2)$ is a Prüfer sequence on 7 numbers. As we can see, S is allowed to have repetitions of numbers in $N = \{1, 2, \ldots, 7\}$. It is obvious that the number of Prüfer sequences that are possible on n symbols are $\underbrace{n \times n \times n \times \cdots \times n}_{n-2 \text{ times}}$ which equals n^{n-2}.

THEOREM 3.2.2 (Cayley's formula) *For a positive integer n, $\tau(K_n) = n^{n-2}$.*

Proof. Since the number of terms in a Prüfer sequence is n^{n-2}, it is obvious that if we can establish a bijection between a labeled spanning tree and a Prüfer sequence, we can conclude that there are n^{n-2} spanning trees in K_n. This bijection is established using a simple algorithm to create a Prüfer sequence from a labeled spanning tree and vice-versa.

Labeled spanning tree to Prüfer sequence: The steps to be followed for determining a Prüfer sequence S from a labeled spanning tree T labeled by the number $\{1, 2, \ldots, n\}$ are as follows.

(i). Find a leaf of the tree which has the smallest label.

(ii). Add the label of the neighbor to a (initially empty) sequence S.

(iii). Delete the leaf from the tree T.

(iv). Repeat the process until only an edge remains in T.

By using these steps, we will obtain a Prüfer sequence which encodes the labeled tree. In T, the corresponding vertices of the members of the sequence has degree at least two because no leaf of T is appended to the Prüfer sequence. Since a vertex v with degree $deg(v)$ gets appended exactly $deg(v) - 1$ times, The number of terms in S is $\sum_{v \in V(T)} (deg(v) - 1) = 2|E(T)| - |V(T)|$, by fundamental theorem of graph theory. Note that, by Theorem 3.1.4, the number of edges of a tree is $|V(T)| - 1$. Therefore the number of terms in S is $n - 2$.

Conversely, the process of obtaining a labeled tree T from a Prüfer sequence S of length $n-2$, can be done by using the following algorithm.

Prüfer sequence to labeled spanning tree: Fix $N = \{1, 2, \ldots, n\}$.

(i). Initialize the Prüfer sequence $S = (a_1, a_2, \ldots, a_{n-2})$ where $a_i \in N$.

(ii). Find the smallest element $x \in N$ such that $x \notin S$.

(iii). Draw an edge connecting x to the first element in S. This is the first edge in T.

(iv). Delete a_1 from S and x from N. Now the first element in S is a_2 and N is $N \setminus \{x\}$.

(v). Repeat steps (ii), (iii) and (iv) until only two elements remain in N.

(vi). The remaining two elements in N are connected by an edge and added to the tree T.

Following these steps, we can reconstruct a labeled spanning tree from a Prüfer sequence. The spanning tree, thus obtained may be oriented differently when compared to the original tree, but it is clear that the degrees of the vertices or their labels are not different from the original tree.

If T_n represents the set of all labeled spanning trees in K_n and U_n represents the set of all Prüfer sequences on n numbers, then using the algorithm defined above, it is clear that a bijection exists between these two sets. Hence they have the same number of elements and thus $\tau(K_n) = |T_n| = |U_n| = n^{n-2}$. $\qquad\square$

Illustration for Cayley's formula: We shall demonstrate the algorithm used in Cayley's formula by a labeled tree T given in Figure 3.8. Let us first encode this tree into a Prüfer sequence.

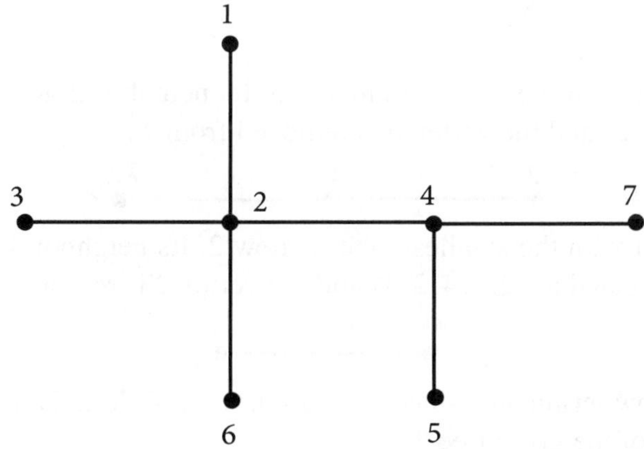

Figure 3.8   A labeled tree T

Iteration 1: The leaf with the smallest label in T is 1. Its neighbor 2 gets added to the sequence S. Thus $S : (2)$ is the sequence after the first iteration. Also 1 is removed from T.

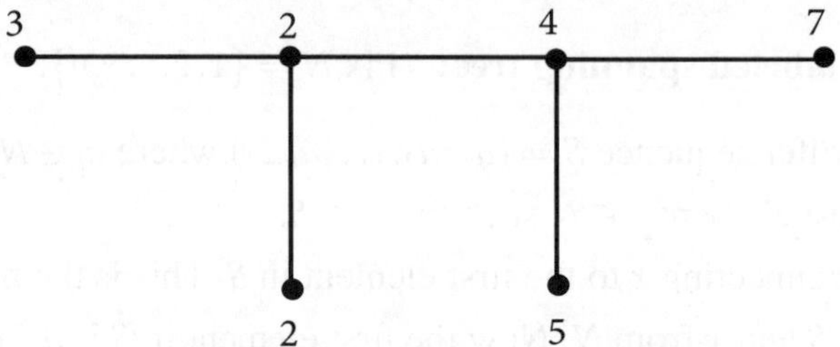

Iteration 2: Now the leaf with the smallest label is 3. Its neighbor 2 gets appended to S. The sequence S is updated to $(2,2)$ and the vertex 3 is removed from T.

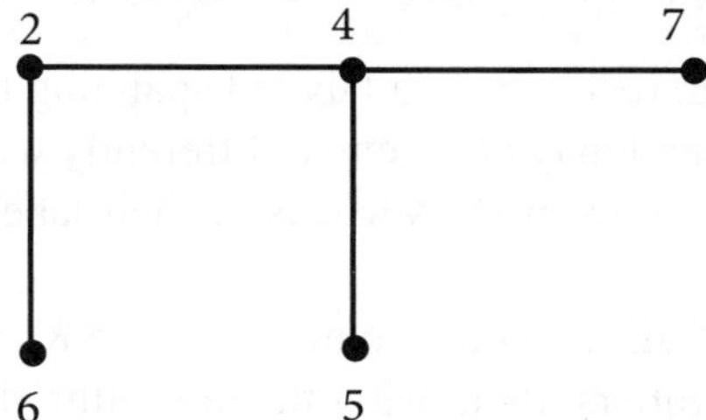

Iteration 3: The leaf with the smallest label is 5. Its neighbor 4 gets appended to S. The sequence S is updated to $(2,2,4)$ and the vertex 5 is removed from T.

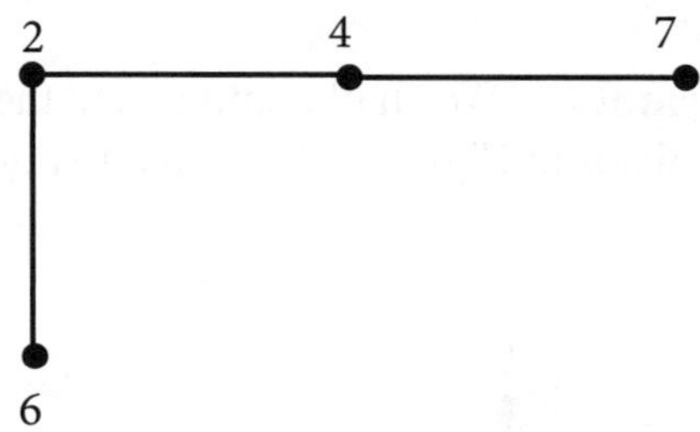

Iteration 4: The leaf with the smallest label is 6. Its neighbor 2 gets appended to S. The sequence S is $(2,2,4,2)$ and the vertex 6 is removed from T.

Iteration 5: The leaf with the smallest label is now 2. Its neighbor 4 gets appended to S. The sequence S is updated to $(2,2,4,2,4)$ and the vertex 2 is removed from T.

The algorithm now terminates since we only have a K_2 left. Thus $S = (2,2,4,2,4)$ is the Prüfer sequence of the given tree T.

Next to check if a labeled tree can be decoded from a Prüfer sequence, it is convenient to use the Prüfer sequence $S : (2,2,4,2,4)$ to obtain the labeled tree associated with it. Now we will use the algorithm to obtain a labeled tree from Prüfer sequence.

Since the number of terms in the Prüfer sequence is 5, we can conclude that $N = \{1,2,3,4,5,6,7\}$.

Iteration 1: Since 1 is the smallest number in N with $1 \notin S$ and 2 is the first member of S, we draw an edge connecting vertex labeled 1 and vertex labeled 2. We remove the number 1 from N and the number 2 from the Prüfer sequence. So $N = \{2,3,4,5,6,7\}$ and $S : (2,4,2,4)$.

Iteration 2: Since 3 is the smallest number in N that is not present in S and 2 is the first member of S, we draw an edge connecting vertex labeled 3 with a vertex labeled 2. We remove the number 3 from N and the number 2 from S. So $N = \{2,4,5,6,7\}$ and $S : (4,2,4)$.

Iteration 3: Now the smallest number in N that is not found in S is 5. We now draw an edge connecting the vertex labeled 5 with a vertex labeled 4. The number 5 is removed from N and the number 4 from S is also removed. So $N = \{2,4,6,7\}$ and $S : (2,4)$.

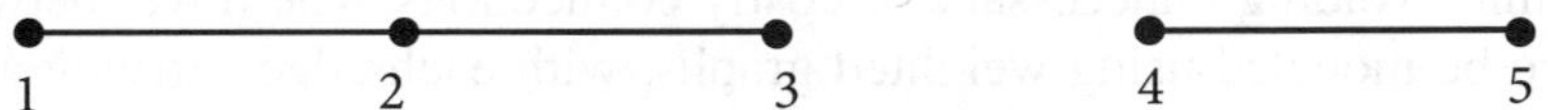

Iteration 4: Once again, we determine the smallest number found in N but not in S to be 6. We draw an edge between 6 and 2. The number 6 is removed from N and the number 2 from S is also removed. So $N = \{2,4,7\}$ and $S : (4)$.

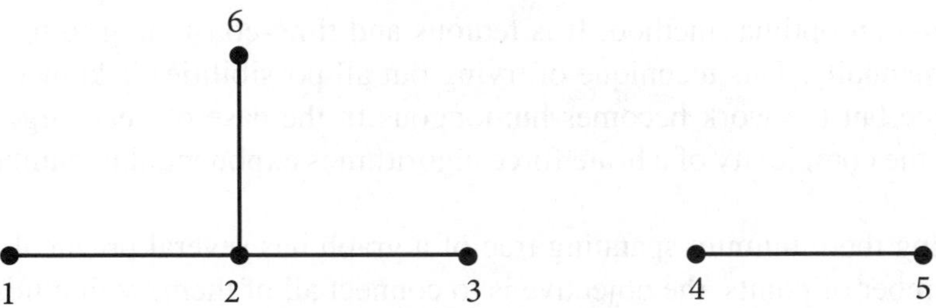

Iteration 5: Now 2 is the smallest number found in N but not in S. Therefore draw an edge connecting 2 and 4. The number 2 is removed from N and the number 4 from S is also removed, leading to an empty sequence. That is $N = \{4,7\}$ and $S : ()$. The algorithm can now terminate after adding the edge between 4 and 7. We will now arrive at a graph of a tree T that has been decoded from the Prüfer sequence $(2,2,4,2,4)$.

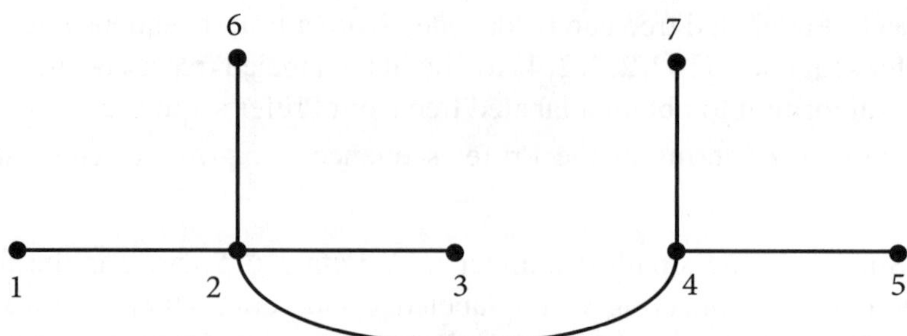

At first sight, the labeled tree obtained from the Prüfer sequence does not look like Figure 3.8, but a simple rearrangement is enough to show that they are indeed the same labeled tree. It has now been established beyond any doubt that it is possible to encode a tree in a Prüfer sequence and also reconstruct a tree from the Prüfer sequence. The 1-1 correspondence between a Prüfer sequence and a labeled tree has been clearly demonstrated.

3.3 Algorithms to determine the minimum spanning tree

In the world of computer networks, telecommunication networks and electrical grids, there is a "cost" to each connection. These networks can be modeled using weighted graphs, with weights assigned to edges to indicate the cost of connections. Hence it is of interest to determine a minimum spanning tree (MST) that will help to connect all the nodes, with the least cost, while avoiding unnecessary or costly connections. The travel between various cities can also be modeled using weighted graphs, with each edge carrying a weight that has been assigned to measure the cost of travel either in distance or time.

For weighted graphs, we wish to determine a *minimum spanning tree* which is a spanning tree whose sum of the individual weights of the edges is the least. Although one can try out different possibilities and choose the best depending on the least total weight, it is not an optimal method. It is tedious and time-consuming to try out different possibilities manually. This technique of trying out all possibilities is known as the brute-force technique but the work becomes humongous in the case of very large networks or graphs. Also, the complexity of a brute force algorithm is exponential in number of vertices and edges.

Determining the minimum spanning tree of a graph has several practical applications. Given any number of points, the objective is to connect all of them, with minimal cost. We can represent the points by the vertices of a graph, the possible connections by the graph's edges and the connection costs by the weights.

In the case of determining a minimum spanning tree, we have a few algorithms that serve the purpose, however we will be discussing only two in this book.

3.3.1 Kruskal's algorithm

Kruskal's algorithm is named after *Joseph Kruskal* who discovered the algorithm when he was a second year graduate student. Kruskal published the algorithm in 1956. Kruskal's algorithm works by constructing a minimum spanning tree for a weighted connected graph $G = (V(G), E(G))$ as an acyclic subgraph with $|V(G)| - 1$ edges for which the sum of the edge weights is the smallest. Kruskal's algorithm belongs to the class of algorithms known as *greedy algorithms*.

DEFINITION 3.3.1 A *greedy technique* of solving a problem requires the use of the following three conditions to construct the optimal solution through a sequence of steps, making the "best" choice at each level, in order to build a complete solution to the problem. The conditions are

- the choice should satisfy the conditions or constraints laid down by the problem.
- the choice should be locally optimal, that is, the best local choice among all feasible choices.
- the choice should not be changed in subsequent steps.

The algorithm begins by sorting all the edge weights in non-decreasing order. An empty forest is initialized and throughout the execution of the algorithm, an edge is added to this forest only if it is a safe edge.

DEFINITION 3.3.2 A *safe edge* is one that does not violate the two following important conditions for building the minimum spanning tree.

- Addition of the edge to the forest does not create a cycle and
- the choice of edge should be greedy, which, in this case, means selecting the one with the least weight among all suitable options.

Usually the greedy technique which allows only for locally optimal choice at each step, might not lead to a solution that is globally optimal. But Kruskal's algorithm works to make a locally optimal choice at every step, which results in a globally optimal minimum spanning tree every time. The time complexity of Kruskal's algorithm is $O(mlogm)$, where m is the number of edges. For a sparse graph, where m is less than $nlogn$, the complexity can also be expressed as $O(mlogn)$.

Algorithm 1: Kruskal's Algorithm

Data: A weighted connected graph $G = (V(G), E(G))$
with $V(G) = \{v_1, v_2 \ldots v_n\}$ and $E(G) = \{e_1, e_2 \ldots e_m\}$
Result: $E_T(G)$, the set of edges of the minimum spanning tree of G
$E_T(G) \leftarrow \emptyset$; /* The set of edges in the tree is initialized as an
  empty set */
$counter \leftarrow 0$; /* The number of edges in $E_T(G)$ is initialized as 0
  */
$k \leftarrow 0$;   /* The number of processed edges is initialized as 0 */
$[w(e_i)] \leftarrow$ a non-decreasing order of weights of edges using a sorting algorithm
while $counter < |V(G)| - 1$ **do**
  $\quad k \leftarrow k+1$ **if** $E_T(G) \cup \{e_k\}$ *is acyclic* **then**
  $\quad\quad E_T(G) \leftarrow E_T(G) \cup \{e_k\}$
  $\quad\quad counter \leftarrow counter + 1$
  $\quad$**end**
end

A *sparse graph* is a type of graph in which the number of edges is significantly less than the maximum possible number of edges. Sparse graphs can be contrasted with dense graphs. A *dense graph* is a type of graph in which the number of edges is close to the maximum possible number $\frac{n(n-1)}{2}$ for the given n number of vertices.

The correctness of the algorithm as given by the statement, "Kruskal's algorithm produces a minimum spanning tree in a non-trivial connected graph" can be proved by a theorem for which one may refer to [5, Theorem 3.12].

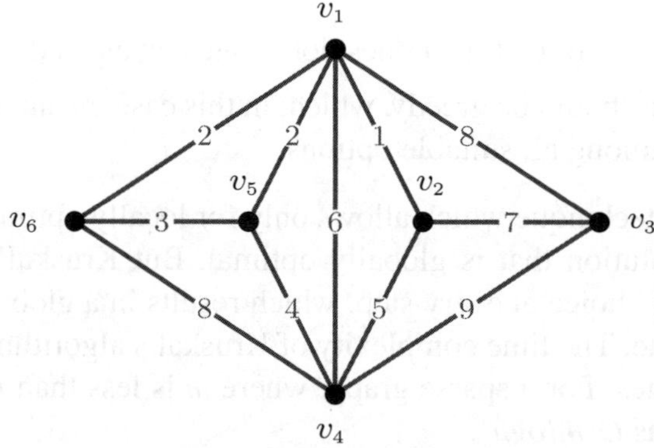

Figure 3.9   A weighted graph G

EXAMPLE 3.3.3 We will demonstrate the iterations of Kruskal's algorithm by finding a minimum spanning tree for the Figure 3.9. The algorithm initializes all the vertices and edges of the weighted graph G. All the edge weights are sorted in non-decreasing order and added to the array $w[e_i] = [1, 2, 2, 3, 4, 5, 6, 7, 8, 8, 9]$.

Iteration 1: Initially all the vertices of the graph are present as a disconnected graph, while edges are added at each iteration after careful examination. Since the edge $v_1 v_2$ is the one with the least weight, we include the edge.

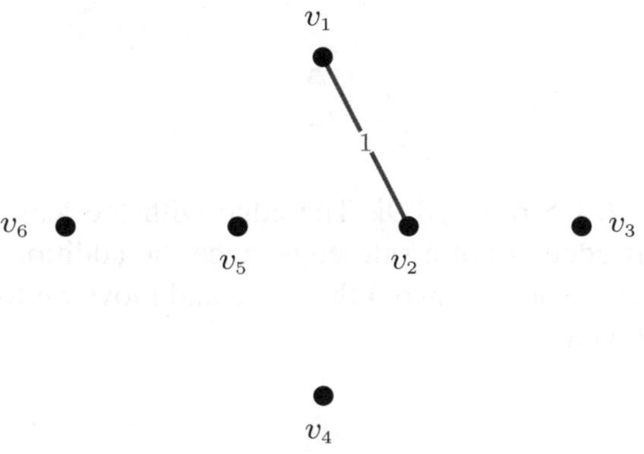

Iteration 2: $w[e_i] = [2, 2, 3, 4, 5, 6, 7, 8, 8, 9]$. Since the edge with weight 1 has been included in the previous iteration, we now include the edge with least weight, which is 2. But there are two edges with the same weight, namely $v_1 v_5$ and $v_1 v_6$. Without loss of generality, we may choose to add the edge $v_1 v_5$ since it is a safe edge.

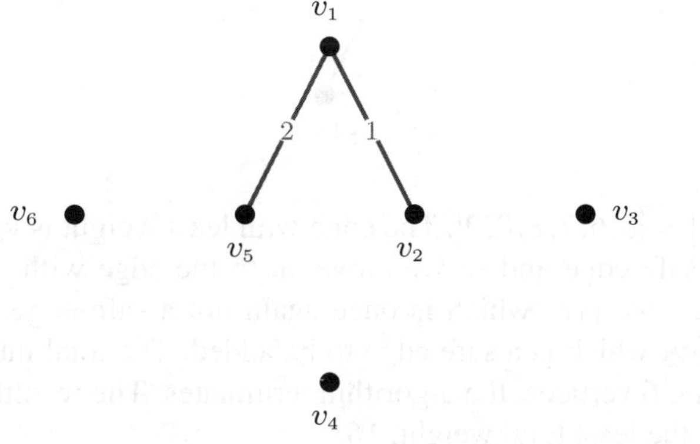

Iteration 3: $w[e_i] = [2, 3, 4, 5, 6, 7, 8, 8, 9]$. Since 2 is the weight of least value among all edges of G, and also because it is a safe edge, we choose to include the edge $v_1 v_6$.

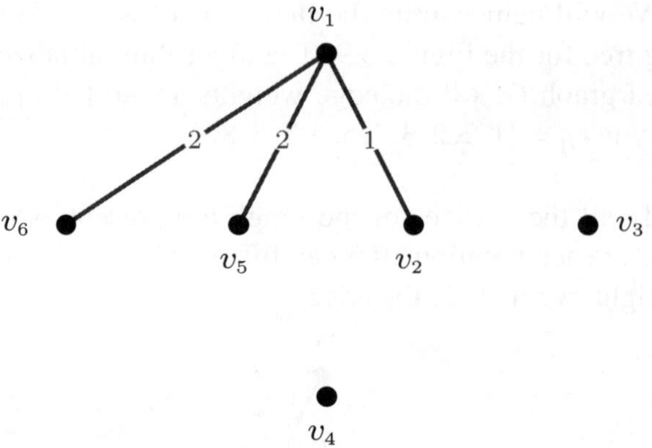

Iteration 4: $w[e_i] = [3,4,5,6,7,8,8,9]$. The edge with the least weight in this is v_5v_6 with weight 3. But this edge is not a safe edge since the addition of this edge results in the formation of a cycle. Hence we avoid this edge and move on to the next edge of least weight namely, v_4v_5 of weight 4.

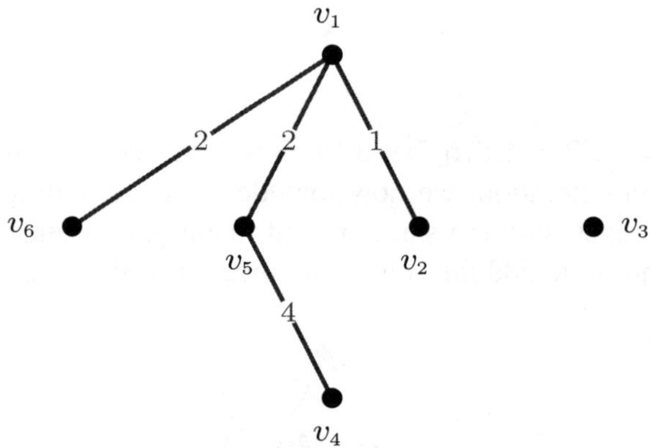

Iteration 5: $w[e_i] = [5,6,7,8,8,9]$. The edge with least weight is v_2v_4 with weight 5. But this edge is not a safe edge and so we move on to the edge with weight 6. This weight corresponds to the edge v_1v_4 which is once again not a safe edge. Hence the next edge with weight 7 is v_2v_3 which is a safe edge to be added. The total number of edges is now 5 and since there are 6 vertices, the algorithm terminates. The resulting tree is a minimum spanning tree with the least total weight, 16.

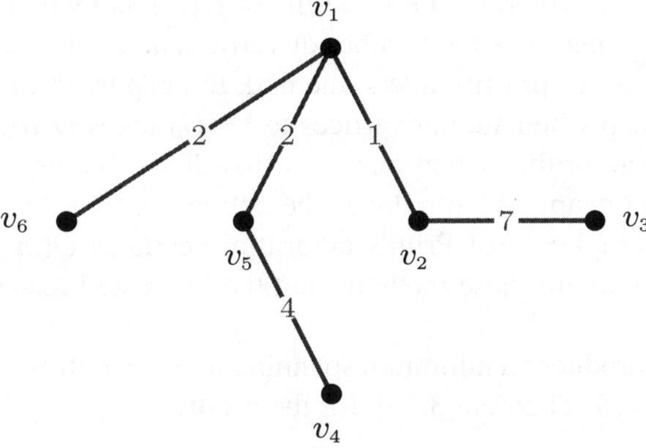

3.3.2 Prim's algorithm

This algorithm was discovered by Czech mathematician *Vojtěch Jarník* in 1930 but was later republished by computer scientists *Robert C. Prim* in 1957 and *Edsgar Dijkstra* in 1959. This algorithm is also known by other names such as Jarník's algorithm or the Prim-Jarník algorithm. Prim's algorithm also yields a minimum spanning tree of a weighted connected graph, but it follows a different approach. Kruskal's algorithm works on disconnected acyclic graphs while Prim's algorithm works only on a connected component, such as a tree. Prim's algorithm also belongs to the class of greedy algorithms, where each edge added is the best option among locally available options.

Algorithm 2: Prim's Algorithm

Data: A weighted connected graph $G = (V(G), E(G))$
with $V(G) = \{v_1, v_2 \ldots v_n\}$ and $E(G) = \{e_1, e_2 \ldots e_m\}$
Result: $E_T(G)$, the set of edges of the minimum spanning tree of G
$V_T(G) \leftarrow \{v_0\}$;   /* The set of vertices in the tree is initialized
 with any vertex */
$E_T(G) \leftarrow \emptyset$;   /* The set of edges in the tree is initialized as an
 empty set */
for ($i = 1$; $i < |V(G)|$; $i = i+1$) {
  Find a minimum weight edge $e* = (v*, u*)$ among all edges (v, u) such that v is
   in $V_T(G)$ and u is in $V(G) - V_T(G)$
  $V_T(G) \leftarrow V_T(G) \cup \{u*\}$ $E_T(G) \leftarrow E_T(G) \cup \{e*\}$
}

The complexity of Prim's algorithm depends on the kind of data structure we use. If we were to use adjacency matrix, with no priority queue to model the graph, we would have to scan all the edges connected to $V_T(G)$ in each iteration, to find the minimum weight edge. Since each iteration would then take a time of $O(m)$ and when done $n-1$ times, the complexity is $O(n.m)$, where n is the number of vertices and m is the number of edges.

If we use binary heap or priority queue method, to keep track of the minimum weight edges, updating the heap when adding vertices to $V_T(G)$ takes $O(logn)$ time. Overall, the complexity of Prim's algorithm when one uses this data structure is $O(mlogn)$. A third option is the Fibonacci heap, which reduces the extraction time for the minimum weight edge and hence the complexity of Prim's algorithm becomes $O(m+nlogn)$. The reader might not be familiar with all these methods, but the interested reader might refer [9] and learn them.

"Prim's algorithm produces a minimum spanning tree in a non-trivial connected graph". The reader can refer to [5, Theorem 3.13], for the proof.

EXAMPLE 3.3.4 We will determine a minimum spanning tree for the weighted connected graph in Figure 3.9 using Prim's algorithm.

Iteration 1: Since we can start with any random vertex in G as v_0, let us start with vertex v_6. Now V_T contains v_6 and therefore $V_T - \{v_6\}$ contains all of the remaining vertices namely $\{v_1, v_2, \ldots, v_5\}$. We need to find the weight of the edges connecting each of these vertices to v_6. We see that $N(v_6) = \{v_1, v_5, v_4\}$ and the weights of edges v_1v_6, v_5v_6 and v_4v_6 are 2, 3 and 8 respectively. The remaining vertices v_2 and v_3 in $V_T - \{v_6\}$ are not connected by an edge to v_6 and their weights are assigned as ∞ to signify the lack of a connecting edge. Now the least weight among 2, 3, 8 and ∞ is 2 and hence the edge v_1v_6 is added to $E_T(G)$ and vertex v_1 is added to $V_T(G)$.

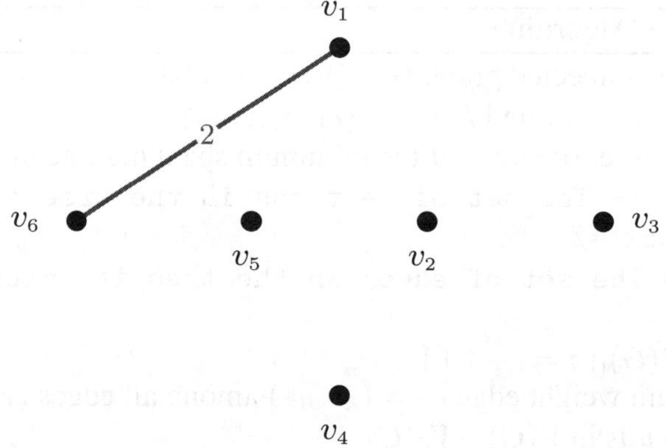

Iteration 2: We can see that the neighbors of v_1 and v_6 are $\{v_2, v_3, v_4, v_5\}$. According to Figure 3.9, the weights of the edges connecting v_1 to v_2, v_3, v_4 and v_5 are 1, 8, 6 and 2 respectively, while weights of edges v_6v_5 and v_6v_4 are 3 and 8 respectively. We choose the edge with least weight, namely v_1v_2 and since it is a safe edge, we add it to $E_T(G)$, while v_2 is added to $V_T(G)$.

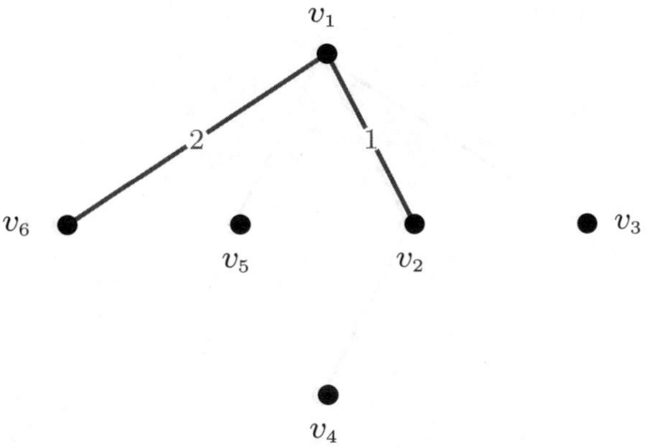

Iteration 3:   In this iteration, $V(G) - V_T(G) = \{v_3, v_4, v_5\}$. We are now required to find edges of least weight to connect these vertices to the vertices belonging to $V_T(G) = \{v_1, v_2, v_6\}$. From Figure 3.9, v_3 and v_4 are connected to the vertex $v_2 \in V_T(G)$ by edges of weight 7 and 5 respectively. The vertices v_4 and v_5 are connected to vertex v_6 by edges of weight 8 and 3 respectively, while v_1 is connected to vertices v_3, v_4 and v_5 through edges of weight 8, 6 and 2 respectively. Since we follow a greedy technique, we find that edge $v_1 v_5$ with weight 2 has the least weight and is also a safe edge. Hence this edge is added to $E_T(G)$ while v_5 is added to $V_T(G)$.

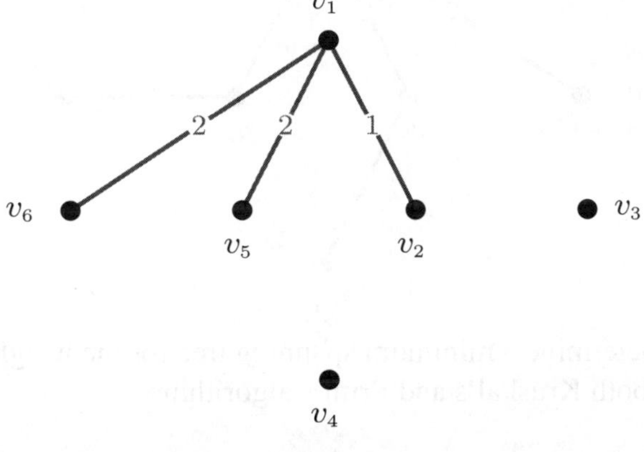

Iteration 4:   Clearly, the vertices in $V(G) - V_T(G) = \{v_3, v_4\}$. The edge with least weight that can help add any vertex in $V_T(G)$ to those in $V(G) - V_T(G)$ is $v_4 v_5$ with weight 4. Hence v_4 is added to $V_T(G)$ and the edge $v_4 v_5$ is added to $E_T(G)$.

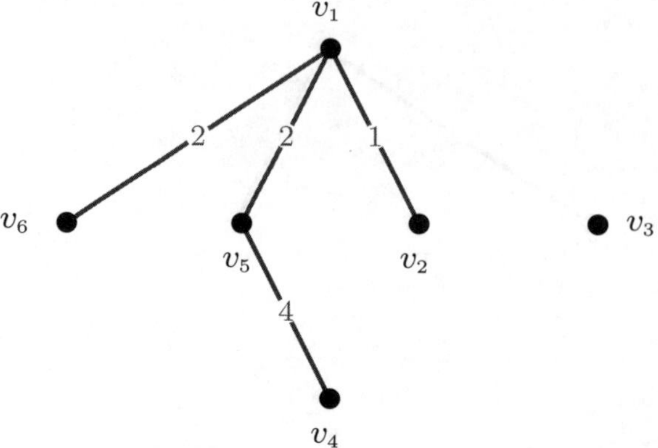

Iteration 5:   We only need to add the vertex v_3 to $V_T(G)$ and we are done. The edge of least weight among all possible options is v_2v_3 with weight 7. Hence v_3 is added to $V_T(G)$ and edge v_2v_3 is added to $E_T(G)$. The algorithm terminates here, because it runs for only $|V(G)| - 1$ times, and we have had 5 iterations. The tree that we have obtained through this algorithm is a minimum spanning tree of total weight 16.

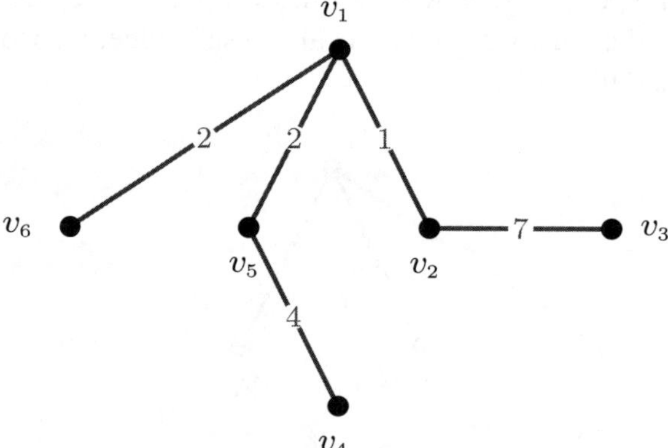

EXAMPLE 3.3.5 Determine a minimum spanning tree for the weighted connected graph in Figure 3.10 using both Kruskal's and Prim's algorithms.

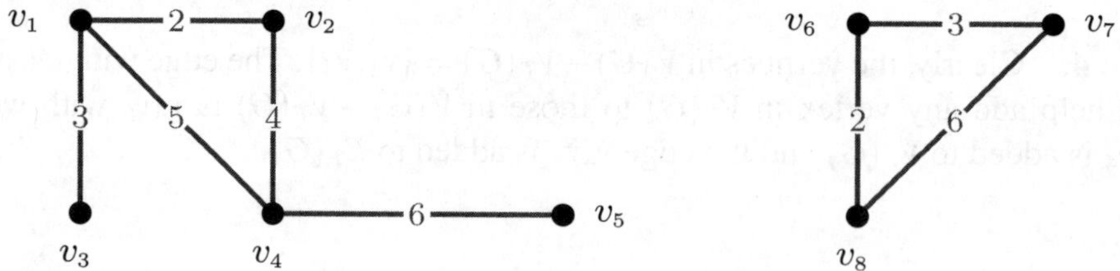

Figure 3.10   A weighted disconnected graph

Since Kruskal's algorithm determines a minimum spanning tree by building forests, it works for the given disconnected graph, while Prim's algorithm fails. The reader is encouraged to try and verify the claim.

Summary

In this chapter, we have introduced the reader to an important class of graphs, called the trees. Different types of trees, their properties and characteristics have been explored. Cayley's formula which helps to determine the number of spanning trees that can be constructed on n vertices has been discussed in detail. Additionally the concept of Prüfer code is demonstrated as a method to prove Cayley's theorem. The process of encoding a labeled tree in a Prüfer code and decoding a Prüfer code to construct a labeled tree have been demonstrated with an example. Also in this chapter are two algorithms, namely Kruskal's and Prim's algorithms, whose objective is to determine the minimum spanning tree of a weighted graph. The two algorithms have been explained in detail through use of pseudocode and examples.

3.4 Exercises

Section 3.1: Characterization of trees and spanning trees

1. Draw a tree T whose complement is also a tree.

2. Draw a graph G with at least one cycle whose complement is also a tree.

3. Draw all trees of order 5.

4. Prove that a graph of order n and size $n - 1$ need not be a tree.

5. Prove or disprove: Every graph with fewer edges than vertices has a component that is a tree.

6. Provide an example of a tree of order 6, having four vertices of degree 1 and two vertices of degree 3.

7. A tree T of order 13 has vertices with degrees of $1, 2,$ and 5. If T has exactly three vertices of degree 2, how many end vertices does it have?

8. Provide an example of a tree of order 6 with four vertices of degree 1 and two vertices of degree 3. (Hint: only one tree has these properties.)

9. A tree T with 50 end-vertices has an equal number of vertices of degree 2, 3, 4, and 5 and no vertices of degree greater than 5. Determine the order of T.

10. Suppose G be a connected graph and H be a connected spanning subgraph of G with as few edges as possible. Show that H is a spanning tree of G.

11. Let T be a tree of order n and size m having n_i vertices of degree i for $i \geq 1$. Then $n = \sum_i n_i$ and $2(n-1) = 2m = \sum_i i n_i$. Prove that $n_1 = 2 + n_3 + 2n_4 + 3n_5 + 4n_6 + \dots$.

Section 3.2: Cayley's formula

12. Given a sequence of positive integers $(d_1, \dots, d_n)$ such that $\sum_{i=1}^n d_i = 2(n-1)$, then prove that there exists a tree T whose vertex set is $\{v_1, \dots, v_n\}$, and the degree of each vertex v_i is d_i for $1 \leq i \leq n$.

13. Let $\{v_1, \dots, v_n\}$ represent a set of $n \geq 2$ vertices, and let $\{d_1, \dots, d_n\}$ be a sequence of positive integers such that $\sum_{i=1}^n d_i = 2(n-1)$. Then prove that the total number of trees with $\{v_1, \dots, v_n\}$ as the vertex set, where vertex v_i has degree d_i for $1 \leq i \leq n$, is given by:

$$\frac{(n-2)!}{(d_1 - 1)!(d_2 - 1)! \cdots (d_n - 1)!}.$$

14. Write an algorithm in pseudocode for encoding a labeled spanning tree into a Prüfer sequence.

15. Write an algorithm in pseudocode for decoding a Prüfer sequence into a labeled spanning tree.

16. Is it possible to graphically realize a degree sequence S of a graph G, S: $(3,3,2,2,2)$ into a tree? Explain.

17. Create a Prüfer sequence for the tree shown in Figure 3.11.

18. Convert the following Prüfer sequences into labeled spanning trees:

    (i). $(1, 1, 1, 2, 4, 6)$

    (ii). $(2, 2, 1, 1, 3, 5, 7)$

    (iii). $(4, 1, 1, 2, 3)$.

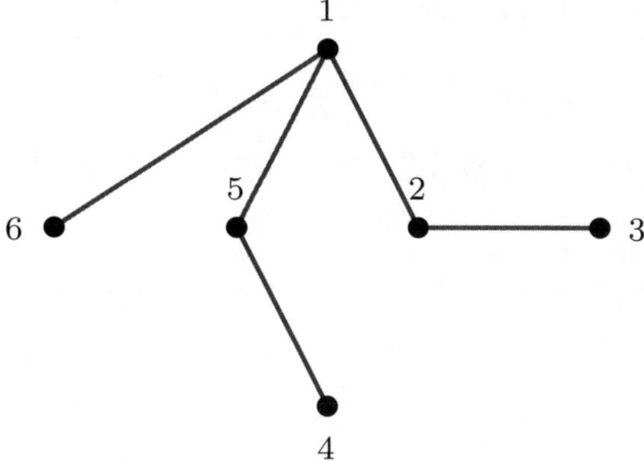

Figure 3.11 Convert the tree into a Prüfer sequence

Section 3.3: Algorithms to determine minimum spanning tree

19. Determine the minimum spanning tree of the weighted connected graph given in Figure 3.12 using

   (i). Kruskal's algorithm

   (ii). Prim's algorithm.

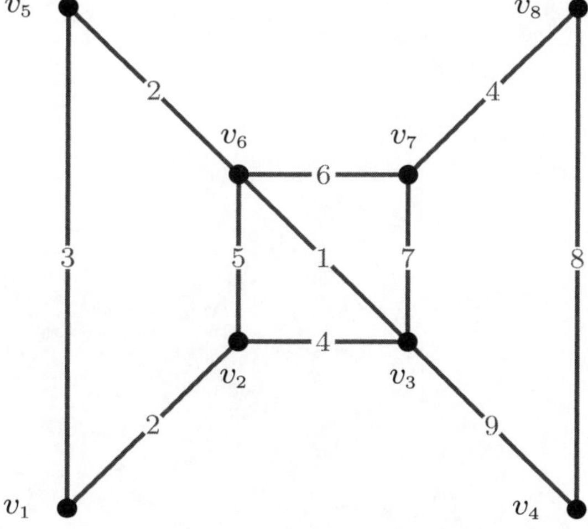

Figure 3.12 A weighted connected graph

4

Connectivity

We are familiar with the concept of connectedness of a graph and how it decides whether we can traverse the graph, starting from any vertex and reaching another vertex. It is possible for a connected graph to lose its connectedness by the removal of a single vertex. There are also other connected graphs in which the removal of several vertices does not disconnect the graph. Hence we have stumbled upon a parameter that can give a "sense" of the connectedness of a graph. It may answer important questions about the nature of its connectedness. In this chapter, we will deal with connectivity, which measures the connectedness of a graph using some specific parameters.

4.1 Blocks, cut vertices and bridges

It is important to define a cut-vertex and a cut-edge before we begin to study connectivity. One must note that when a vertex is removed from a graph, the graph is cleared of that specific vertex and its incident edges. In a graph, an edge removal refers to the removal of just one specific edge.

4.1.1 Cut-vertex and bridge

DEFINITION 4.1.1 A *cut-vertex* or *cut-edge* of a connected graph is a vertex or an edge whose removal results in a disconnected graph.

A cut-edge is also called a *bridge*. In Figure 4.1, the vertices u_2 and u_7 are the cut-vertices. The edge e is the only cut-edge or bridge of the graph.

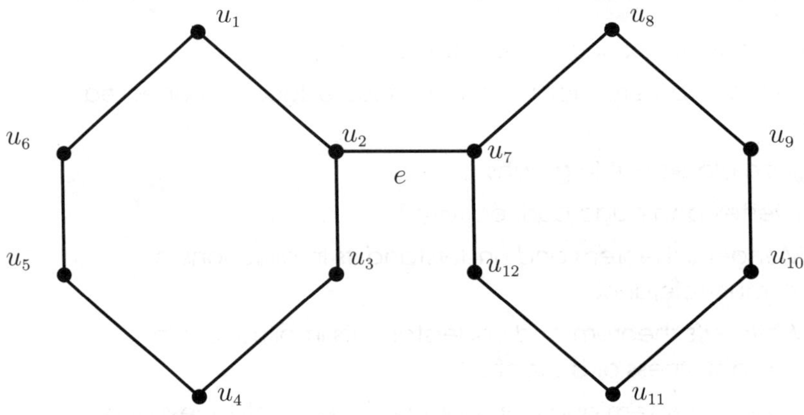

Figure 4.1   Graph with cut-vertex and bridge

REMARK 4.1.2 It is interesting to note that $G - e$ contains exactly two components if G is a connected graph and e is one of its cut-edges, since eliminating an edge yields a maximum of two components. However, this is not the case when G's cut-vertex is involved. Take the star graph $K_{1,n}$ as an example, and remove the central vertex. Afterwards, there are n isolated vertices and so n components in the resulting graph.

There is a characterization of a cut-edge that we will see in the following theorem that helps us to identify a cut-edge easily.

THEOREM 4.1.3 *An edge in a connected graph is a bridge if and only if it belongs to no cycle.*

Proof. Let $e = xy$ be any edge in a graph G. Let us also assume that e is a bridge of G. We claim that e is not a part of any cycle in G. Alternatively, we can prove that e is a part of a cycle in G, only if it is not a cut-edge or a bridge. Assume $G - \{e\}$ is connected, which can also be understood as e not being a bridge. Due to this assumption, $G - \{e\}$ has an $x - y$ path that does not involve e. Therefore the $x - y$ path, together with e, completes a cycle, which naturally leads to the conclusion that e is part of a cycle in G.

Conversely, suppose that e is part of a cycle in G. We shall prove that $G - \{e\}$ is connected, which in turn implies that e is not a bridge. Let e lie on a cycle C of G. Choose any two arbitrary vertices $u, v \in V(G)$. Since G is connected, there exists a $u - v$ path P in G. If P does not contain edge e, then clearly the path P exists in $G - \{e\}$, thus proving the connectivity of $G - \{e\}$. Suppose P contains the edge e. Note that $G - \{e\}$ contains a $u - x$ path along P, an $x - y$ path along C and a $y - v$ path along P; refer to Figure 4.2. Therefore a $u - v$ walk exists in $G - \{e\}$. Now, by Theorem 1.2.5, we have a $u - v$ path in $G - \{e\}$. Since u and v are arbitrary, $G - \{e\}$ is connected.   □

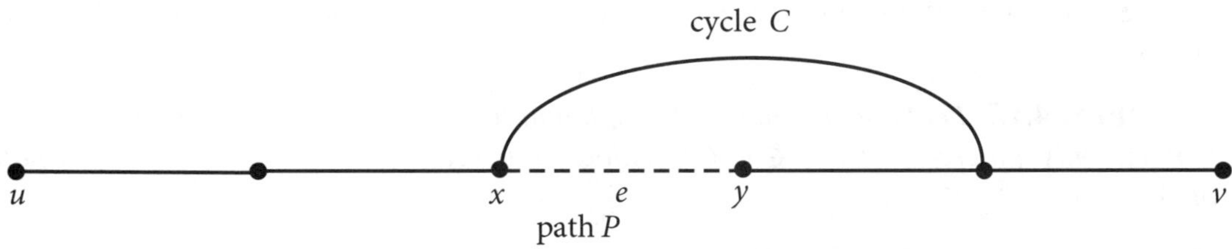

Figure 4.2   Graph $G - e$ with an alternate $u - v$ path

Note that no tree contains a cycle. So we have the following result.

COROLLARY 4.1.4 *A graph G is a tree if and only if every edge of G is a bridge.*

REMARK 4.1.5 Let G be a connected graph of order 3 or more. If G contains a bridge $e = xy$, then the edge e is not available in the graph $G - \{x\}$ and so x is a cut-vertex of G. Thus, if G contains a bridge, then G contains a cut-vertex. But the converse is not true. For instance, consider the graph G in Figure 4.3. Clearly, $G - \{u_2\}$ is disconnected, whereas $G - \{e\}$ is connected for any $e \in E(G)$. So G contains a cut-vertex but not a bridge.

THEOREM 4.1.6 *Let v be a cut-vertex in a connected graph G, and let u and w be vertices in distinct components of $G - v$. Then v lies on every $u - w$ path in G.*

Proof. If u and w are distinct components of $G - v$, then u and w are not connected in $G - v$. On the other hand, u and w are necessarily connected in G. Therefore every $u - w$ path in G must contain the vertex v.   □

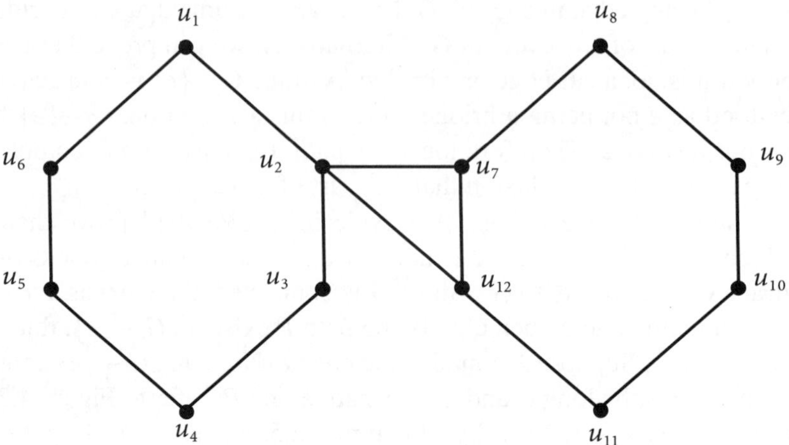

Figure 4.3   Graph with cut-vertex and but not bridge

The following result gives a necessary and sufficient condition for a vertex to be a cut-vertex.

THEOREM 4.1.7 *Let G be a connected graph and let $v \in V(G)$. Then v is a cut-vertex of G if and only if there exists $u, w \in V(G)$ distinct from v such that v lies on every $u - w$ path of G.*

Proof. If v is a cut-vertex of G, then $G - \{v\}$ is disconnected. Choose $u, w \in V(G)$ with u and w in different components of $G - \{v\}$. Then, by Theorem 4.1.6, every $u - w$ path in G contains v.

Conversely, assume that there exists $u, w \in V(G)$ such that every $u - w$ path in G must contain the vertex v. Then there is no $u - w$ path in $G - \{v\}$ so that $G - \{v\}$ is disconnected. Thus v is a cut-vertex of G. $\qquad\square$

Identifying a connected graph, when the number of vertices is a much larger number can be a hassle. If you can imagine, a graph of say 1000 vertices, can you determine if the graph is connected or not? You might think that it is easy because you can look at a graph's structure and see if it is connected. But that is not always so. It would be time consuming to draw the graph and check it component wise. A much more efficient way to check if a graph is connected or not, is to be able to identify two different vertices, whose separate removal still results in connected graphs.

THEOREM 4.1.8 *Let G be a graph of order 3 or more. Then G is connected if and only if G contains two distinct vertices u and v such that $G - \{u\}$ and $G - \{v\}$ are connected.*

Proof. Consider a graph G of order 3 or more. Suppose there exists two vertices $u, v \in G$ such that $G - \{u\}$ and $G - \{v\}$ are connected. We claim that G itself is connected.

Choose any two distinct vertices $x, y \in V(G) \setminus \{u, v\}$. Since $G - \{u\}$ is connected, there exists an $x - y$ path P in $G - \{u\}$. Consequently, G also has the $x - y$ path P in $G - \{v\}$ and therefore x and y are connected in G.

Choose any two distinct vertices x, y in G such that $x, y \in \{u, v\}$. It suffices to prove that u and v are connected by a path in G. Recall that G is a graph of order 3 or more. So we can guarantee that there exists a vertex $w \in V(G) \setminus \{u, v\}$. Since $G - \{v\}$ is connected, there is at least one $u - w$ path P' in $G - \{v\}$. Since $G - \{u\}$ is connected, it contains a $w - v$ path P''. So P' followed by P'' produces a $u - v$ walk. We know that every $u - v$ walk contains a $u - v$ path (by Theorem 1.2.5). Thus u and v are connected in G.

Conversely, assume that G is connected with $|V(G)| \geq 3$. Choose u and v in G such that u and v are connected by a path of the longest length in G. This is possible because we just need to find u, v such that it puts a path of the greatest length in the graph between them. We claim that $G - \{u\}$ and $G - \{v\}$ are also connected. Assuming that is not the case, it follows that at least one of $G - \{u\}$ and $G - \{v\}$ is disconnected. Let us assume that $G - \{v\}$ is disconnected. This implies that $G - \{v\}$ contains two vertices x, y that are not connected in $G - \{v\}$. Clearly $x \neq y \neq v$. Since G is connected, let P' be the shortest $x - u$ path and P'' be the shortest $u - y$ path in G. Since the $u - v$ path P, is a path of the longest length in G together with the fact that $x \neq y \neq v$, the vertex v cannot lie on P' nor on P''. Thus P' and P'' are paths that are available in $G - \{v\}$. The path P' followed by P'' produces an $x - y$ walk W in $G - \{v\}$. This implies that $G - \{v\}$ has an $x - y$ path, and that leads to the implication that x and y are connected in $G - \{v\}$, which is a contradiction. $\qquad \square$

4.1.2 Blocks

We have discussed in detail about connected graphs with cut-vertices. We shall now turn our attention to graphs with no cut-vertices. Graphs that do not disconnect despite the removal of vertices, or at least have subgraphs that have the desired property of being immune to the removal of vertices are of great interest in the study of fault-tolerant networks.

DEFINITION 4.1.9 A non-trivial connected graph with no cut-vertex is called a *non-separable* graph. A maximal non-separable subgraph of a graph is called a *block*.

Note that if the graph itself is non-separable, then the graph itself, by definition, becomes a block.

In the graph G, given in Figure 4.4, B_1, B_2, B_3, B_4 and B_5, being the maximal connected subgraphs, are the blocks of G. We may notice that in graph G, u and v are the cut-vertices.

THEOREM 4.1.10 *A graph of order at least 3 is non-separable if and only if every two vertices lie on a common cycle.*

Proof. First, suppose that G is a graph of order at least 3, such that every two vertices of G lie on a common cycle. Consequently, G is connected. Suppose to the contrary, that G is

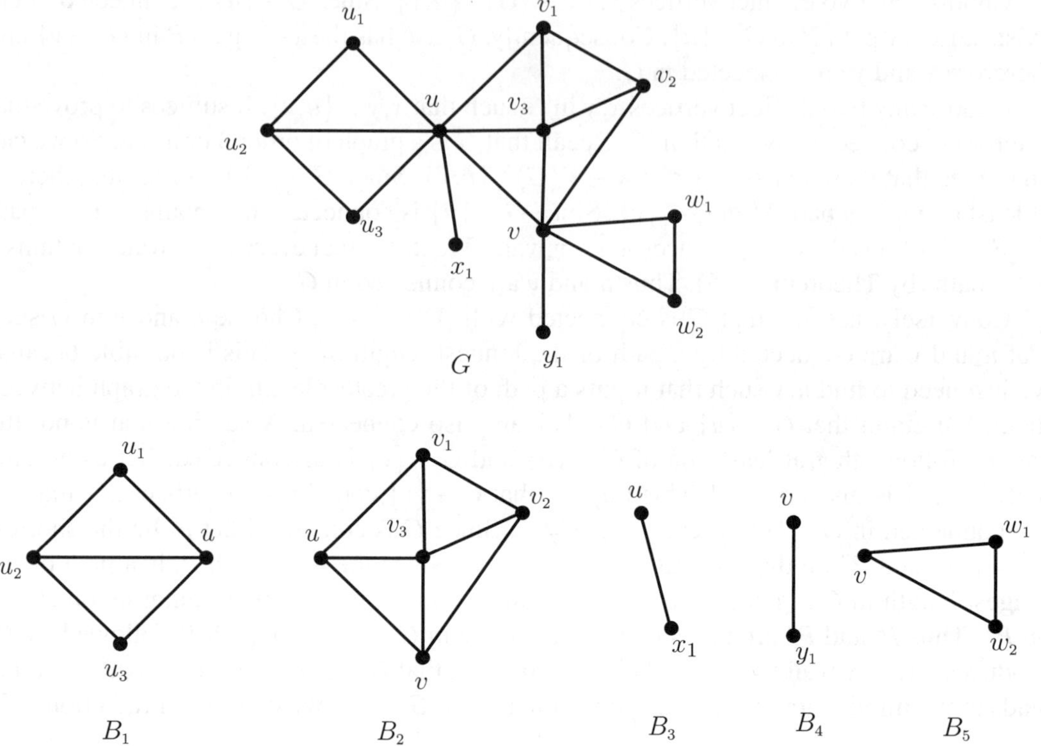

Figure 4.4   A graph with its blocks

not non-separable. That is G must contain a cut-vertex, say v. Let u and w be two vertices that belong to different components of $G - v$. By assumption, u and w lie on a common cycle C on G. Then C determines two distinct $u - v$ paths of G at least one of which does not contain v, which is a contradiction to Theorem 4.1.6. Thus G contains no cut-vertices and so G is non-separable.

Conversely, let G be a non-separable graph of order at least 3. That is G has no cut-vertex. Let us assume, to the contrary, that there are pairs of vertices of G, that do not lie in a common cycle. Among all such pairs, let u, v be a pair for which the length of a path P between u and v is minimum. If the length of the path P is 1, then $uv \in E(G)$. Since uv belongs to no cycle in G, by Theorem 4.1.3, uv is a bridge in G. Therefore $G - \{u\}$ is disconnected and so u is a cut-vertex of G, a contradiction. Thus the length of the path P is at least 2.

Let $P : u = v_0, v_1, \ldots, v_{k-1}, v_k = v$ be a $u - v$ path of length k in G. Since the length of the path $u, v_1, \ldots, v_{k-1}$ is $k - 1 < k$, there is a cycle C in G containing u and v_{k-1}. By assumption, v is not on C. By the necessary part, v_{k-1} is not a cut-vertex of G. Since v_{k-1} is not a cut-vertex of G and u and v are different from v_{k-1}, we have by Theorem 4.1.7 that

v_{k-1} is not part of a $u - v$ path in G, say Q. Let x be the last vertex common to Q and C; existence is guaranteed by u; see Figure 4.5. Let Q' be the $v - x$ sector of Q and let P' be a $v_{k-1} - x$ path on C that contains u. Now the edge vv_{k-1} together with $v_{k-1} - x$ path P' and $x - v$ path Q' form a cycle which contains both u and v, a contradiction. Thus every two vertices lie on a common cycle.

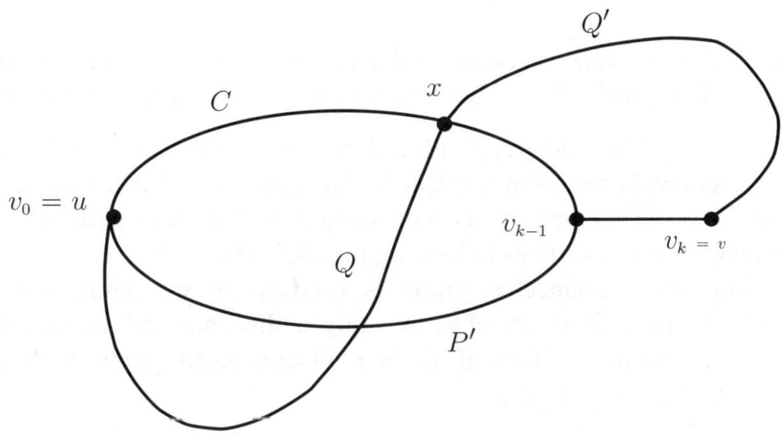

Figure 4.5   Diagrammatic representation for Theorem 4.1.10

$\square$

COROLLARY 4.1.11 *Every two distinct blocks, B_1 and B_2 in a non-trivial connected graph G have the following properties:*

(i). *The blocks B_1 and B_2 are edge disjoint.*

(ii). *The blocks B_1 and B_2 have at most one vertex in common.*

(iii). *If B_1 and B_2 have a vertex v in common, then v is a cut-vertex of G.*

4.2 Vertex connectivity

In some of the graphs we have encountered so far, the strategic deletion of a single vertex or an edge was enough to disconnect a graph. So it is quite possible for the reader to labor under the misapprehension that a single deletion or maybe two, is enough to result in an increase in the component size. It is not so all the time. So we may ask the question, what is the least possible number of removals of vertices in a graph G that disconnects it? This is where the concept of connectivity comes into play.

DEFINITION 4.2.1 The set of vertices whose removal will disconnect a connected graph is called a *cut-set*, a *separating set*, or a *vertex cut*. That is, a *vertex cut* of a connected graph

G is a set $S \subseteq V(G)$ such that $G - S$ has more than one component. The *vertex connectivity* of a graph G is the minimum number of vertices that must be removed in order for the graph to be disconnected or a single vertex graph (K_1). The vertex connectivity of a graph G is denoted by $\kappa(G)$ and it is the cardinality of the minimum vertex cut of G.

A vertex cut S of a connected graph G is *minimal* if no proper subset of S is a vertex cut of G.

DEFINITION 4.2.2 A graph G is *ℓ-connected*, where $\ell \in \mathbb{N}$, if $G - S$ is connected for every set of vertices $S \subseteq V(G)$ with $|S| < \ell$. In other words, G is ℓ-connected for every $\ell < \kappa(G)$.

It is useful to recall that a subgraph induced by a set S is the same as the one obtained through a vertex removal process of a graph G. The graph $G - S$ is the induced subgraph $G[V(G) \setminus S]$ which is the subgraph of G with vertices that do not belong to S along with edges that are incident upon vertices belonging to $V(G)$, but not S.

Keep in mind that a connected graph is used while we define the concept of connectivity. In any case, it is possible to study a disconnected graph's connectivity component by component. In fact, if G is a disconnected graph with components $G_1, G_2, \ldots, G_\ell$, then $\kappa(G) = \min\limits_{1 \leq i \leq \ell} \kappa(G_i)$.

We shall compute the vertex connectivity of some graphs that are familiar to us.

The path P_n: For the $u - v$ path of n vertices, denoted by P_n, the removal of end vertices u or v does not do the trick. A vertex cut consisting of any one internal vertex in the path P_n is sufficient to increase the components of P_n. Thus $\kappa(P_n) = 1$.

The cycle C_n: For the n-cycle C_n with $n \geq 4$, the vertex connectivity $\kappa(C_n) = 2$. The choice of any vertex x and a vertex that is not adjacent to x for the vertex cut results in a disconnected graph. In fact, the removal of any vertex from C_n results in a path P_{n-1} which is a connected graph. So we require a minimum of two vertices in a vertex cut to disconnect a cycle C_n.

The complete graph K_n: For the complete graph K_n, it is obvious that all the induced subgraphs of K_n are complete. The removal of any $n - 1$ vertices from K_n results in a singleton vertex K_1. Thus we obtain $\kappa(K_n) = n - 1$. Note that we consider the resulting K_1 as disconnected in this case. By the same reason, we have $\kappa(C_3) = 2$.

The complete bipartite graph $K_{m,n}$: In the case of the complete bipartite graph $K_{m,n}$, consider a bipartition X, Y of $K_{m,n}$. Every induced subgraph that has at least one vertex from X and from Y is connected. Therefore every vertex cut of $K_{m,n}$ contains X or Y. Since $\kappa(K_{m,n})$ is the cardinality of the minimum vertex set together with the fact that X and Y themselves are separating sets, we have $\kappa(K_{m,n}) = \min\{m, n\}$.

Petersen graph: The connectivity of the Petersen graph is not too difficult to compute. The structure of the Petersen graph itself lets us know that there are 2 cycles in it, namely the inner cycle and an outer cycle, see Figure 4.6. Since we need to disconnect either one of the cycles and since each of these cycles is C_5, we have to remove two vertices from one cycle. The removal of two vertices alone does not disconnect the graph since we have another cycle that is still connected. So we strategically remove one vertex from the other cycle, and this results in a disconnected graph. In fact, removal of three vertices in the neighborhood set $N(v)$ for any vertex v disconnects the Petersen graph. Thus $\kappa(G) = 3$ where G is the Petersen graph.

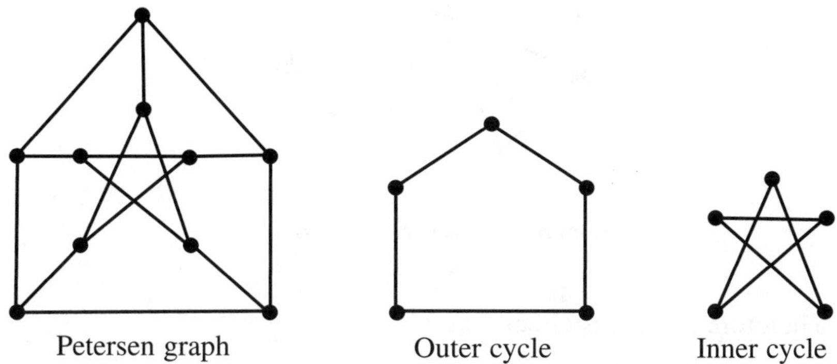

Figure 4.6 Petersen graph

Harary Graphs $H_{k,n}$: We have already encountered the Harary graphs earlier; see Theorem 2.1.8. But we shall discuss their construction anyway. Given $k < n \in \mathbb{N}$, place n vertices around a circle, equally spaced. If both k and n are even, $H_{k,n}$ is constructed by making each vertex adjacent to the nearest $k/2$ vertices in each direction around the circle. If k is odd and n is even, construct $H_{k,n}$ by making each vertex adjacent to the nearest $\frac{(k-1)}{2}$ vertices and to the diametrically opposite vertex. When k and n are both odd, index the vertices by the integers modulo n. Construct $H_{k,n}$ from $H_{k,n-1}$ by adding the edges i to $i + \frac{(n-1)}{2}$ for $0 \leq i \leq \frac{n-1}{2}$.

But $H_{k,n}$ is k-regular only if at least one of k or n is even; refer Theorem 2.1.8. For instance, Harary graph $H_{3,8}$ is given in Figure 4.7.

The following theorem explicitly gives the connectivity of the Harary Graphs.

THEOREM 4.2.3 *For every two integers k and n with $2 \leq k \leq n$, the vertex connectivity* $\kappa(H_{k,n}) = k$.

The proof of this theorem is beyond the scope of this book. However, we can discuss a proof for smaller values of k. If we consider $H_{2,n}$, it is merely the cycle C_n with some additional edges. We know that the connectivity of C_n is 2. It follows that $\kappa(H_{2,n}) = 2$.

We will discuss the Harary graph $H_{3,n}$ in the following example.

EXAMPLE 4.2.4 Consider the Harary graph $H_{3,n}$ for $4 \leq n \in \mathbb{N}$. Clearly $H_{3,n}$ is 3-regular graph. Since v is an isolated vertex in the graph $H_{3,n} - N(v)$ for any $v \in V(H_{3,n})$, we have $\kappa(H_{3,n}) \leq 3$. Since $H_{3,n}$ contains a spanning n-cycle $C : v_1, v_2, \ldots, v_n, v_1$, it follows that

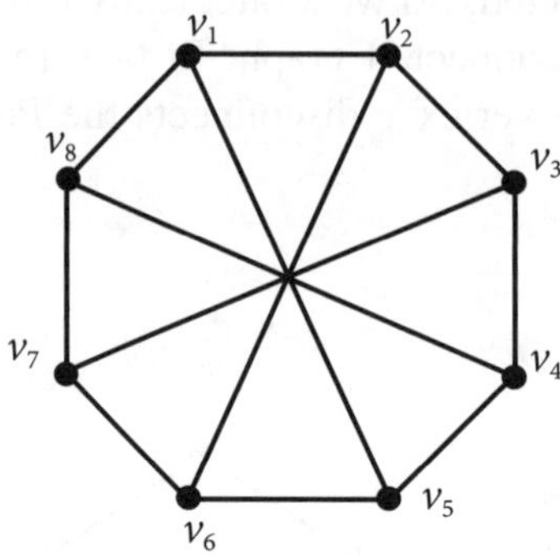

Figure 4.7   Harary graphs $H_{3,8}$

$\kappa(H_{3,n}) \geq 2$. Therefore $\kappa(H_{3,n})$ is either 2 or 3.

Case 1: Suppose $n = 2\ell \geq 4$ is even. Let $U = \{u, v\}$ be a set of two vertices of $H_{3,n}$. We show that $H_{3,n} - U$ is connected. Note that there is a $u - v$ path P on C of length at most ℓ. Without loss of generality, we may assume that $u = v_1$ and $v = v_j$ where $2 \leq j \leq \ell + 1$. If $v = v_2$, then $H_{3,n} - U$ contains the spanning path $v_3, v_4, \ldots, v_n$ and so $H_{3,n} - U$ is connected. Thus $3 \leq j \leq \ell + 1$. However then, $H_{3,n} - U$ contains the paths $P' : v_2, v_3, \ldots, v_{j-1}$ and $P'' : v_{j+1}, v_{j+2}, \ldots, v_n$ as well as the edge $v_2 v_{\ell+2}$. Since $v_{\ell+2}$ belongs to P'', the graph $H_{3,n} - U$ is connected. Hence $\kappa(H_{3,n}) = 3$.

Case 2: Suppose that $n = 2\ell + 1 \geq 5$ is odd. Let $U = \{v_i, v_j : 1 \leq i < j \leq n\}$ be a set of two vertices.

If $j = i + 1$, then $v_{j+1}, v_{j+2}, \ldots, v_n, v_1, \ldots, v_{j-1}$ is a path in $H_{3,n} - U$ which contains all the vertices of $H_{3,n} - U$. Therefore $H_{3,n} - U$ is connected.

Assume that $j \neq i + 1$. Then either the path $P : v_i, v_{i+1}, \ldots, v_j$ or the path $Q : v_j, v_{j+1}, \ldots, v_i$ has length at most ℓ. If the length of P is at most ℓ, then $P' : v_{i+1}, v_{i+2}, \ldots, v_{j-1}$ and $Q' : v_{j+1}, v_{j+2}, \ldots, v_{i-1}$ as well as the edge $v_{i+1} v_{i+\ell+1}$, where $v_{i+\ell+1}$ belongs to Q' and so $H_{3,n} - U$ is connected. If the length of Q is at most ℓ, then the paths P' and Q', as well as the edge $v_{j+2} v_{j+\ell+2}$, where $v_{j+\ell+2}$ belongs to P', are contained in $H_{3,n} - U$. Thus $\kappa(H_{3,n}) = 3$ because U is not a cut-vertex.

4.3   Edge connectivity

Edge connectivity is a measure of the minimum number of edge removals that results in a disconnected graph.

DEFINITION 4.3.1 The set of edges $X \subseteq E(G)$, whose removal will disconnect a graph, is called an *edge-cut*. The *edge connectivity* of a connected graph G is the minimum number of edges that must be removed in order for the graph to be disconnected. The edge connectivity of a graph is denoted by $\lambda(G)$ and it is the cardinality of the minimum edge-cut of G.

An edge-cut X of a graph G is called a *minimal edge cut* if no proper subset of X is an edge-cut of G.

A graph G is *ℓ-edge connected*, where $\ell \in \mathbb{N}$ if $G - X$ is connected for every set of edges X with $|X| < \ell$. That is, if $\ell < \lambda(G)$, then G is ℓ-edge connected graph.

If G is a disconnected graph with components $G_1, G_2, \ldots, G_\ell$, then $\lambda(G) = \min_{1 \leq i \leq \ell} \lambda(G_i)$. Note that $\lambda(G) = 1$ if and only if G contains a bridge.

EXAMPLE 4.3.2 Consider the graph G in Figure 4.8. The removal of edge e_3 is not enough to disconnect the graph, nor is the removal of $\{e_3, e_4\}$. But the removal of $X_1 = \{e_2, e_3, e_4\}$ results in a disconnected graph. Similarly $X_2 = \{e_4, e_6\}$ is an edge-cut of G and so is $X_3 = \{e_1, e_4, e_6\}$. Among these three edge-cuts, X_2 is a minimal edge-cut of G. Since the cardinality of X_2 is the least, we have $\lambda(G) = 2$.

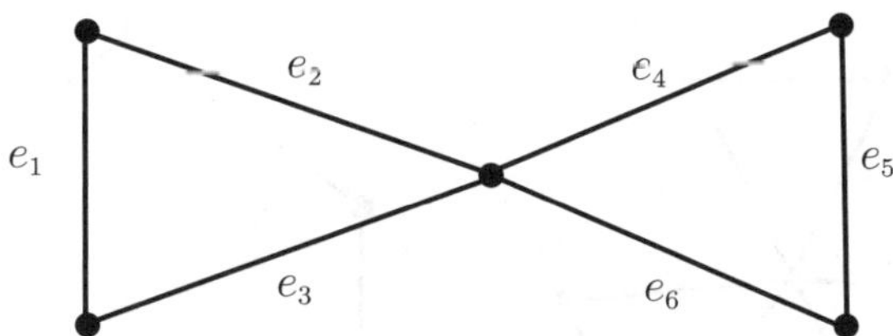

Figure 4.8 Edge connectivity of a graph G

It should be noted if G is a connected graph and X is a minimum edge-cut of G, then $G - X$ contains exactly two components. Because the removal of the last edge in X results in two components, refer to Remark 4.1.2. However, this is not the case when we consider a minimum vertex cut of G.

We will calculate the edge connectivity of a few well-known graphs.

The path P_n: For the path of length n, the removal of any edge disconnects the graph and so $\lambda(P_n) = 1$.

The cycle C_n: For the n-cycle C_n with $n \geq 3$, C_n becomes disconnected if any two edges are removed, but a connected graph arises if any edge is removed. Therefore $\lambda(C_n) = 2$.

Complete graph K_n: The edge connectivity of the complete graph K_n is $n - 1$. Clearly $\lambda(K_1) = 0$. Assume that $n \geq 2$. Since every vertex has degree $n - 1$, the removal of $n - 1$ edges incident with a vertex results in a disconnected graph. Thus $\lambda(K_n) \leq n - 1$. Let us consider a minimum edge-cut X of K_n. So $|X| = \lambda(K_n)$. Then $K_n - X$ has exactly two components, G_1 and G_2 where G_1 has order k, say, and G_2 has order $n - k$. Since X consists of all edges joining G_1 and G_2 and G is complete, it follows that $|X| = k(n - k)$. Because

$k \geq 1$ and $n - k \geq 1$, we have $(k-1)(n-k-1) \geq 0$ and so $(k-1)(n-k-1) = k(n-k) - n + 1 \geq 0$. Therefore $\lambda(G) = |X| = k(n-k) \geq n - 1$. Thus $\lambda(K_n) = n - 1$.

Complete bipartite graph $K_{m,n}$: The edge connectivity of the complete bipartite graph $K_{m,n}$ is $min\{m,n\}$. Clearly $\lambda(K_{m,n}) \leq min\{m,n\}$ because each vertex has degree m or n. In order to show that $\lambda(K_{m,n}) \geq min\{m,n\}$, we use Theorem 4.3.4, which states that it is sufficient to show that $\kappa(K_{m,n}) \geq min\{m,n\}$, by the previous section, we know that $\kappa(K_{m,n}) = min\{m,n\}$. Therefore $\lambda(K_{m,n}) = min\{m,n\}$.

Petersen graph: The edge connectivity $\lambda(G)$ of the Petersen graph G is the same as its vertex connectivity $\kappa(G)$, which equals 3. The proof for this assertion can be found in Theorem 4.3.6. Note that the Petersen graph is a 3-regular graph.

Recall that for a non-negative integer ℓ, a graph G is ℓ-edge connected if $\lambda(G) \geq \ell$. Consequently, every ℓ-edge connected graph is t-edge connected for every integer $0 \leq t \leq \ell$. It follows that every 1-edge connected graph is non-trivial and connected and every 2-edge connected graph is a connected graph of order 3 or more that contains no bridges. In Figure 4.9, G_1, G_2 and G_3 are 2-edge connected graphs which are not 1-edge connected.

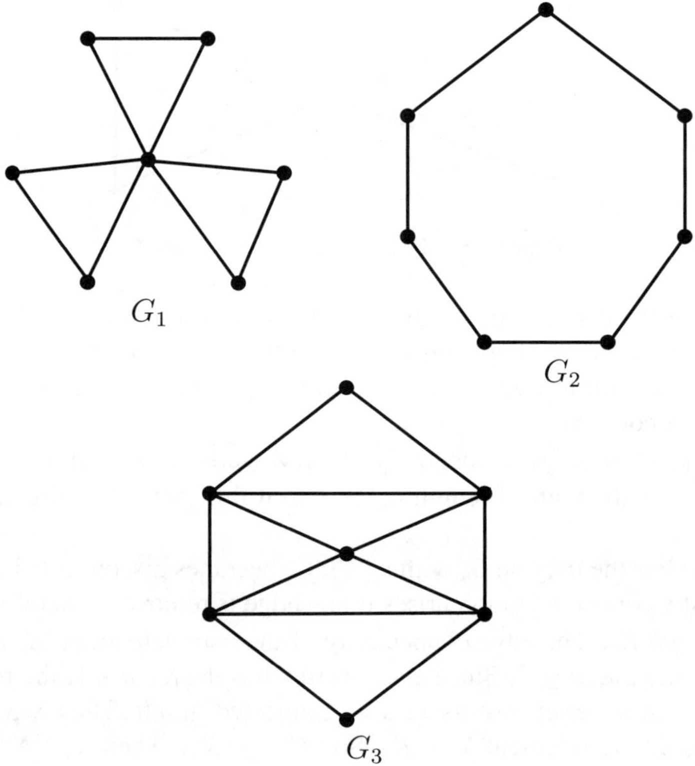

Figure 4.9   2-edge connected graphs

Before exploring the connectivity of a cubic graph, one must be familiar with the inequality that governs the vertex and edge connectivity of any graph. We will also study how connectivity and the degree of a vertex, as concepts are themselves connected. The degree of a graph and its connectivity are inter-twined concepts in the sense that the maximum and minimum degree of a graph helps to decide how many vertices should be removed in order for a graph to increase its components.

REMARK 4.3.3 Both the vertex connectivity number and the edge connectivity number of a graph are bounded above by the the minimum degree of the graph. For, let $v \in V(G)$ such that $deg(v) = \delta(G)$, the minimum degree of G. If we remove all the vertices in the neighbors of v or if we remove all the edges incident with v, then v is an isolated vertex in the resulting graph. So $\kappa(G) \leq \delta(G)$ and $\lambda(G) \leq \delta(G)$.

THEOREM 4.3.4 For every graph G, $\kappa(G) \leq \lambda(G) \leq \delta(G)$.

Proof. If G is trivial, that is a graph with no edges, then $\kappa(G) = \lambda(G) = 0$ and so the inequalities hold. If $G \cong K_n$ for some integer $n \geq 2$, then $\kappa(G) = \lambda(G) = \delta(G) = n-1$ and so the inequalities hold. The study of connectivity of a disconnected graph depends on its connected components, we may assume that G is a connected graph of order $n \geq 3$, that is not complete. Hence $\delta(G) \leq n-2$.

By Remark 4.3.3, it follows that $\lambda(G) \leq \delta(G) \leq n-2$. We only need to show that $\kappa(G) \leq \lambda(G)$.

Let X be a minimum edge-cut of G. Then $|X| = \lambda(G) \leq n-2$. Necessarily, $G-X$ contains exactly two components, G_1 and G_2. Suppose that the order of G_1 is k. So the order of G_2 is $n-k$. Therefore $k \geq 1$ and $n-k \geq 1$. Consequently every edge in X joins a vertex of G_1 and a vertex of G_2. We consider two cases.

Case 1: Every vertex of G_1 is adjacent to every vertex of G_2 in G. It is clear that $|X| = k(n-k)$. Since $(k-1)(n-k-1) \geq 0$, it follows that $(k-1)(n-k-1) = k(n-k)-n+1 \geq 0$, and so $\lambda(G) = |X| = k(n-k) \geq n-1$. However $\lambda(G) = |X| = k(n-k) \geq n-1$. But $\lambda(G) \leq n-2$. This is not possible and so this case cannot occur.

Case 2: There exist vertices u in G_1 and v in G_2 such that u and v are not adjacent in G. We now define a set U of vertices of G. For each $e \in X$, we select a vertex for U in the following way. If u is incident with e, then choose the vertex in G_2 that is incident with e as an element of U; otherwise, select the vertex that is incident with e and belongs to G_1 as an element of U. By the choice of U, we have $|U| \leq |X|$ and $u,v \notin U$. Since one end vertex of each edge of X is in U, every edge in X is removed in $G-U$. Therefore there is no $u-v$ path in $G-U$, it follows that $G-U$ is disconnected and so U is a vertex cut in G. Hence $\kappa(G) \leq |U| \leq |X| = \lambda(G)$ as desired. $\square$

Recall that $\kappa(C_n) = \lambda(C_n) = \delta(C_n) = 2$ for $3 \leq n \in \mathbb{Z}$. Also the graph G in Figure 4.10 is a graph with $\kappa(G) = 1, \lambda(G) = 2$ and $\delta(G) = 3$. These graphs provide evidence that there is a graph G where strict inequality and equality hold.

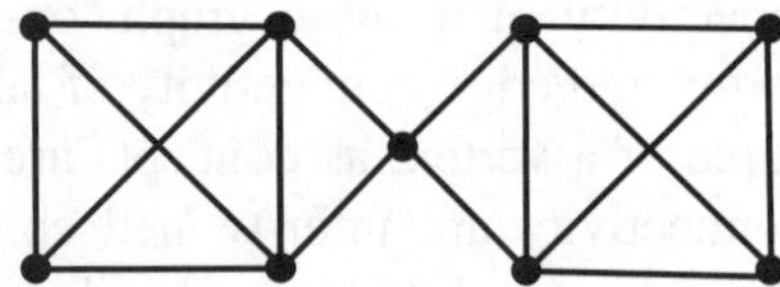

Figure 4.10　　A graph with $\kappa(G) = 1$, $\lambda(G) = 2$ and $\delta(G) = 3$

COROLLARY 4.3.5 If G is a graph of order n and size m, then $\kappa(G) \leq \lfloor \frac{2m}{n} \rfloor$.

Proof. We know that the sum of the degrees of the vertices of G is $2m$. It follows that the average degree of the vertices of G is $\frac{2m}{n}$ and so $\delta(G) \leq \frac{2m}{n}$. Since $\delta(G)$ is an integer, $\delta(G) \leq \lfloor \frac{2m}{n} \rfloor$. We know that $\kappa(G) \leq \lambda(G) \leq \delta(G)$. Hence $\kappa(G) \leq \lfloor \frac{2m}{n} \rfloor$. □

In what follows, the cubic graph is a 3-regular graph.

THEOREM 4.3.6 If G is a cubic graph, then $\kappa(G) = \lambda(G)$.

Proof. For a cubic graph G, it follows that $\kappa(G) \neq 0$ and $\lambda(G) \neq 0$. If $\kappa(G) = 3$, then by Theorem 4.3.4, we have $\lambda(G) = 3$. So we need to discuss only two cases, namely $\kappa(G) = 1$ or $\kappa(G) = 2$. Let U be a minimum vertex cut of G. Then $|U| = 1$ or $|U| = 2$. So $G - U$ is disconnected. Let G_1 and G_2 be two components of $G - U$. Since $deg(u) = 3$ for each $u \in U$, at least one of G_1 or G_2 contains exactly one neighbor of u.

Case 1: Suppose $\kappa(G) = |U| = 1$. Thus U consists of a cut-vertex u of G. Since some component of $G - U$ contains exactly one neighbor of u, say w, the edge uw is a bridge of G and so $\lambda(G) = \kappa(G) = 1$.

Case 2: Suppose $\kappa(G) = |U| = 2$. Let $U = \{u, v\}$. Assume that each of u and v has exactly one neighbor, say u' and v' respectively, in the same component of $G - U$, say G_1. Refer to Figure 4.11(a), when u and v are not adjacent. Let $X = \{uu', vv'\}$. Then there is no $u - u'$ path in $G - X$ and so X is an edge-cut of G. Therefore $\lambda(G) = \kappa(G) = 2$. We may assume that u has one neighbor u' in G_1 and two neighbors in G_2; while v has two neighbors in G_1 and one neighbor v' in G_2. For better understanding, the reader may refer to Figure 4.11(b). Since $deg(u) = deg(v) = 3$, we have $uv \notin E(G)$. Then $X = \{uu', vv'\}$ is an edge-cut of G. So $\lambda(G) = \kappa(G) = 2$. □

The concept of vertex connectivity or edge connectivity plays an important role in determining if a network is fault-tolerant. In large networks and network flow, the concept of connectivity helps to determine if the loss of a vertex or edge can still allow network flow without disconnecting the network. This measure is known as fault-tolerance and it is an important parameter for judging the quality of a network.

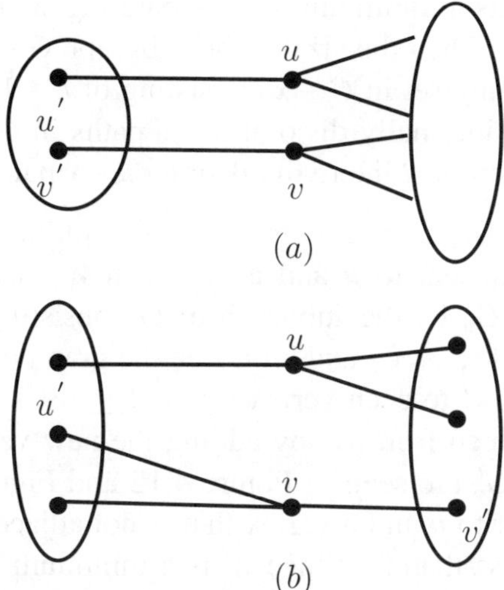

Figure 4.11   Diagrammatic representation for Theorem 4.3.6

4.4 Menger's, Whitney's and Dirac's theorems

What we have discussed so far is connectivity in terms of the degrees of the vertices or the number of vertices or edges to be removed in order to disconnect a graph. But there is another perspective on connectivity which will be explored through Menger's theorem.

DEFINITION 4.4.1 A set of vertices S of a graph G is said to *separate* two vertices u and v of G if $G - S$ is disconnected and u and v belong to different components of $G - S$. A $u - v$ separating set of minimum cardinality is called a *minimum $u - v$ separating set*.

DEFINITION 4.4.2 A collection $\{P_1, P_2, \ldots, P_k\}$ of $u - v$ paths in a graph is said to be *internally disjoint* if every pair of these paths has no vertices in common other than u and v.

THEOREM 4.4.3 (Menger's theorem) *Let u and v be non-adjacent vertices in a graph G. Then the minimum $u - v$ separating set equals the maximum number of internally disjoint $u - v$ paths in G.*

Proof. We proceed by induction on the size of the graph. Clearly, the result is trivially true for an empty graph. Assume that the result holds for all graphs of size less than m, a positive integer. Let G be a graph of size m. Let u and v be two vertices of G that are not adjacent. Assume that a minimum $u - v$ separating set has k vertices. Undoubtedly, G is limited to a maximum of k internally disjoint $u - v$ paths. Our claim is that G contains k internally disjoint $u - v$ paths. Three cases are to be examined.

Case 1: Suppose there exists a minimum $u-v$ separating set U in G containing a vertex x that is adjacent to u and v. Then the size of the subgraph $G-x$ is less than m and $U-x$ is a minimum $u-v$ separating set in $G-x$ consisting of $k-1$ vertices. By the induction hypothesis, there are $k-1$ internally disjoint $u-v$ paths in $G-x$. These paths together with the path $u-x-v$ constitute k internally disjoint $u-v$ paths in G.

Case 2: Suppose there exists a minimum $u-v$ separating set W in G containing a vertex in W that is not adjacent to u and a vertex in W that is not adjacent to v. Let $W = \{w_1, w_2, \ldots, w_k\}$. Let G_u be the subgraph of G consisting of all $u-w_i$ paths in G, where $w_i \in W$ for each $i \in \{1, \ldots, k\}$ and let G'_u be the graph obtained from G_u by adding a new vertex v' and joining v' to each vertex w_i for $1 \le i \le k$. Let G_v and G'_v be defined similarly where G'_v is obtained from G_v by adding the new vertex u'. Representations of the graphs G_u, G'_u, G_v and G'_v are seen in Figure 4.12 and Figure 4.13. Since W contains a vertex that is not adjacent to u and a vertex that is not adjacent to v, the size of each of the graphs G'_u and G'_v is less than m. Since W is a minimum $u-v$ separating set in G'_u, it follows by induction hypothesis that G'_u contains k internally disjoint $u-v'$ paths, each consisting of a $u-w_i$ path P_i followed by the edge w_iv'. Similarly there are k internally disjoint $u'-v$ paths in G'_v, each consisting of a w_i-v path Q_i preceded by the edge $u'w_i$. Since W is a $u-v$ separating set in G, the two graphs G_u and G_v have only the vertices of W in common. Therefore the k-paths obtained by following P_i by Q_i for each $i \in \{1, \ldots, k\}$ are internally disjoint paths in G.

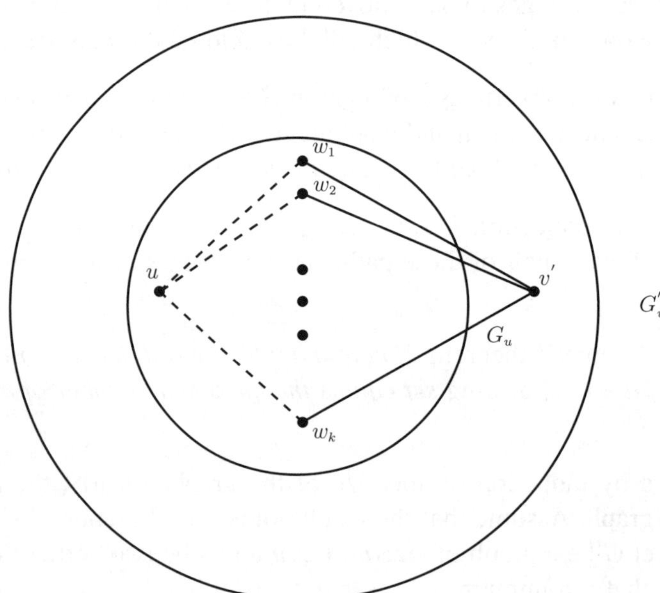

Figure 4.12　Diagrammatic representation of Menger's theorem on G'_u

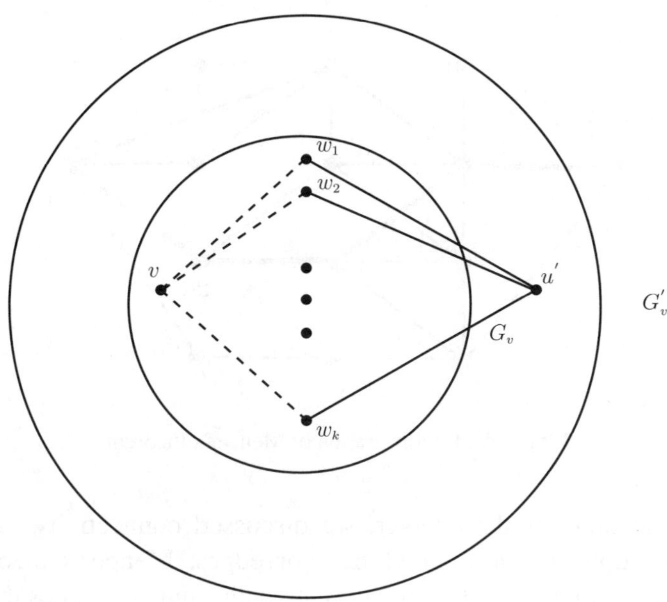

Figure 4.13   Diagrammatic representation of Menger's theorem on G_v'

Case 3: For each minimum $u - v$ separating set S in G, either every vertex of S is adjacent to u and not adjacent to v, or every vertex of S is adjacent to v and not adjacent to u. Let $P : u, x, y, \ldots, v$ be a $u - v$ shortest path in G and let $e = xy \in E(G)$. Consider the subgraph $G - e$ in G. Certainly, every minimum $u - v$ separating set in $G - e$ contains at least $k - 1$ vertices.

We claim that a minimum $u - v$ separating set in $G - e$ contains k vertices. Assume to the contrary that $G - e$ contains a minimum $u - v$ separating set $Z = \{z_1, z_2, \ldots, z_{k-1}\}$. Then $Z \cup \{x\}$ is a minimum $u - v$ separating set in G. Since x is adjacent to u, each vertex z_i, where $1 \le i \le k - 1$ is also adjacent to u and not adjacent to v. On the other hand, $Z \cup \{y\}$ is a minimum $u - v$ separating set in G. Since each z_i $(1 \le i \le k - 1)$ is adjacent to u, y is adjacent to u. Now $p - x$ is a $u - v$ path in G, which contradicts the fact that P is the shortest $u - v$ path in G. Therefore, as claimed, k is the minimum number of vertices in a $u - v$ separating set in $G - e$.

Now, by the induction hypothesis, there are k internally disjoint $u - v$ paths in $G - e$. Hence there are k internally disjoint $u - v$ paths in G as well. $\qquad\square$

An illustration of Menger's theorem is given in Figure 4.14. In the graph of Figure 4.14, $U = \{w_1, w_2, w_3\}$ is a $u - v$ separating set. Since there is no $u - v$ separating set with fewer than three vertices, U is a minimum $u - v$ separating set. By Menger's theorem, there are three internally disjoint $u - v$ paths in G, which are indicated by dotted, bold and dashed edges.

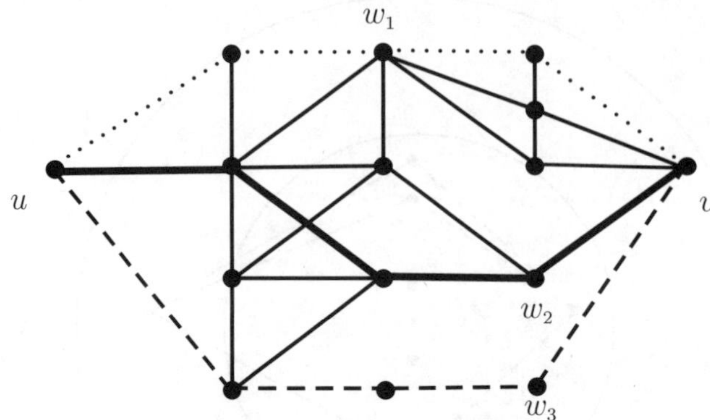

Figure 4.14   Illustration of Menger's theorem

In the earlier sections of the chapter, we discussed connectivity in terms of being invulnerable to multiple deletions of vertices or edges. Menger's theorem talks about connectivity in terms of having multiple alternate paths and also states that both concepts are essentially the same.

We are already familiar with the concept of a k-connected set. A set is k-connected if every vertex cut has at least k vertices where $k \geq 2$. In continuation of Menger's theorem, we will explore graph connectivity in terms of multiple $u - v$ paths and how they are related to a k-connected graph where $k \geq 2$.

THEOREM 4.4.4 (Whitney's theorem) *A non-trivial graph G is ℓ-connected for some integer $\ell \geq 2$ if and only if for each pair u and v of distinct vertices of G, there are at least ℓ internally disjoint $u - v$ paths in G.*

Proof. Assume that G is an ℓ-connected graph where $\ell \geq 2$. Let u and v be two distinct vertices of G. Suppose first that u and v are not adjacent. Let U be a minimum $u - v$ separating set. Then $|U| \geq \kappa(G) \geq \ell$. By Menger's theorem (Theorem 4.4.3), G contains at least ℓ internally disjoint $u - v$ paths. Next, suppose that u and v are adjacent, where $e = uv$. Then $G - e$ is $(\ell - 1)$-connected. Let W be a minimum $u - v$ separating set in $G - e$. Thus $|W| \geq \kappa(G - e) \geq \ell - 1$. By Menger's theorem, $G - e$ contains at least $\ell - 1$ internally disjoint $u - v$ paths, and so G contains at least ℓ internally disjoint $u - v$ paths.

Conversely, assume that G is a graph containing at least ℓ internally disjoint $u - v$ paths for any pair of distinct vertices u and v of G. Let U be a minimum vertex cut of G. Then $|U| = \kappa(G)$. Let x and y be vertices in distinct components of $G - U$. So U is an $x - y$ separating set of G. Since there are at least ℓ internally disjoint $x - y$ paths in G, by Menger's theorem, we have $\kappa(G) = |U| \geq \ell$. Thus G is ℓ-connected. $\qquad\square$

COROLLARY 4.4.5 *Let G be a k-connected graph and let S be any set of k vertices. If a graph H is obtained from G by adding a new vertex w and joining w to the vertices of S,* *then H is also k-connected.*

Proof is left as an exercise to the reader.

COROLLARY 4.4.6 *If G is a k-connected graph and $u, v_1, v_2, \ldots, v_k$ are $k+1$ distinct vertices of G, then there exists internally disjoint $u - v_i$ paths for all $i \in \{1, \ldots, k\}$ in G.*

Proof is left as an exercise to the reader.

We have "edge"versions of Menger's and Whitney's theorems. Interested readers may try to prove the following two results.

THEOREM 4.4.7 *For distinct vertices u and v in a graph G, the minimum number of edges of G that separate u and v equals the maximum number of pairwise edge-disjoint $u - v$ paths in G.*

THEOREM 4.4.8 *A non-trivial graph G is k-edge connected if and only if G contains k pairwise edge-disjoint $u - v$ paths for each pair u and v of distinct vertices of G.*

Before closing this chapter, let us discuss a sufficient condition for a graph to be k-connected graph. The following theorem is due to Gabriel Dirac.

THEOREM 4.4.9 (Dirac's theorem on k-connected graph) *If G is a k-connected graph where $k \geq 2$, then every set of k vertices of G lie on a common cycle of G.*

Proof. Let $S = \{v_1, v_2, \ldots, v_k\}$ be a set of k vertices of G. We show that there exists a cycle in G containing every vertex of S. Among all cycles in G, let C be the one containing the maximum number ℓ of vertices of S. We claim that $\ell = k$. Assume to the contrary that $\ell < k$. Since G is k-connected with $k \geq 2$, it follows that G is 2-connected. Then by Theorem 4.1.10, we have $2 \leq \ell$. Without loss of generality, we may assume that C contains the vertices $v_1, v_2, \ldots, v_\ell$ of S and that the vertices of S on C appear in the order $v_1, v_2, \ldots, v_\ell$ as we proceed cyclically about C, refer Figure 4.15. In particular, there is a $v_1 - v_2$ path P on C that contains no internal vertices belonging to S. Since $\ell < k$, there is a vertex $u = v_j \in S$, where $\ell + 1 \leq j \leq k$ that does not belong to C. Since $2 \leq \ell < k$, the graph G is ℓ-connected. Now applying Corollary 4.4.6 to the vertices $u, v_1, v_2, \ldots, v_\ell$, we see that G contains internally disjoint $u - v_i$ paths P_i for all $i \in \{1, \ldots, \ell\}$). Now the $u - v_2$ path P_2 together with $v_2 - v_1$ sector of C via v_ℓ and the $v_1 - u$ path P_1 produces a cycle C' containing the vertices $u = v_j, v_1, v_2, \ldots, v_l$. This is a contradiction to the fact that C contains the maximum number of vertices of S. Thus there exists a cycle in G which contains all the vertices of S.

$\square$

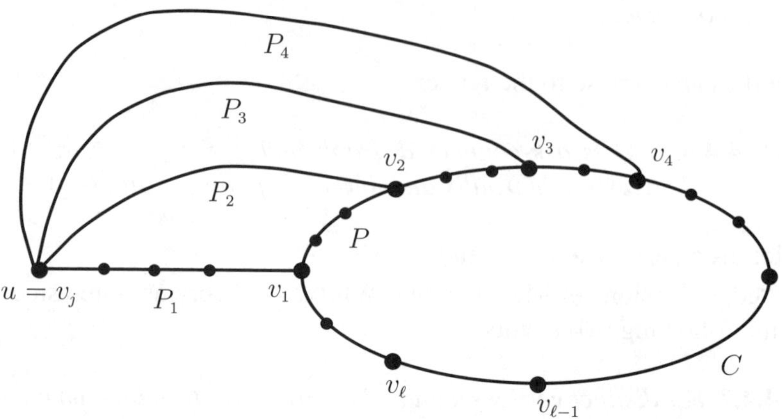

Figure 4.15 Diagrammatic representation for Dirac's Theorem

Summary

In this chapter, the concept of connectivity of a graph is discussed through different perspectives. We initially defined the cut-vertex and cut-edge of a graph. The properties of connectivity and how connectivity is perceived in a graph are clarified through various theorems. We discuss a type of graph called block, which does not disconnect despite removals of vertices and edges. In parallel, the concept of edge connectivity is also explored. The relationship between connectedness and connectivity is further analyzed using Menger's, Whitney's and Dirac's theorems which also provide a different perspective on how connectivity can be visualised. Detailed proofs of these theorems have been discussed.

4.5 Exercises

Section 4.1: Blocks, cut vertices and bridges

1. Provide an example of a graph that

    (i). contains more bridges than cut-vertices.

    (ii). contains more cut vertices than bridges.

2. Suppose a simple graph G on $n \geq 2$ vertices has at least $\frac{(n-1)(n-2)}{2} + 1$ edges. Then show that G is connected.

3. Suppose a general graph G has exactly two odd-degree vertices, v and w. Let G' be the graph created by adding an edge joining v to w. Prove that G' is connected if and only if G is connected.

4. Let G be a simple graph with degree sequence $d_1 \leq d_2 \leq \ldots \leq d_n$. For $k \leq n - d_n - 1$ and $d_k \geq k$, prove that G is connected.

5. A graph is k-regular if all the vertices have the same degree say, k. If G is a simple k-regular graph with n vertices, then what is the smallest integer k such that G is connected?.

6. Prove that a vertex v of a simple graph G can never be a cut-vertex of both G and its complement.

7. Show that the blocks of a connected graph is $k-$edge connected if and only if the graph is $k-$edge connected.

Section 4.2: Vertex connectivity

8. Suppose that G is a connected graph with at least three vertices. Form G' from G by adding an edge between every pair of vertices that are at a distance 2 apart in G. Prove that G' formed in this way is 2-connected.

9. Show that if G is a k-connected graph, then $G + K_1$ is $(k+1)$-connected.

10. Prove or disprove that if H is a subgraph of G, then $\kappa(H) \leq \kappa(G)$.

11. For a simple graph G with $\Delta(G) \leq 3$, show that $\kappa(G) = \kappa(G)'$.

Section 4.3: Edge connectivity

12. Provide a simple graph with

      (i). $\kappa(G) < \lambda(G) < \delta(G)$.

      (ii). $\kappa(G) = \lambda(G) < \delta(G)$.

      (iii). $\kappa(G) < \lambda(G) = \delta(G)$.

      (iv). $\kappa(G) = \lambda(G) = \delta(G) = 4$.

13. Show that if G is a k-edge connected graph, then $G + K_1$ is $(k+1)$-edge connected.

14. Determine $\kappa(G)$ and $\lambda(G)$ for the graph G given in Figure 4.16.

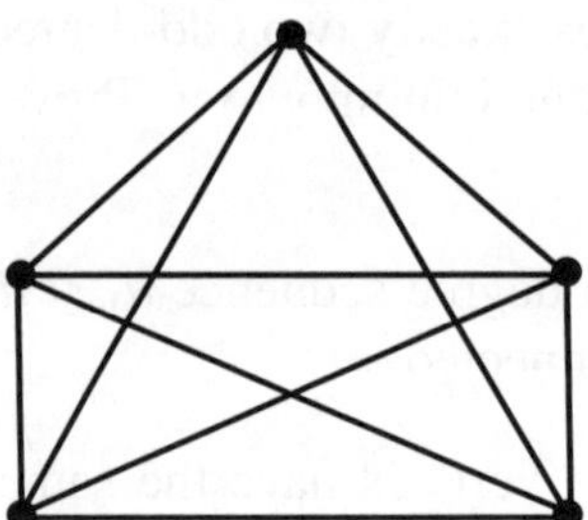

Figure 4.16   Graph G

15. Assume that $\lambda(G) = k > 0$. Prove that there are sets of vertices U and V, that partition the vertex set of G such that there are exactly k edges with one end point in U and one end point in V.

16. Find the vertex connectivity and edge connectivity of the $4-$cube Q_4. [The hypercube graph Q_n with $n = 4$ consists of vertices $\{(0,0,0,0),(0,0,0,1)\ldots(1,1,1,1)\}$. An edge exists between two vertices only if they are different by one co-ordinate. The Q_4 is also called a *tesseract*.]

17. Show that we obtain an isolated vertex by deleting an edge-cut of size 3 in the Petersen graph.

Section 4.4: Menger's, Whitney's and Dirac's theorem

18. Find the minimum number of vertices required to disconnect $K_{5,5}$. Explain how this relates to Menger's theorem.

19. Assume that G is a 5-connected graph and suppose u, v, and w be three distinct vertices of G. Establish that G contains two cycles C and C' that have only u and v in common but neither of which contains w.

20. Prove or disprove the converse of Dirac's theorem. Provide an example to support your answer.

5

Distance in Graphs

LEARNING OBJECTIVES

- Define distance and calculate distance between two vertices in a graph.
- Define eccentricity and calculate eccentricities of different vertices.
- Define radius and diameter of a graph.
- Discuss alternative definitions of distance such as the detour.
- Apply Dijkstra's algorithm to calculate shortest paths.
- Apply depth-first search to some graphs and record its traversal order.
- Apply breadth-first search to some graphs and document its traversal order.

As we discuss the topic of traversal within a graph, it becomes necessary to define the concept of distance in a graph. The definition of distance within a graph follows the same style of definition of distance between two cities or from one's location to one's destination. Just as it is logical to seek shortest paths in one's day-to-day life, we discuss the distance between vertices, in terms of the shortest path from vertex u to vertex v. In a connected graph G, the distance between two vertices u and v denoted as $d(u,v)$ is the shortest path between two vertices in a graph. In a disconnected graph, there are no paths between two vertices which lie in different components and so it follows that if u and v lie in different components of a disconnected graph, then $d(u,v) = \infty$. In this chapter, we will also be introduced to shortest path algorithms and graph traversal algorithms, that help one to navigate a graph while also paying attention to distance and connectivity.

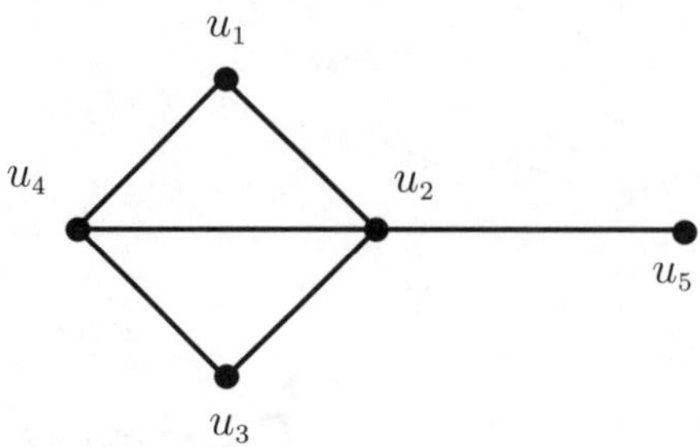

Figure 5.1   A connected graph G

In Figure 5.1, there are many ways to get from u_1 to u_5, like through the path P_1 : u_1, u_4, u_3, u_2, u_5 as well as the path P_2 : u_1, u_2, u_5. However it is P_2 that wins as the shortest path from u_1 to u_5. The reader can write down other paths between u_1 and u_5 and see for oneself that our claim is correct. It therefore follows that $d(u_1, u_5) = 2$. Since we are leaning towards defining the shortest path as the distance between two vertices in a connected graph, the definition follows.

DEFINITION 5.0.1 The *distance* between two vertices u and v, denoted by $d(u, v)$, in a graph is defined as the length of the shortest $u - v$ path. The $u - v$ path of length $d(u, v)$ is also called a $u - v$ *geodesic*. Also $d(u, v) = \infty$ if there is no $u - v$ path in a graph.

The concept of distance between two vertices does satisfy the following properties of a relation:

(a) $d(u, v) \geq 0$ for all $u, v \in V(G)$.

(b) $d(u, v) = 0$ if and only if $u = v \in V(G)$.

(c) $d(u, v) = d(v, u)$ for all $u, v \in V(G)$ [the symmetry property].

(d) $d(u, w) \leq d(u, v) + d(v, w)$ for all $u, v, w \in V(G)$ [the triangle inequality].

It is clear that the distance definition satisfies the property of equivalences in relation which the reader can verify as an exercise. We will verify the triangle inequality. Let P_1 be a $u - v$ geodesic and let P_2 be a $v - w$ geodesic in a graph G. Hence $P_1 \cup P_2$ is a $u - w$ path in G but it need not be a $u - w$ geodesic. Since $d(u, w)$ is the length of the $u - w$ geodesic, it follows that $d(u, w) \leq d(u, v) + d(v, w)$ for all $u, v, w \in V(G)$. The properties satisfied by distance $d(u, v)$ makes it a metric space and therefore opens some interesting doors to explore.

5.1   Eccentricity of a vertex

As we defined distance as a geodesic, it makes sense also to identify the farthest vertex from a given vertex. The concept called the eccentricity of a vertex v in a graph G is the distance between v and the vertex farthest from v in G.

DEFINITION 5.1.1 Let G be a connected graph and let $v \in V(G)$. Then, the *eccentricity* $e(v) = \max\{d(u,v) : u \in V(G)\}$.

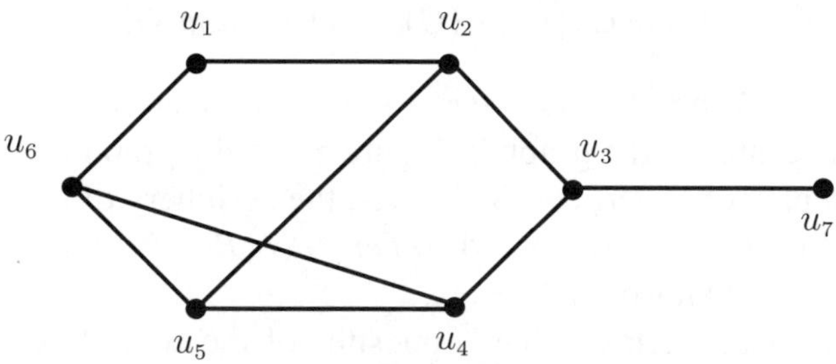

Figure 5.2   A connected graph G

We will use the graph in Figure 5.2 to determine the eccentricities of different vertices. For example, the vertex u_1 has eccentricity $e(u_1) = 3$. This is determined by calculating the distances, $d(u_1, u_i)$ for all $i \in \{2, 3, \ldots, 7\}$ and choosing the maximum among them as $e(u_1)$. Note that $d(u_1, u_2) = 1$, $d(u_1, u_3) = 2$, $d(u_1, u_4) = 2$, $d(u_1, u_5) = 2$, $d(u_1, u_6) = 1$ and $d(u_1, u_7) = 3$. Therefore $e(u_1) = 3$. Similarly one can find the eccentricities of all the vertices in this manner. The reader can determine the eccentricities of the rest as an exercise.

5.1.1   Radius and diameter of a graph

As mentioned in the previous section, it is possible to identify the one or more vertices which are the most distant from a given vertex using its eccentricity. While we are on the topic, we can ask an interesting question. Since eccentricities, as a concept, helps to visualize the most distant vertex from itself, thus imagining itself to be the "center", can there actually be a "center" of a graph, in terms of distance from the other vertices? The answer to the question is yes. We can identify a vertex that is central to the graph, in terms of distance. The "central" vertex (or vertices) is the one with the least eccentricity among all vertices, which in turn, translates to being "closer" in distance to all other vertices.

DEFINITION 5.1.2 In a graph G, the least eccentricity is called the *radius* and it is denoted by $rad(G)$ while the maximum eccentricity is called the *diameter* and denoted by $diam(G)$. That is $rad(G) = \min\limits_{v \in V(G)} e(v)$ and $diam(G) = \max\limits_{v \in V(G)} e(v)$.

DEFINITION 5.1.3 The *central vertex* of a graph G is the vertex v whose eccentricity is $rad(G)$. The *center* of G, denoted by $C(G)$, is the subgraph induced by central vertices. That is $C(G) = G[\{v \in V(G) : e(v) = rad(G)\}]$. Also if $C(G) = G$, then the graph is said to be *self-centered*.

DEFINITION 5.1.4 The vertex whose eccentricity equals $diam(G)$ is called a *peripheral vertex* of the graph G. The *periphery* of a graph G, denoted by $Per(G)$, is the graph induced by peripheral vertices. That is

$$Per(G) = G\left[\{v \in V(G) : e(v) = diam(G)\}\right].$$

The reader may verify that the diameter of the graph given in Figure 5.2 is 3 while its radius is 2. Also, the center of the graph in Figure 5.2 is the graph induced by $\{u_2, u_3, u_4\}$ and so $C(G) = P_3$, a path on 3 vertices. Moreover, the periphery of the graph in Figure 5.2 is the graph induced by $\{u_1, u_5, u_6, u_7\}$ and so $Per(G) = P_3 \cup K_1$, which is a disjoint union of a path on 3 vertices and an isolated vertex.

In a way the peripheral vertex is the "opposite" of the central vertex. The use of the words "radius " and "diameter" is intentional since we are envisioning the graph as a sort of circular structure with a "center". In conventional notation, the radius of a circle is half of its diameter. We are led to the question if the same is true for graphs as well. For the graph given in Figure 5.2, we saw that $rad(G) = 2$ and $diam(G) = 3$ which goes against the diameter being twice that of the radius. However it is possible that $diam(G) = 2rad(G)$, for some connected graphs. For instance, consider the path on 5 vertices P_5. Then $rad(P_5) = 2$ and $diam(P_5) = 4$. Furthermore, for the 4 cycle C_4, $rad(C_4) = diam(C_4) = 2$. So we can conclude that there is no equation that connects radius and diameter of a graph G. But we have the following inequality.

THEOREM 5.1.5 *For a non-trivial connected graph G, $rad(G) \leq diam(G) \leq 2rad(G)$.*

Proof. The inequality $rad(G) \leq diam(G)$ follows immediately from the fact that the smallest eccentricity cannot exceed the largest eccentricity. Let $diam(G) = d(u, v)$ for some $u, v \in V(G)$ and let w be a central vertex of G. By applying the triangle inequality to the distances between the vertices u, v and w, we have $d(u, v) \leq d(u, w) + d(w, v)$ which leads to $diam(G) \leq rad(G) + rad(G) = 2rad(G)$. $\square$

THEOREM 5.1.6 *For every two adjacent vertices u and v in a connected graph, the difference between their eccentricities do not exceed 1.*

Proof. Let u and v be any two adjacent vertices in a graph G and let $e(u)$ and $e(v)$ be their eccentricities. Without loss of generality, let us assume that $e(u) \geq e(v)$. Let x be a vertex, that is farthest from u. So $d(u, x) = e(u)$. Using the triangle inequality, $e(u) = d(u, x) \leq d(u, v) + d(v, x) \leq 1 + e(v)$. Hence $e(u) \leq 1 + e(v)$ which implies that $0 \leq e(u) - e(v) \leq 1$. Therefore $|e(u) - e(v)| \leq 1$. $\square$

The following result is a clear observation from Theorem 5.1.6.

COROLLARY 5.1.7 *If u and v are adjacent vertices in a graph G and $x \in V(G) \setminus \{u, v\}$, then $|d(u, x) - d(v, x)| \leq 1$.*

THEOREM 5.1.8 *Every graph is the center of some graph.*

Proof. Let G be any graph. We claim that G is the center of some graph G_1. Let u and v be any two new vertices that are added to the graph G such that u and v are made adjacent to all the vertices of G except themselves. Let u_1 and v_1 be two more vertices which are added to the newly constructed graph such that u_1 is adjacent to only u and v_1 is adjacent to only v. Let us name the resulting graph as G_1. Then in G_1, $e(u_1) = e(v_1) = 4$, $e(u) = e(v) = 3$ and $e(x) = 2$ for all $x \in V(G)$. It follows that all the vertices in G have the least eccentricity of 2. Therefore $C(G_1) = G$. $\square$

5.1.2   Detour

Is it possible to define distance in graphs in an alternate way, or perhaps in the opposite way? What if we defined distance in terms of the longest path between two vertices? That is precisely, where the concept of *detour* comes in. The concept of detour was successfully used in the *channel assignment problem* where FM radio stations located within a certain proximity of one another must be assigned distinct channels. Two FM stations which are on 101.3 Mhz and 101.5 Mhz, which are adjacent frequencies, are required to be atleast 72 kilometers apart. This distance might differ according to the classes of stations. But if the stations are closer to each other than the minimum required 72 kilometers, they must be assigned channels which have a bigger difference in their frequencies. This problem is tackled using graph coloring, which we will discuss in forthcoming chapters and the concept of detour, which we will define now.

DEFINITION 5.1.9 For two vertices u and v in a connected graph G of order n, the *detour* $D(u,v)$ from u to v is defined as the length of the longest $u - v$ path in G. A path of length $D(u,v)$ is called a $u - v$ detour.

For the graph G in Figure 5.2, we have $D(u_1, u_2) = 5$. The reader can determine the detour distance for other pairs of vertices. It is easy to see that, as in the case of geodesic, the detour distance also forms a metric.

DEFINITION 5.1.10 The *detour eccentricity* $e_D(v)$ of a vertex v is the maximum detour distance from a vertex v to any vertex of the graph G.

DEFINITION 5.1.11 The minimum detour eccentricity among all the vertices v of graph G is the *detour radius* and the maximum detour eccentricity is the *detour diameter*.

We can see that the definitions of detour works just like the usual definitions of distance, as in the case of geodesic. The reader can, as a fun exercise, think of other plausible ways of defining distance in graphs or even between graphs. There is a plethora of definitions that define or rather re-define distance, to suit certain specifications. But for now, let us think of distance to be the shortest path between two vertices as the normative definition.

5.2   Dijkstra's algorithm

When discussing simple connected graphs of a smaller order or size, it is easy to manually determine the shortest paths. But as graphs grow in size or order, it is not feasible to do so. We must rely on algorithms to find a shortest path between two vertices. The same argument holds for a directed graph or a weighted graph.

When discussing about a weighted connected graph, it is practical to think of shortest paths between two vertices of a weighted graph in terms of distance. Although the concept of distance is based on shortest path or geodesic between two vertices, it is not the same in the case of weighted graphs where the weight of each edge determines the path. We will discuss a famous algorithm in 1959 due to *Edsgar W. Dijkstra*, a Dutch computer scientist used to determine the shortest path between two vertices in a weighted graph.

Dijkstra's algorithm falls within the category of algorithms known as greedy algorithms. As discussed in the earlier chapter, the greedy technique works by constructing a solution through a sequence of steps such that each step builds on a partially constructed solution until a complete solution is obtained.

It is also a *single-source shortest path algorithm* meaning that, given a start or source vertex, the algorithm helps to determine the shortest path to another vertex from the source vertex in a weighted graph with non-negative weights.

Process of the algorithm: The algorithm works by first creating two sets of vertices, say S and T to represent the visited and unvisited vertices of a weighted graph G respectively. Let us assume that we need to determine the shortest path between two vertices, say u and v in G. In that case u is the *source vertex* while v is the *stop vertex* .

Initially S is empty while T contains all of the vertices. We initialize an empty queue called the priority queue and a minimum weight vertex, in each iteration is enqueued. We are also initializing a weight function $W(x_i)$ for all vertices $x_i \in V(G)$, which is based on distance. Initially the source vertex u has $W(u) = 0$, while the rest have weights $w(x_i) = \infty$. Just like the priority vertex, the weight function for each vertex keeps changing as we move through the graph. Let us keep in mind that in the weighted graph, the weights of the edges between the vertices x_i and x_j have already been given as $e(x_i, x_j)$ and this is different from the weight function $W(x_i)$.

We start at u which has an assigned weight of $W(u) = 0$ and we assign u to P, the priority vertex. We determine which of the vertices in the neighborhood of P is the next vertex to be visited based on some rules.

Rule 1: Update the weight function of all the vertices in the neighborhood of P except the vertices belonging to S. If $x_i \in N(P) - S$, then $W(x_i) = min\{W(x_i), W(P) + e(P, x_i)\}$.

Rule 2: Update the priority vertex by choosing the vertex with least weight among all the vertices in the neighborhood of P, except those belonging to S. Determine $W(x_k) =$

Algorithm 1: Dijkstra's Algorithm

Data: A weighted connected graph with $V(G)$ and non-negative weights $e(x_i, x_j)$ between each pair of vertices $x_i, x_j \in V(G)$.

Result: $W(P)$

```
;       /* The distance from source vertex u to the stop vertex v */
P ← 0;                      /* The priority queue, initially empty */
S ← 0;                          /* The set of visited vertices */
T ← V(G);                         /* The set of all vertices */
```

for (*Each vertex x_i in T*) {

```
    W(x_i) ← ∞;                 /* Initialize the distances as ∞ */
    W(u) ← 0;            /* The distance to the source vertex is 0 */
    P ← {(u,0)};        /* Add source vertex u to priority queue with
     distance 0 */
```

}

while $T \neq \emptyset$ **do**

  $P \leftarrow$ vertex in T with $minW(x_i)$

  if $P = v$ **then**

```
    | Output W(v);              /* Destination vertex v is reached */
```

  end

```
    Add P to S;                           /* Mark P as visited */
    Remove P from T; /* Remove P from the set of unvisited vertices
     */
    ;                              /* For each neighbor of P */
```

  for ($x_i \in N(P)$) {

```
        ;                       /* A neighbor of P not yet visited */
```

    if $x_i \notin S$ **then**

      if $W(P) + e(P, x_i) < W(x_i)$ **then**

        $W(x_i) \leftarrow W(P) + e(P, x_i)$

```
        Add or update (x_i, W(x_i)) in P; /* Priority queue is updated
         with new distances */
```

      end

    end

    Output $W(v)$

  }

end

$min\{W(x_i)/x_i \in N(P) - S\}$. Since x_k is the vertex with the least weight in $N(P) - S$, the vertex which was originally the priority vertex is now moved to set S of visited vertices, while x_k from T becomes the new priority vertex P.

Rule 3: Repeat Rules 1 and 2 until S contains the stop vertex v.

The complexity of Dijkstra's algorithm is dependent on the data structures used to implement the minimum priority queues. The complexities of Dijkstra's algorithm are given in Table 5.1.

If we were to use a *simple linear search* with no priority queue, it represents a data structure that does not proritize elements or complex search paths: it is a straighforward search based only on a sequence.

A *binary heap* is a complete binary tree data structure that maintains a specific order property and is commonly used to implement priority queues.

A *Fibonacci heap* is an advanced heap data stucture that consists of a collection of trees and provides a highly efficient way to manage dynamic sets for operations required in priority queues.

For the correctness of Dijkstra's algorithm given by "Dijkstra's algorithm runs on a weighted, directed graph $G = (V,E)$ with non-negative function w and source s terminates with a distance $d(s,u)$ for all vertices $u \in V$", the reader may refer to [9, Theorem 24.6].

Table 5.1   Complexities of Dijkstra algorithm

Data structure	Complexity	Suitable for
Simple linear search (No priority queue)	$O(n^2)$	Dense graphs
Binary Heap	$O((n+m)logn)$	Sparse to medium density graphs
Fibonacci Heap	$O(m+nlogn)$	Large, dense graphs

EXAMPLE 5.2.1 We will see a demonstration of Dijkstra's algorithm using the weighted graph in Figure 5.3.

Iteration 0: We need to find the shortest path from u to v in graph G. We initialize S with all visited nodes as containing u since u is the only vertex with least weight and also being the source vertex. T is initialized with unvisited nodes as $\{x_1, x_2, x_3, x_4, x_5, v\}$. Refer Table 5.2.

Iteration 1: In Iteration 1, the neighbors of u are x_1 and x_4 and the edge weights from u to x_1 and x_4 are revised from ∞ to 2 and 8 respectively. This is due to the fact that the weights $W(P) + e(P,x_i)$ for x_1 and x_4 are $0+2$ and $0+8$, since the current priority vertex is u and the weight function W_i for vertex u (which is currently P) has been initialized as 0. Now the least weight among the two is due to the vertex x_1 and so vertex x_1 becomes the

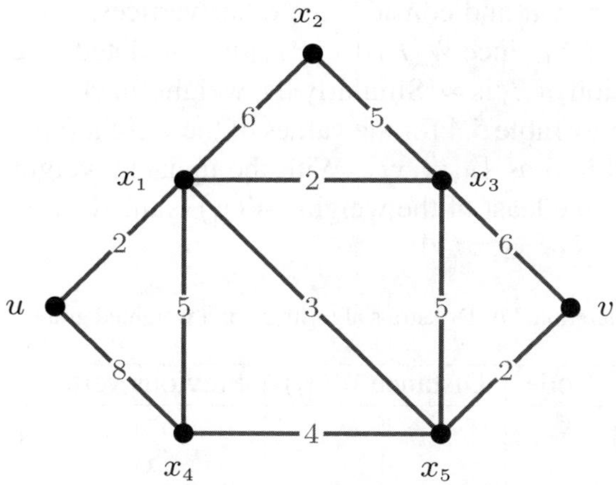

Figure 5.3   A connected weighted graph G

Table 5.2   Iteration 0 of Dijkstra's algorithm for a weighted graph in Figure 5.3

Node	Distance $W(x_i)$	Previous vertex
u	0	-
x_1	∞	-
x_2	∞	-
x_3	∞	-
x_4	∞	-
x_5	∞	-
v	∞	-

next priority vertex. The visited nodes in S are updated to $\{u, x_1\}$ and unvisited nodes to $T = \{x_2, x_3, x_4, x_5, v\}$. Table 5.3 gives more clarity to the iteration.

Table 5.3   Iteration 1 of Dijkstra's algorithm for a weighted graph in Figure 5.3

Node	Distance $W(x_i)$	Previous vertex
x_1	2	u
x_2	∞	-
x_3	∞	-
x_4	8	u
x_5	∞	-
v	∞	-

Iteration 2:   Since the vertex x_1 is currently the priority vertex P, we look into the weight function of all the neighbors of x_1. Now $N(x_1) = \{x_2, x_3, x_4, x_5, u\}$. Since we only consider

$N(P) - S$, we ignore vertex u and consider the other vertices. The weight function of x_2, $W(x_2)$ is now $min\{\infty, 2+6\}$ since $W(P) + e(P, x_i)$ is calculated as 2+6 and the previously initialized weight function of x_3 is ∞. Similarly the weight function of x_3 is now $min\{\infty, 2+2\}$. The reader can refer to Table 5.4 for the values of the weight function $W(x_i)$. The visited nodes are now updated in S as $\{u, x_1, x_3\}$. With the updated weight function in hand, the vertex corresponding to the least of the weights $W(x_i)$ is the next priority vertex, which in this case is x_3. Also $T = \{x_2, x_4, x_5, v\}$.

Table 5.4　　Iteration 2 of Dijkstra's algorithm for a weighted graph in Figure 5.3

Node	Distance $W(x_i)$	Previous vertex
x_2	8	x_1
x_3	4	x_1
x_4	7	x_1
x_5	5	x_1
v	∞	-

Iteration 3:　In Iteration 3, the visited nodes are $S = \{u, x_1, x_3, x_5\}$ and the next priority vertex, based on the least distance from u is x_5. Also the weight function of v has been updated to 10, in keeping with the conditions in the algorithm for updating the distances. Table 5.5 sheds light on the distance update for this iteration.

Table 5.5　　Iteration 3 of Dijkstra's algorithm for a weighted graph in Figure 5.3

Node	Distance $W(x_i)$	Previous vertex
x_2	8	x_1
x_4	7	x_1
x_5	5	x_1
v	10	x_3

Iteration 4:　Since the priority vertex is now x_5, we look at the vertices belonging to $N(x_5) - S$. Since this is $\{x_4, v\}$, we will now update the weight function of these two vertices. Clearly, from the table $W(x_4)$ and $W(v)$ are both 7. Now the choice of the priority vertex between these two vertices can be done in two ways. We can choose x_4 as a priority vertex and in the next iteration, terminate at v otherwise we can choose v as the priority vertex in this iteration, and move it to S, thus terminating the algorithm. Since our objective is a shortest path from u to v, there is no point in prolonging the process, and we choose the latter. The visited nodes are $S = \{u, x_1, x_3, x_5, v\}$. The vertex x_3 is omitted because the weight function of V is altered in the last iteration to 7. The change is due to the fact that x_5 is the previously visited vertex. Once v enters the set of visited nodes in S, the algorithm stops and yields us the shortest path as $u \rightarrow x_1 \rightarrow x_5 \rightarrow v$ with the least total weight of 7

as calculated by adding the weight functions of the vertices in the shortest path. Refer to Table 5.6 for more clarity.

Table 5.6   Iteration 4 of Dijkstra's algorithm for a weighted graph in Figure 5.3

Node	Distance	Previous vertex
x_2	8	x_1
x_4	7	x_1
v	7	x_5

Table 5.7 gives a detailed picture of the iterations of Dijkstra's algorithm to obtain a shortest path for the graph G in Figure 5.3

Table 5.7   Iterations of Dijkstra's algorithms for graph G in Figure 5.3

Vertex	Shortest distance					Previous vertex
	$W(x_i)$	I iteration	II iteration	III iteration	IV iteration	
u	0					
x_1	∞	$min\{\infty,2\}$				u
x_2	∞		$min\{\infty,8\}$	$min\{8,9\}$		
x_3	∞		$min\{\infty,4\}$			x_1
x_4	∞	$min\{\infty,8\}$	$min\{8,7\}$		$min\{7,9\}$	
x_5	∞		$min\{\infty,5\}$	$min\{5,9\}$		x_3
v	∞			$min\{\infty,10\}$	$min\{10,7\}$	x_5

REMARK 5.2.2 It is interesting to note that Dijkstra's algorithm works well for directed graphs with weights too. Dijkstra's algorithm works for a disconnected weighted graph, a non-weighted simple graph and also the directed graph. But Dijkstra's algorithm does not work for the weighted graph with negative weights. Instead, the Bellman–Ford algorithm determines the shortest path for a directed weighted graph with negative weights. Since we are yet to study directed graphs, the reader can look forward to the study of the Bellman–Ford algorithm in the last chapter. Notably, Dijkstra's algorithm has also been proven to be universally optimal in 2024 by a group of computer scientists from Zurich, Copenhagen and the USA [14]. It essentially means that Dijkstra's algorithm performs better than other shortest path algorithms in any graph.

5.3   Algorithms for graph traversal

The study of distance as a concept is incomplete without understanding the working of BFS (breadth-first search) and DFS (depth-first search). These two algorithms are used in a versatile manner, for various objectives. For example, we may use the BFS to determine the shortest path in a simple graph (without weights), the DFS for determining if a graph is connected or not. However, much more can be done using BFS or DFS. One can use the BFS or DFS to determine the vertices that are reachable from a designated root vertex in a graph. In other words, it traverses the graphs in levels, and creates a stack/queue of vertices, which have been "visited" during the traversal at each level. The word "level" is used to signify the vertices which are at a certain distance from a specified root vertex. For example, all the vertices which are at the distance of 2 from the root vertex may belong to one level, while those at a distance of 3 belong to the next level. These forests shed light on how the graph G, after a BFS or DFS may be reconstructed to trees (in connected graphs) or forests (in disconnected graphs).

As the name suggests, the BFS covers the "breadth" of the graph thoroughly at each level, before moving on to the next level. But the DFS works differently by going as "deeply" as possible, crossing different levels of a graph. Upon reaching the very end of the thread, (encountering a leaf) it backtracks to previous vertices, trying to find different paths or neighbors that were not "visited" earlier. Depending on one's needs and objectives, we can choose either algorithm.

5.3.1   Depth-first search

We shall now determine the DFS of a simple graph. The algorithm behind DFS works by merely following the 6 steps outlined below:

  (i) Mark all the vertices of the graph G as "unvisited".

  (ii) Choose any vertex v which has been marked as "unvisited".

 (iii) Change the status from "unvisited" to "visited".

 (iv) Move to another unvisited vertex $u \in N(v)$ and change its status to "visited".

  (v) Take $v = u$ and repeat Step (iv) until $N(v) = \emptyset$.

 (vi) Go to predecessor of v and repeat (ii), (iii) and (iv)

(vii) Output a list of vertices in the order in which they were visited.

Although it seems like the DFS is a simple sequence of steps, with the objective of marking a vertex as "visited ", there is more to the algorithm than what meets the eye. We must define the "backtracking" procedure, where the algorithm explores new vertices, by going back to a vertex, where there is a possibility of undiscovered vertices in its neighborhood. This is precisely, the reason for the creation of a Depth-First forest, composed of several Depth-First trees [9]. It is therefore, a necessity to color code the "visits" themselves, so that the back tracking procedure is clearly defined. We may use a particular color, say white, to mark all vertices as "unvisited ". The first time, a vertex v is encountered by the DFS, its color is altered to gray to mark a DFS-visit. But once its neighbors are all encountered and "grayed", v is now colored black, to indicate that all paths, from the vertex v have been explored. The reason we use a stack [12] is because it makes it easier to trace the operation of the DFS and the stack is a last-in-first-out (LIFO) data structure. We push a vertex on to the stack when the vertex is reached for the first time and we pop it off the stack, when it becomes a dead-end.

Algorithm 2: Depth-First Search Algorithm

DFS(G)

Data: A graph G with vertex set $V(G)$ and edge set $E(G)$ and a root vertex or start vertex u

Result: Stack S

```
;   /* All the vertices reachable from u are visited and added to
  a Stack S */
```

for (*each vertex* $v \in V(G)$) {

$\quad$ $color[v] \leftarrow$ WHITE ; `/* All unvisited vertices are colored WHITE */`

}

$S \leftarrow \emptyset$

for (*each vertex* $u \in V(G)$) {

$\quad$ **if** $color[u] = WHITE$ **then**

$\quad\quad$ | DFS-VISIT(u,S)

$\quad$ **end**

}

DFS-VISIT(u,S)

$color[u] \leftarrow GRAY$;   `/* When DFS encounters a WHITE vertex */`

$S \leftarrow S \cup \{u\}$; `/* A visited vertex is added to the top of the stack */`

for (*Each vertex* $v \in N(u)$) {

$\quad$ **if** $color[v] = WHITE$ **then**

$\quad\quad$ | DFS-VISIT (v,S)

$\quad$ **end**

$\quad$;   `/* Recursively visit all the neighbors */`

}

$color[u] \leftarrow BLACK$; `/* u is blacked to indicate it has been fully explored, and popped off the stack */`

We will now construct a table to understand how the DFS traverses the graph G in Figure 5.4.

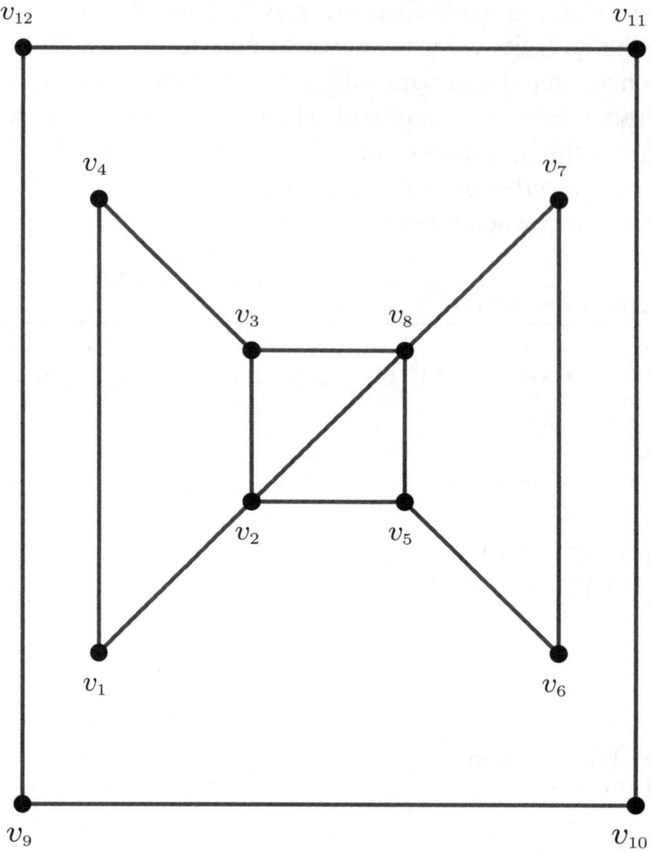

Figure 5.4   A simple graph G

In Table 5.8, we can see how the DFS is conducted. The **Visit** column contains the queue S in the order in which the vertices are visited. The adjacent column **COLOR=GRAY** gives a list of vertices whose color is changed to GRAY after being visited once. As per the algorithm, they are colored BLACK only if all the different paths have been explored. The **already visited** column contains vertices in Stack S, which when compared with $N(v)$ gives us the next vertex to be visited because we ignore vertices which are already visited in $N(v)$. The last column consists of vertices to which the DFS backtracks in order to explore a new path because there are two connected components.

There are two DFS trees that can be built from the graph G in Figure 5.4. As seen from Table 5.8, the vertex v_1 is the root of one DFS tree while v_9 is the root of the other DFS tree. After reaching v_7, the DFS cannot proceed further and it backtracks to v_6, v_5 and v_8 looking for vertices in its neighborhood which have not yet been visited. Since all vertices have been visited, the vertices are assigned the color BLACK and the DFS visit ends. But there are vertices in $V(G)$ which are still not visited and therefore the DFS picks a new

Table 5.8 Depth-first search of graph G in Figure 5.4

Visit	COLOR=GRAY	$N(v)$	Already visited	Backtrack
v_1	v_1	$\{v_2, v_4\}$	-	-
v_2	v_2	$\{v_1, v_3, v_5, v_8\}$	$\{v_1\}$	-
v_3	v_3	$\{v_2, v_4, v_8\}$	$\{v_2\}$	-
v_4	v_4	$\{v_1, v_3\}$	$\{v_1, v_3\}$	v_3
v_8	v_8	$\{v_2, v_3, v_5, v_7\}$	$\{v_2, v_3\}$	-
v_5	v_5	$\{v_2, v_6, v_8\}$	$\{v_2, v_8\}$	-
v_6	v_6	$\{v_5, v_7\}$	$\{v_5\}$	-
v_7	v_7	$\{v_8, v_6\}$	$\{v_8, v_6\}$	v_6, v_5, v_8, END.
v_9	v_9	$\{v_{10}, v_{12}\}$	-	-
v_{10}	v_{10}	$\{v_9, v_{11}\}$	v_9	-
v_{11}	v_{11}	$\{v_{10}, v_{12}\}$	v_{10}	-
v_{12}	v_{12}	$\{v_9, v_{11}\}$	$\{v_9, v_{11}\}$	v_{11}, v_{10}, v_9, END.

vertex v_9 and continues with the DFS visit. Backtracking occurs in v_{12} to v_{11} and later to v_{10}, v_9. Since the root has been reached and there are no further paths to explore, the DFS visit ends and $S = [v_1, v_2, v_3, v_4, v_8, v_5, v_6, v_7, v_9, v_{10}, v_{11}, v_{12}]$. A DFS forest helps to give a clearer picture of the graph G as seen in Figure 5.5. There are two trees with root vertex v_1 and v_9 with backtracking occurring at v_4, v_7 and v_{12}.

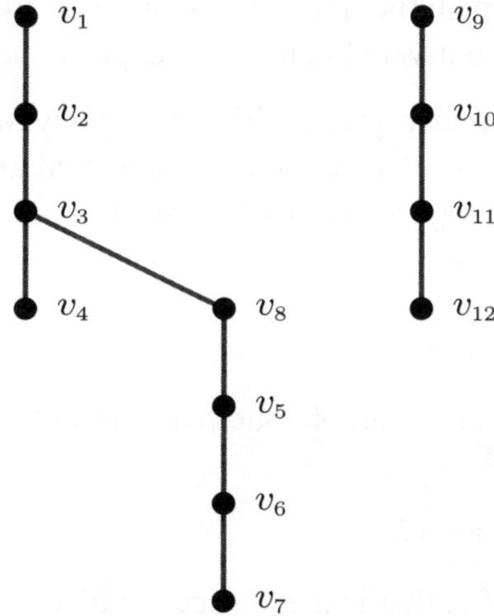

Figure 5.5 DFS forest of graph G in Figure 5.4

In the DFS algorithm, the vertex operation (initial coloring, gray coloring, adding to the stack, black coloring) is performed $O(n)$ times. Each edge uv is visited exactly once while exploring the neighbors of each vertex u. Combining the complexities of vertices and edges, the overall complexity is $O(n) + O(m) = O(n + m)$, where n is the number of vertices and m is the number of edges of the graph G. The linear complexity reflects that each vertex and edge is processed exactly once, thus making the DFS an efficient algorithm for traversing graphs.

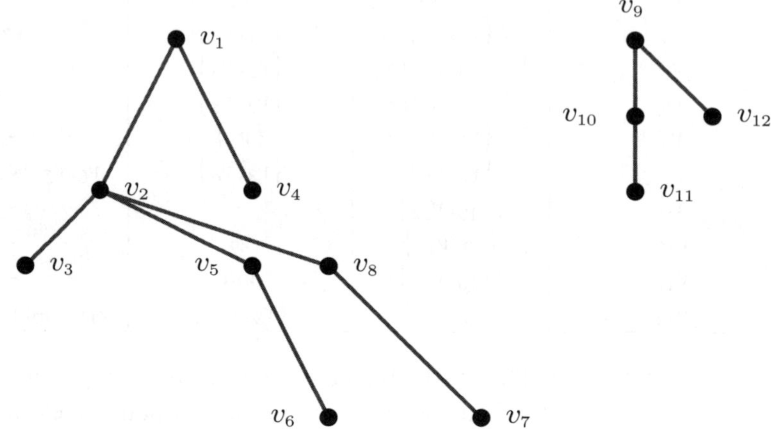

Figure 5.6   BFS forest of the graph G in Figure 5.4

For the correctness of the DFS, we direct the reader to the Paranthesis theorem [9, Theorem 22.7] and the white-path theorem [9, Theorem 22.9]. These theorems are beyond the scope of the book but the motivated reader can explore them further.

NOTE 5.3.1 A DFS forest can also give an idea of the acyclic nature of a graph. It helps one to visualize the different components of a disconnected graph, and also as a test to see if a graph is connected or disconnected. It can also be used to identify the cut-vertex or cut-edge of a graph.

5.3.2   Breadth-first search

The BFS also helps one to traverse a graph, but in a different manner. It works by following the simple steps outlined below:

(a) Begin at the start vertex $v \in G$.

(b) Visit all the neighbors of v, that is the vertices in $N(v)$.

(c) Once all vertices in this level are visited, move to the next level and repeat the process.

(d) Continue this process, until all the vertices of G are visited or a target vertex is found.

As mentioned earlier, the BFS [12] works through a graph by covering the breadth of a graph, which in a way is like traversing a graph in concentric circles unlike DFS, where we go as far as the path takes us. To initialize, a start vertex is identified and marked as "visited". In subsequent iterations the algorithm identifies all the unvisited vertices in the neighborhood of the start vertex and adds them to the queue after marking them as visited. At this point, the start vertex is removed from the queue. The neighborhood vertices of the next vertex in the queue(which is now the start vertex) is explored and if unvisited, are added to the queue. A BFS forest can be constructed by connecting each unvisited vertex in the neighborhood of a start vertex v, by means of a tree-edge to v, thus signifying that it is the child of v.

Algorithm 3: Breadth-First Search Algorithm

Data: A graph G with vertex set $V(G)$ and edge set $E(G)$ and a root vertex or start
       vertex u

Result: Queue Q

;   /* All the vertices reachable from u are visited and added to
a Queue Q */

Search(G)

for (*each vertex $v \in V(G)$*) {

    | $mark[v] \leftarrow$ NOT VISITED; /* All unvisited vertices are marked so
    | */

}

for (*each vertex $v \in V(G)$*) {

    | **if** $mark[v] == NOT\ VISITED$ **then**

    | | BFS(v)

    | **end**

}

BFS(v)

$Q \leftarrow \emptyset$

$mark[v] \leftarrow$ VISITED

ENQUEUE v into Q

while $Q \neq \emptyset$ **do**

    | $u \leftarrow FIRST(Q)$

    | DEQUEUE u from Q

    | **for** (*Each node $w \in N(u)$*) {

    | | **if** $mark[w] == NOT\ VISITED$ **then**

    | | | $mark[w] \leftarrow$ VISITED

    | | | ENQUEUE w into Q

    | | **end**

    | }

end

The overall time complexity of the BFS algorithm is $O(n+m)$. We will now discuss the BFS of graph G in Figure 5.4.

Table 5.9   BFS of the graph G in Figure 5.4

Start vertex	$N(v)$	Mark as VISITED	Dequeue	Queue
v_1	-	$\{v_1\}$	-	$[v_1]$
v_1	$\{v_2,v_4\}$	$\{v_2,v_4\}$	v_1	$[v_2,v_4]$
v_2	$\{v_3,v_5,v_8\}$	$\{v_3,v_5,v_8\}$	v_2	$[v_4,v_3,v_5,v_8]$
v_4	$\{v_1,v_3\}$	-	v_4	$[v_3,v_5,v_8]$
v_3	$\{v_2,v_8\}$	-	v_3	$[v_5,v_8]$
v_5	$\{v_2,v_6,v_8\}$	$\{v_6\}$	v_5	$[v_8,v_6]$
v_8	$\{v_2,v_3,v_5,v_7\}$	$\{v_7\}$	v_8	$[v_6,v_7]$
v_6	$\{v_5,v_7\}$	-	v_6	$[v_7]$
v_7	$\{v_8,v_6\}$	-	v_7	$[\,]$
v_9	-	$\{v_9\}$	-	$[v_9]$
v_9	$\{v_{10},v_{12}\}$	$\{v_{10},v_{12}\}$	v_9	$[v_{10},v_{12}]$
v_{10}	$\{v_9,v_{11}\}$	$\{v_{11}\}$	v_{10}	$[v_{12},v_{11}]$
v_{12}	$\{v_9,v_{11}\}$	-	v_{12}	$[v_{11}]$
v_{11}	$\{v_{10},v_{12}\}$	-	v_{11}	$[\,]$

As seen in Table 5.9, the order of visit of the BFS is $[v_1,v_2,v_4,v_3,v_5,v_8,v_6,v_7, v_9,v_{10},v_{12},v_{11}]$.

The start vertex represents the vertex whose neighbors are considered and marked as visited if they have not already been done so. The $N(v)$ column gives the complete list of the neighbors of the start vertex. Once they are marked as visited, they are added to the queue Q, and the start vertex is dequeued. This procedure is repeated until all the vertices of the graph G have been visited. The BFS tree is a representation of the breadth-first search conducted on a graph G and can be found in Figure 5.6. The correctness of the BFS algorithm is given by [9, Theorem 22.5].

"Let $G = (VE)$ be a directed or undirected graph and suppose that BFS is run on G from a given source vertex $s \in V$. Then, during its execution, BFS discovers every vertex $u \in V$ that is reachable from the source s."

REMARK 5.3.2 To summarize the two traversal algorithms, the DFS is used when one needs to travel very far and explore a path to its very end. A BFS is used when one needs to move through different levels of a graph, while thoroughly covering ground at each level, before moving on to the next level. To simply state it, the DFS is an algorithm for the brave, while the BFS is an algorithm for the cautious [12], and both of them are extremely useful for traversal in a graph.

Summary

The concept of distance as the length of the shortest path between any two vertices is initially defined. Auxiliary definitions such as eccentricity of a vertex, radius, diameter and periphery of a graph are illustrated with examples and their rationale is clarified. An alternative definition of distance as the longest path between the two vertices is also introduced as detour. Dijkstra's algorithm, a single-source shortest path algorithm is discussed in detail. We have also explored two important algorithms which are the corner stones of graph traversal. The DFS and the BFS have been investigated with the help of detailed explanations, pseudocodes and example problems.

5.4   Exercises

Section 5.1: Eccentricity of a vertex

1. Consider the graph in Figure 5.7. Find

    (i). $d(v_1, v_4)$.

    (ii). $d(v_3, v_7)$.

    (iii). $d(v_1, v_7)$.

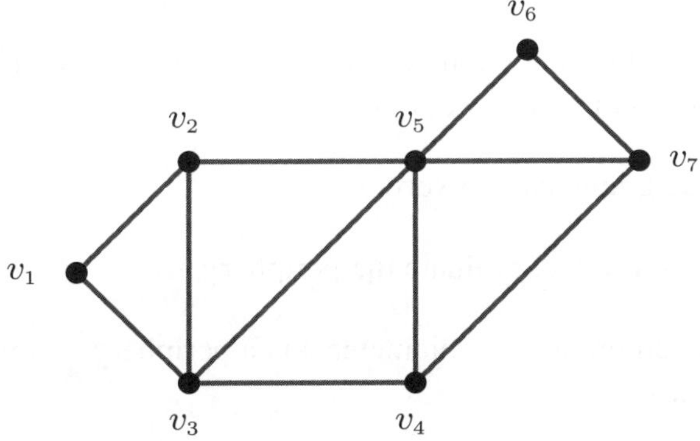

Figure 5.7   Graph G

2. Consider the graph G in Figure 5.7. Find

    (i). eccentricity $e(v_i)$ for all $v_i \in V(G)$.

    (ii). the center of G.

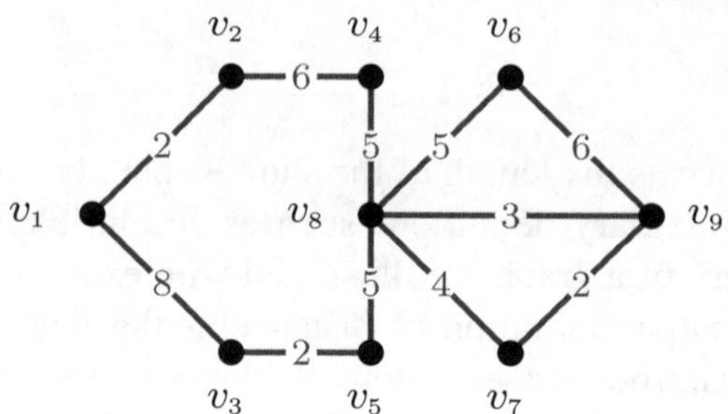

Figure 5.8   A connected weighted graph G

3. Determine the radius and diameter of $K_{s,t}$ for $1 \leq s \leq t$. Find the center of $K_{s,t}$.

4. Provide an example of a connected graph G where $C(G)$ is disconnected.

5. Suppose that G is a connected graph with a cut-vertex. If $u, v \in V(G)$ with $d(u,v) = diam(G)$, then prove or disprove that no block of G contains both u and v.

6. Show that $diam(\overline{G}) \leq 2$ when G is a disconnected graph.

7. Suppose that G is a connected graph and k is an integer with $rad(G) < k < diam(G)$. Show that there is a vertex v of G with $e(v) = k$.

8. Show that the center of every connected graph G is a subgraph for some block of G.

9. Suppose that G is a connected graph of order $n \geq 3$ that is not complete. Show that G is the periphery of some graph if and only if $\Delta(G) \leq n - 2$.

10. Draw a self-centered graph using 5 vertices.

11. For the graph G in Figure 5.7, evaluate the periphery.

12. Prove that a connected graph G of diameter 2 is a periphery of some graph if and only if G is self-centered.

Section 5.2: Dijkstra's algorithm

13. Find the shortest path from vertex x_2 to x_5 in Figure 5.3.

14. Find the shortest path from vertex v_1 to v_9 in Figure 5.8.

Section 5.3: Algorithms for graph traversal

15. Construct the DFS forest for the graph in Figure 5.2 by conducting a DFS on the graph.

16. Construct the BFS forest for the graph in Figure 5.2 by conducting a BFS on the graph.

17. Plot the DFS forest for the $C_5 + K_3$.

18. Plot the BFS forest for the Petersen graph.

6

Eulerian Graphs and Hamiltonian Graphs

- Define Eulerian graphs and distinguish between Eulerian paths and circuits.

- Recognize conditions under which a graph becomes Eulerian.

- Apply Fleury's algorithm to a graph.

- Apply Hierholzer's algorithm to a graph.

- Define Hamiltonian graphs and distinguish between Hamiltonian paths and cycles.

- Recognize conditions under which a graph becomes Hamiltonian.

- Discuss the Chinese postman problem (CPP) and construct the shortest Eulerian circuit for a graph.

- Discuss the traveling salesman problem (TSP) and construct the shortest Hamiltonian path for a graph.

In this chapter we explore some powerful tools for efficient traversal across graphs; the Eulerian and the Hamiltonian modes of traversal. These traversals are inspired by the efficiency of traversing an edge or a vertex exactly once. In this chapter, we will discuss two algorithms, namely Fleury's and Hierholzer's algorithms to determine an Eulerian circuit in a graph. This chapter also discusses the application of efficient traversals to two well known problems in network optimization: the Chinese postman problem (CPP) and the traveling salesman problem (TSP).

6.1   Eulerian graph

We are familiar with the bridges of Königsberg problem, which later became a question about the existence of Euler circuits in a graph. Let us now define the Euler graph in a formal manner.

DEFINITION 6.1.1 A trail in a graph G is called an *Euler trail* if it includes every edge of G. Thus a Euler trail is a walk through the graph that uses every edge exactly once.

EXAMPLE 6.1.2 Consider the graph G in Figure 6.1. We can observe that the graph G has Euler trail $v_2, e_1, v_1, e_5, v_5, e_6, v_4, e_{10}, v_1, e_9, v_6, e_8, v_4, e_7, v_3, e_4, v_2, e_2, v_6, e_3, v_3$.

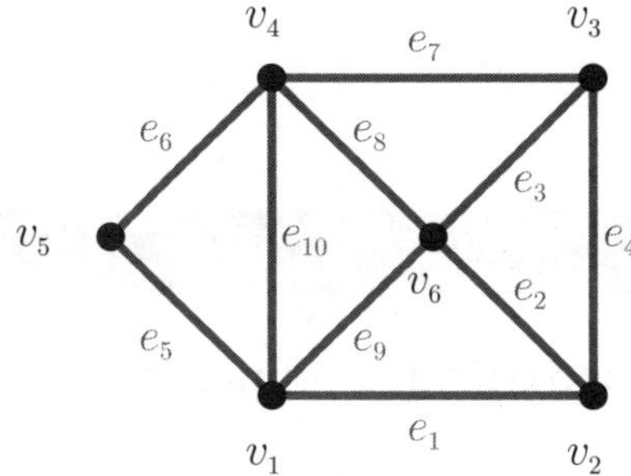

Figure 6.1   A graph with Euler trail

DEFINITION 6.1.3 An *Euler Circuit* of a graph G is a circuit that includes each edge of G exactly once. Thus an Euler circuit is just a closed Euler trail.

DEFINITION 6.1.4 A graph G is called an *Eulerian* or Euler graph if it has an Euler circuit.

EXAMPLE 6.1.5 Consider the graph G in Figure 6.2. Clearly G is Eulerian because it has an Eulerian circuit $v_2, e_1, v_1, e_5, v_5, e_6, v_4, e_{10}, v_1, e_9, v_6, e_8, v_4, e_7, v_3, e_4, v_2, e_2, v_6, e_3, v_3,$ e_{12}, v_7, e_{11}, v_2.

Note that the graph in Figure 6.1 is not Eulerian as it does not have any Euler circuit.

EXAMPLE 6.1.6 The graph G in Figure 6.3 representing the seven Königsberg bridges is not Eulerian as it does not contain any Euler circuit. The reader may refer to Chapter 1 for historical details of the Königsberg bridge.

At this point, it may not be straightforward to see why the graphs given in Figure 6.1 and Figure 6.3 are not having Eulerian circuits. But in the next section, we will give characterization of the Eulerian graph in terms of the degree of the vertices through which we can easily verify that the graphs given in Figure 6.1 and Figure 6.3 are not Eulerian graphs.

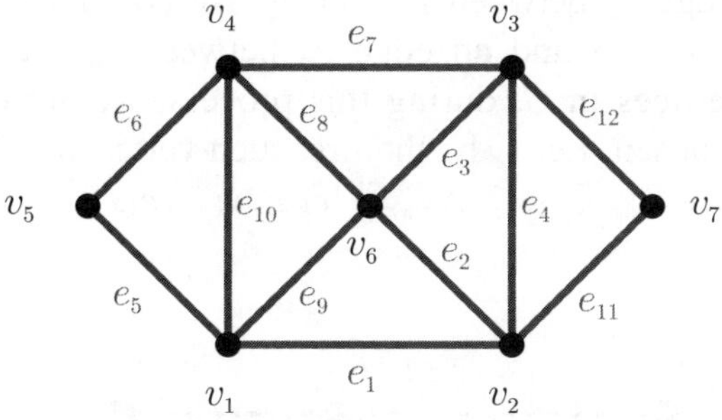

Figure 6.2 An Eulerian graph

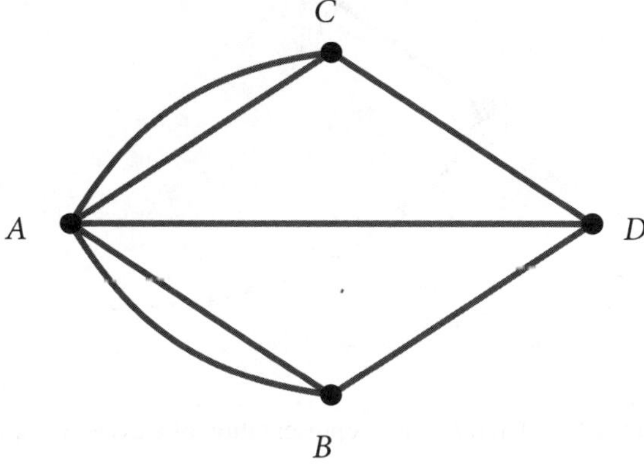

Figure 6.3 Graph representing Königsberg bridge problem

6.2 Characterization of Eulerian graphs

It is usually not hard to find an Euler trail in graphs of smaller order. Our objective is to determine whether a graph, regardless of its size, has an Euler trail or circuit. In this section, we are going to discuss about the necessary and sufficient conditions for a graph to be an Eulerian graph. Let us start with a result dealing with the existence of a cycle in a graph.

THEOREM 6.2.1 *Let G be a graph in which the degree of every vertex is at least two. Then G contains a cycle.*

Proof. Suppose G is not simple. Then G contains either a loop or a pair of parallel edges. In either case, G has a cycle of length 1 or a cycle of length 2. Hence in this case the result holds true.

Now, assume that G is simple. Let v_0 be any vertex of G. Since $deg(v_0) \geq 2$, there exists a vertex v_1 and an edge e_1 between v_0 and v_1. As $deg(v_1) \geq 2$, there exists a vertex $v_2 \neq v_0$ and an edge e_2 between v_1 and v_2. By continuing this process n times, there exists a vertex $v_n \neq v_{n-2}$ and an edge e_n between v_{n-1} and v_n. Since there are only finite number of vertices in G, during this process, we are compelled to choose a vertex which is already chosen. Let v_k be the first such vertex which has been chosen two times. Then $v_k, e_{k+1}, v_{k+1}, e_{k+2}, v_{k+2}, \ldots, v_{k+r-1}, e_{k+r}, v_{k+r}, e_{k+r+1}, v_k$ is a cycle as shown in the Figure 6.4.

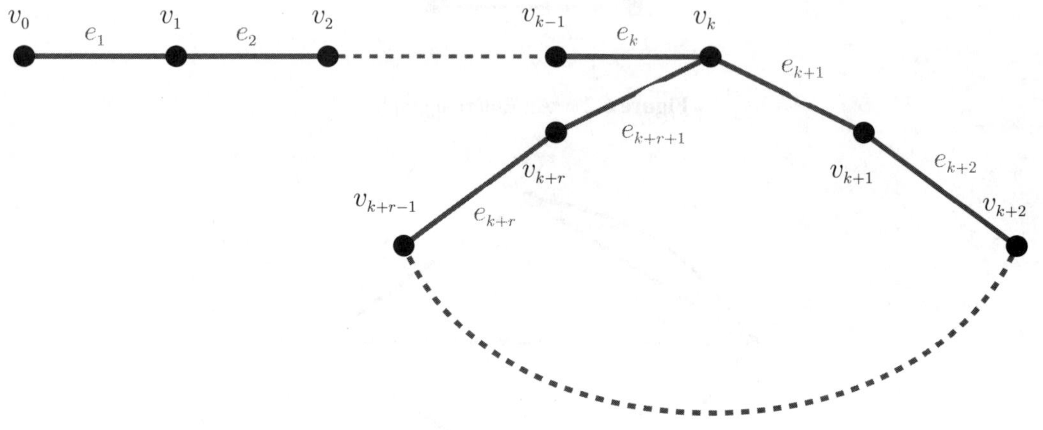

Figure 6.4   Diagrammatic representation of a cycle in a graph

$\square$

The following result is a characterization for the graph to be an Eulerian graph in terms of the degree of its vertex.

THEOREM 6.2.2 *A connected graph G is Eulerian if and only if the degree of every vertex in G is even.*

Proof. Assume that G is Eulerian. Then there exists an Eulerian circuit C in G. Let u be the initial and terminal vertex of C. Let $v \in V(G) \setminus \{u\}$. Since G is connected and the circuit C contains all the edges of G, v must be a vertex in C. Since each edge of G appears exactly once in C, each time v appears in C, there are two distinct edges incident with v. Therefore, if v appears k times in C, then the degree of v is $2k$, which is even. For the initial vertex u, if the vertex u appears ℓ times internally in C, then $deg(u) = 2\ell + 2$ because C begins and ends with u. Thus the degree of every vertex of G is even.

Conversely, assume that the degree of each vertex is even in the connected graph G. We prove that G is Eulerian by the induction on the number of edges of G. Initially, suppose G does not have any edges. Since G is connected, G should have only one vertex. Then G is trivially Eulerian. Using the induction hypothesis, assume that the result holds for all graphs whose number of edges is less than the number of edges of G. Since G is connected, there is no vertex of degree 0 in G. Since the degree of every vertex is even, we have $deg(v) \geq 2$ for all $v \in V(G)$. Then by Theorem 6.2.1, G contains a cycle C. Now if C contains all the edges of G, then C is an Euler circuit and hence G is Eulerian. Suppose that every edge of G is not in C. It is then possible to construct the subgraph H from G by deleting all the edges of C. The H, thus constructed, is a proper spanning subgraph of G. Note that the degree of each vertex in H is either same as that of G or decreased by two, due to the removal of edges of C. Therefore the degree of each vertex in H is also even. Thus, H with lesser edges than G, satisfies the induction hypothesis and therefore each component of H is Eulerian. Also each component of H has at least one vertex from C as these components are formed by deleting the edges of C. Refer to Figure 6.5, for the structural representation of the graph H. Starting at any vertex u of C, we will now construct an Euler circuit in G by going through the edges of C until we reach a vertex v_1 of a nonempty component of H. We can now traverse all of the edges of this particular component and return to v_1 because every component of H is Eulerian. We then keep going around C until the next non-empty component of H appears. We repeat this process until we reach the vertex u. This construction contains all the edges of C and all the edges of the non-empty components of H. Thus it contains all the edges of G exactly once, forming an Euler circuit. Hence G is an Eulerian graph. $\qquad\square$

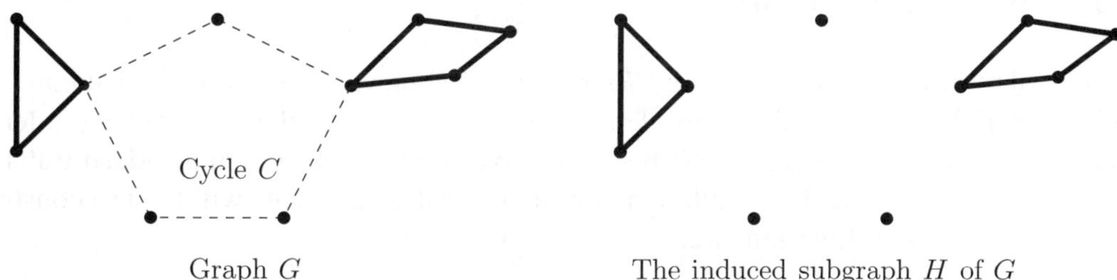

Figure 6.5   Graph G and the graph H after removing the edges of C

Euler stated the above remarkable result and was able to prove only the necessary part of the statement in 1736. The sufficiency was proved more than a century later in 1873 by *Carl Hierholzer*. Surprisingly, Hierholzer didn't publish the result when he was alive. After his death, his colleague with whom Hierholzer had discussed the proof, published the result with Hierholzer as the sole author. We will be seeing another contribution of Hierholzer in the form of the Hierholzer algorithm in the upcoming section.

Using Theorem 6.2.2, we give a characterization of graphs having Euler trails.

THEOREM 6.2.3 *A connected graph G has an Euler trail if and only if G has two vertices of odd degree. Furthermore, each Euler trail of G has these odd vertices as end vertices.*

Proof. Let $T : u, e_1, \ldots, e_m, v$ be an Euler trail of G. Now construct a new graph H from G' by adding a new vertex w and joining w with u and v. Denote the edge between w and u by e_{m+1} and the edge between w and v by e_{m+2}. Then G' has an Eulerian circuit $C : w, e_{m+1}, u, e_1, \ldots, e_m, v, e_{m+2}, w$. Then by Theorem 6.2.2, every vertex of G' has even degree. Since w is adjacent only to u and v in G', we get that u and v are the only two odd degree vertices in $G = H - w$.

Conversely, assume that G is a connected graph with exactly two odd vertices, say u and v. Construct a graph H from G by adding an edge e joining u to v. Then the degrees of u and v are increased by one and thus they are of even degree in H. Therefore each vertex in H is of even degree and hence, by Theorem 6.2.2, H has an Euler circuit, say $C : u, e, v, e_1, \ldots, e_m, u$. Now, by deleting e from the circuit C, we get an Euler trail $v, e_1, \ldots, e_m, u$ from v to u in G. $\qquad\square$

REMARK 6.2.4 (a) The graph given in Figure 6.1 has an Euler trail because the vertices v_2 and v_3 are the only vertices with odd degrees. Also, by Theorem 6.2.2, the graph given in Figure 6.1 is not Eulerian.

(b) The Königsberg graph, refer to Figure 6.3, is not Eulerian as all the vertices are of odd degree. Moreover, it does not have an Euler trail.

6.3 Fleury's algorithm

Fleury's algorithm was introduced by *Pierre Henri Fleury*, a French mathematician and engineer in 1883 as a recreational problem. However it is now used to construct an Eulerian circuit within an Eulerian graph. It is necessary to ensure that the graph is indeed Eulerian before we proceed to create an Eulerian circuit. The following rules will help to construct an Eulerian trail or an Eulerian circuit.

Process of the algorithm: The following steps give a brief description of the structure of Fleury's algorithm.

Step 1: Ensure that the graph G with vertex set $V(G)$ and edge set $E(G)$ has an Eulerian circuit or an Eulerian trail. This can be done by checking for the following parameters.

- All the vertices of the graph must have an even degree for an Eulerian circuit to exist.

- For an Eulerian trail to exist, exactly two of the vertices must have an odd degree, while the others have an even degree.

Step 2: If all the vertices are of even degree, start at an arbitrary vertex (say v_0). But if there are exactly two vertices of odd degree start at one of the vertices with an odd degree. Among the edges incident on v_0, select an edge (say v_0v_1) that is not a bridge. We may choose a bridge, if there is no other choice. The reason for this stipulation is to ensure that one does not isolate a part of the graph, unless necessary.

Step 3: Add the selected edge to a queue T.

Step 4: Remove this selected edge from $E(G)$.

Step 5: Move to the next vertex in the queue T (which is v_1, in this case) and check for edges incident upon this vertex that are not bridges. Repeat Steps 3, 4 and 5 until all edges are removed from $E(G)$.

Algorithm 1 gives the pseudocode to practically implement Fleury's algorithm. We initialize the queue T, which will contain the Eulerian circuit or path after the culmination of the algorithm. The *Function Eulerian*(G), consists of an odd-degree check to confirm if the graph has an Eulerian path or circuit by counting the number of vertices with odd degrees. If *Eulerian*(G) is valid, the *Start* vertex is selected based on the number of odd vertices. The *Function isBridge* (G, u, v) determines if an edge is a bridge by temporarily removing it and checking for connectivity of the $G - uv$ by using a breadth-first search (BFS). The *Function BFS* $(G, u, Target)$ helps to identify a target vertex and add it to the queue, while dequeuing the first vertex of the queue and making it the start vertex for the next iteration. The Function $DFS(G, u)$ conducts a depth-first search (DFS) in G for an edge that is not a bridge and adds it to T while removing it from the $E(G)$. The function terminates only if all the edges of $E(G)$ are added to T. Finally, the algorithm is run and the output is the Eulerian trail or circuit T, which contains a list of vertices, which is in fact, the sequence of edges forming the Eulerian circuit or trail.

EXAMPLE 6.3.1 Use Fleury's algorithm to construct an Euler circuit of the Eulerian graph G shown in Figure 6.6.

Iteration 1: Choose $v_0 = a$ and $T_0 = a$. Then choose edge e_1 to form $T_1 : a, e_1, d$ as e_1 is not a bridge of $G_0 = G$.

Iteration 2: Choose edge e_2 to form $T_2 : a, e_1, d, e_2, a$ as e_2 is not a bridge of $G_1 = G_0 - e_1$. Then choose edge e_3 to form $T_3 : a, e_1, d, e_2, a, e_3, b$ as e_3 is not a bridge of $G_2 = G_1 - e_2$.

Iteration 3: Choose edge e_4 to form $T_4 : a, e_1, d, e_2, a, e_3, b, e_4, c$ as e_4 is not a bridge of $G_3 = G_2 - e_3$.

Algorithm 1: Fleury's Algorithm

Data: A connected graph $G = (V(G), E(G))$
with $V(G) = \{v_1, v_2 \ldots v_n\}$ and $E(G) = \{e_1, e_2 \ldots e_m\}$
Result: T, a sequence of edges that form an Eulerian Circuit or Path of G

```
T ← ∅ ;                /* The Queue of edges of the Eulerian trail is
 initialized as an empty set */
```

Function Eulerian (G)
$oddvertices \leftarrow$ list of vertices in G with odd degree
if $length(oddvertices) > 2$ **then**
 | Return "No Eulerian path or circuit exists"
end
if $length(oddvertices){==}2$ **then**
 | $start \leftarrow oddvertices[0]$; /* Start from an odd degree vertex */
else
 | $start \leftarrow$ any vertex $\in V(G)$; /* Start from any vertex to construct
 | Eulerian circuit */
end

Function isBridge (G, u, v)
Remove edge (u, v) from G.
if *u and v are not reachable using BFS* **then**
 | Return TRUE ;   /* (u, v) is a bridge */
else
 | Return FALSE
end

Function BFS $(G, u, \textbf{Target})$
for (*Each* (u, v) *in* $E(G)$, *AND* $v \in N(u)$) {
 | **if** *(u, v) is not a bridge or there is only one incident edge at u* **then**
 | | $T \leftarrow T \cup \{(u, v)\}$ /* If an edge is not a bridge or if the
 | | number of incident edges on vertex u equals one, the edge
 | | is selected */
 | | RemoveEdge(G, u, v)
 | | DFS(G, v)
 | **end**
}
RUN Algorithm
$T \leftarrow$ Eulerian (G)
if $T{==}$*"No Eulerian Path or Circuit exists"* **then**
 | Print T
else
 | Print("Eulerian path or circuit:", T)
end

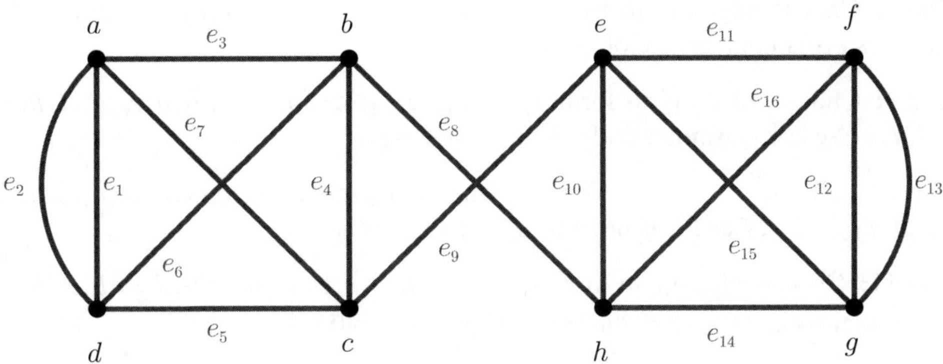

Figure 6.6 An Eulerian graph G

Iteration 4: Choose edge e_5 to form $T_5 : a, e_1, d, e_2, a, e_3, b, e_4, c, e_5, d$ as e_5 is not a bridge of $G_4 = G_3 - e_4$.

Iteration 5: Choose edge e_6 to form $T_6 : a, e_1, d, e_2, a, e_3, b, c_4, c, e_5, d, e_6, b$, though e_6 is a bridge of $G_5 = G_4 - e_5$, but we have no other choice than to take a bridge; see Figure 6.7.

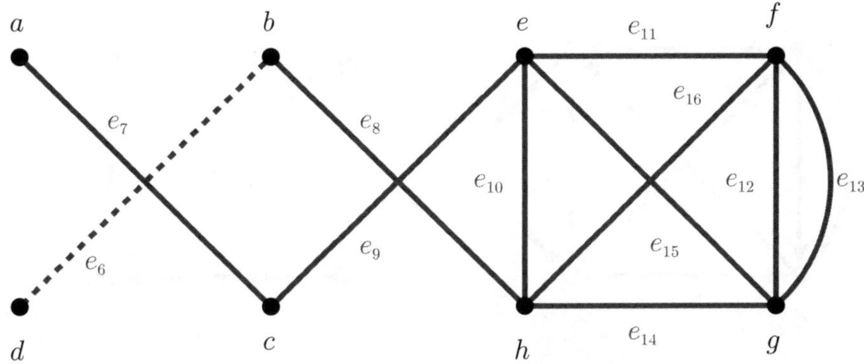

Figure 6.7 e_6 is a bridge of $G_5 = G_4 - e_5$

Iteration 6: Choose edge e_8 to form $T_7 : a, e_1, d, e_2, a, e_3, b, e_4, c, e_5, d, e_6, b, e_8, h$ though e_8 is a bridge of $G_6 = G_5 - e_6$, but we have no other choice than to take a bridge.

Iteration 7: Choose edge e_{10} to form $T_8 : a, e_1, d, e_2, a, e_3, b, e_4, c, e_5, d, e_6, b, e_8, h, e_{10}, e$ as e_{10} is not a bridge of $G_7 = G_6 - e_8$.

Iteration 8: Choose edge e_{11} to form $T_9 : a,e_1,d,e_2,a,e_3,b,e_4,c,e_5,d,e_6,b,e_8,h,e_{10},e,$ e_{11},f as e_{11} is not a bridge of $G_8 = G_7 - e_{10}$.

Iteration 9: Choose edge e_{12} to form $T_{10} : a,e_1,d,e_2,a,e_3,b,e_4,c,e_5,d,e_6,b,e_8,h,e_{10},e,$ e_{11},f,e_{12},g as e_{12} is not a bridge of $G_9 = G_8 - e_{11}$.

Iteration 10: Choose edge e_{13} to form $T_{11} : a,e_1,d,e_2,a,e_3,b,e_4,c,e_5,d,e_6,b,e_8,h,e_{10},e,$ $e_{11},f,e_{12},g,e_{13},f$ as e_{13} is not a bridge of $G_{10} = G_9 - e_{12}$.

Iteration 11: Choose edge e_{16} to form $T_{12} : a,e_1,d,e_2,a,e_3,b,e_4,c,e_5,d,e_6,b,e_8,h,e_{10},e,$ $e_{11},f,e_{12},g,e_{13},f,e_{16},h$ as e_{16} is the only choice left.

Iteration 12: Choose edge e_{14} to form $T_{13} : a,e_1,d,e_2,a,e_3,b,e_4,c,e_5,d,e_6,b,e_8,h,e_{10},e,$ $e_{11},f,e_{12},g,e_{13},f,e_{16},h,e_{14},g$ as e_{14} is the only choice left.

Iteration 13: Choose edge e_{15} to form $T_{14} : a,e_1,d,e_2,a,e_3,b,e_4,c,e_5,d,e_6,b,e_8,h,e_{10},e,$ $e_{11},f,e_{12},g,e_{13},f,e_{16},h,e_{14},g,e_{15},e$ as e_{15} is the only choice left.

Iteration 14: Choose edge e_9 to form $T_{15} : a,e_1,d,e_2,a,e_3,b,e_4,c,e_5,d,e_6,b,e_8,h,e_{10},e,$ $e_{11},f,e_{12},g,e_{13},f,e_{16},h,e_{14},g,e_{15},e,e_9,c$ as e_9 is the only choice left.

Iteration 15: Choose edge e_7 to form $T_{16} : a,e_1,d,e_2,a,e_3,b,e_4,c,e_5,d,e_6,b,e_8,h,e_{10},e,$ $e_{11},f,e_{12},g,e_{13},f,e_{16},h,e_{14},g,e_{15},e,e_9,c,e_7,a$ as e_7 is the only choice left.

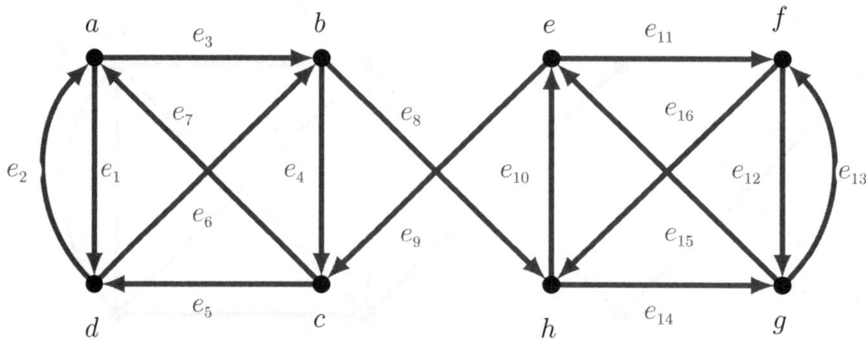

Figure 6.8   An Eulerian circuit T_{16}

Iteration 16: We stop at T_{16} as it contains all edges of G and is the required Eulerian circuit; see Figure 6.8.

Table 6.1 gives a bird's eye view of the various iterations of Fleury's algorithm. The START column, lists all the vertices that are chosen to become the START vertex, from which we try to choose an incident edge that is not a bridge. In case there are no other options, we can choose a bridge, as we see in the choice of e_6 and e_8. The second column provides a list of incident edges on the START vertex of the graph G'. The third column, gives the edge that was chosen and the fourth column TARGET provides the vertex on which the selected edge is incident. The Queue T lists out the trail of vertices and edges and the last column G' is the graph G_i that is created after the removal of the selected edges.

The dominating factor in Fleury's algorithm is the bridge checking process which requires performing a BFS traversal for each edge to determine if it is a bridge. The overall complexity is therefore $O(m \times (n+m))$. This can be further simplified to $O(m \times n)$. Since m is atleast proportional to n in a connected graph, the time complexity of Fleury's algorithm is $O(m \times n)$. The complexity makes Fleury's algorithm less efficient for large graphs, especially when there are many edges. The bridge checking process makes this algorithm slower when compared to other Eulerian path/circuit algorithms.

Table 6.1   Demonstration of Fleury's algorithm for the graph G in Figure 6.6

START	Incident edges	Edge chosen	Target	Queue T	G'
a	-	-	-	$T_0 = [a]$	$G_0 = G$
a	e_3, e_7, e_1, e_2	e_1	d	$T_1 = T \cup [e_1, d]$	$G_1 = G_0 - \{e_1\}$
d	e_2, e_6, c_5	e_2	a	$T_2 = T_1 \cup [e_2, a]$	$G_2 = G_1 - \{e_2\}$
a	e_3, e_7	e_3	b	$T_3 = T_2 \cup [e_3, b]$	$G_3 = G_2 - \{e_3\}$
b	e_4, e_6, e_8	e_4	c	$T_4 = T_3 \cup [e_4, c]$	$G_4 = G_3 - \{e_4\}$
c	e_5, e_7, e_9	e_5	d	$T_5 = T_4 \cup [e_5, d]$	$G_5 = G_4 - \{e_5\}$
d	e_6	e_6	b	$T_6 = T_5 \cup [e_6, b]$	$G_6 = G_5 - \{e_6\}$
b	e_8	e_8	h	$T_7 = T_6 \cup [e_8, h]$	$G_7 = G_6 - \{e_8\}$
h	e_{10}, e_{14}, e_{15}	e_{10}	e	$T_8 = T_7 \cup [e_{10}, e]$	$G_8 = G_7 - \{e_{10}\}$
e	e_9, e_{11}, e_{15}	e_{11}	f	$T_9 = T_8 \cup [e_{11}, f]$	$G_9 = G_8 - \{e_{11}\}$
f	e_{12}, e_{13}, e_{16}	e_{12}	g	$T_{10} = T_9 \cup [e_{12}, g]$	$G_{10} = G_9 - \{e_{12}\}$
g	e_{13}, e_{14}, e_{15}	e_{13}	f	$T_{11} = T_{10} \cup [e_{12}, f]$	$G_{11} = G_{10} - \{e_{13}\}$
f	e_{16}	e_{16}	h	$T_{12} = T_{11} \cup [e_{16}, h]$	$G_{12} = G_{11} - \{e_{16}\}$
h	e_9, e_{14}	e_{14}	g	$T_{13} = T_{12} \cup [e_{14}, g]$	$G_{13} = G_{12} - \{e_{14}\}$
g	e_{15}	e_{15}	e	$T_{14} = T_{13} \cup [e_{15}, e]$	$G_{14} = G_{13} - \{e_{15}\}$
e	e_9	e_9	c	$T_{15} = T_{14} \cup [e_{19}, c]$	$G_{15} = G_{14} - \{e_9\}$
c	e_7	e_7	a	$T_{16} = T_{15} \cup [e_7, a]$	$G_{16} = G_{15} - \{e_7\}$

6.4 Hierholzer's algorithm

In 1873, *Carl Hierholzer*, a German mathematician, created an algorithm for finding the Eulerian circuit in an Euler graph, using a different method of expanding a circuit while adding new edges to an existing circuit. In fact, Hierholzer's algorithm is more efficient than Fleury's algorithm to determine Eulerian circuits in undirected graphs. It works by following these steps in an Eulerian graph.

Step 1: Choose any arbitrary vertex as a start vertex, if the graph has no odd vertices. If the graph has two odd degree vertices, the start vertex is one of the odd degree vertices. Add it to a stack, which has been initialized with the start vertex.

Step 2: Follow the edges of the start vertex, to new vertices and edges. Take care not to traverse previously followed edges. New vertices are added to the stack.

Step 3: Repeat Step 2 until there are no more new edges to follow. If that is the case, we should have now formed a cycle, that is we returned to where we started from. The last vertex which stopped all progress, is pushed to the top of the stack which now holds all the vertices beginning with the start vertex.

Step 4: Backtrack from this vertex to the previous one in the stack. Check if there are unexplored edges. If true, go to Step 2.

Step 5: If there are no further edges left to traverse, the stack now holds the complete Eulerian circuit and we are done. Else, repeat Step 4.

We will see the pseudocode of this algorithm in Algorithm 2. In this pseudocode of Hierholzer's algorithm, we will first check if the graph satisfies the condition of being Eulerian, using the *Function Eulerian* which is the same as Fleury's algorithm. We initialize two data structures, namely *stack*[ ] and *path*[ ]. The start vertex from which we begin to create an Eulerian circuit is initialized using a variable *start*. The vertex that becomes *start* is determined by *Function Eulerian*. If the graph has only even-degree vertices, *start* is any vertex, while if there are two odd-degree vertices, one of them is assigned to *start*.

The array *stack* gets assigned with the *start* vertex. In the first few iterations, since most vertices are unvisited and edges are untraveled, visited vertices are added to *stack* while visited edges are added to *path*. An unmentioned point is that, edges are being removed from $E(G)$ while being added to *path*.

This leads to a trail of vertices and edges being visited until we arrive at a vertex, from which there are no untraveled edges. At this point, the *currentvertex* gets appended to path, while being removed from *stack* signifying the completion of a circuit. We backtrack to the previous vertex in the *stack* checking for unused edges, emanating from previous vertices. If there is any such vertex, it gets pushed onto the stack, making it the *currentvertex* while the *nextvertex* is the vertex to which the unused edge leads us. Thus the edge (*currentvertex*, *nextvertex*) gets appended to *path* while being removed from the set of edges $E(G)$. Now *nextvertex* is assigned to *currentvertex* and the process starts again, looking for a new circuit, if it exists.

Algorithm 2: Hierholzer's Algorithm

Data: A connected graph $G = (V(G), E(G))$
with $V(G) = \{v_1, v_2 \ldots v_n\}$ and $E(G) = \{e_1, e_2 \ldots e_m\}$
Result: *path*, a sequence of edges and vertices that form an Eulerian trail of G
Function Eulerian (G)
oddvertices $\leftarrow$ list of vertices in $V(G)$ with odd degree
if *length(oddvertices)* > 2 **then**
| Return "No Eulerian path or circuit exists"
end
if *length(oddvertices)*$==2$ **then**
| *start* $\leftarrow$ *oddvertices*$[0]$;   /* Start from an odd degree vertex */
else
| *start* $\leftarrow$ any vertex $\in V(G)$;   /* Start from any vertex to construct
| Eulerian circuit */
end

stack $\leftarrow$ *start* ; /* The stack of vertices of the Eulerian circuit is
  initialized with the start vertex */
path $\leftarrow [\,]$;   /* The list of edges of the Eulerian circuit is
  initialized as an empty set */
currentvertex $\leftarrow$ *start*
while *stack* $\neq \emptyset$ **do**
   if *currentvertex has no unused edges* **then**
     path.append(currentvertex); /* The currentvertex with no unused
      edges is added to the path list */
     currentvertex $\leftarrow$ *stack.pop*() ;   /* Remove the vertex from the
      stack and backtrack */
   else
     stack.push(currentvertex); /* The *currentvertex* is pushed onto the
      stack */
     nextvertex $\leftarrow \{v \in N(currentvertex)\}$;   /* Choose an adjacent
      vertex of the *currentvertex*, which is reached by an unused
      edge */
     path $\leftarrow$ *path.append(currentvertex, nextvertex)*
     $E(G) \leftarrow$ *removeedge(currentvertex, nextvertex)*
     currentvertex $\leftarrow$ *nextvertex* ;   /* Move to the next vertex */
   end
end
return *path*

We will see the working of the Hierholzer algorithm on the graph given in Figure 6.6.

EXAMPLE 6.4.1 Determine an Eulerian circuit for the graph in Figure 6.6, using Hierholzer's algorithm.

We have seen how Fleury's algorithm works for the same graph. Since in the graph all vertices with even degrees, we may start at any vertex and hence, *start* gets assigned with vertex a.

First circuit: The *path* array gets $[a,e_1,d,e_5,c,e_4,b,e_3,a,e_2,d,e_6,b,e_8,h,e_{10},e,e_9,c,e_7,a]$ while *stack* consists of the list of vertices $[a,d,c,b,a,d,b,h,e,c,a]$ in the order in which they were encountered. Since a has no more unused edges, it is popped off the stack, while being appended to *path* to indicate the completion of one circuit. The algorithm backtracks to c which is found to have no unused edges. So we backtrack further to e which is set to *currentvertex* since there are unused edges.

Second circuit: We now start a new circuit with e as $[e,e_{11},f,e_{12},g,e_{15},e]$. Since all unused edges have been exhausted, e is appended to the path and popped off the stack. The *path* array is currently $[a,e_1,d,e_5,c,e_4,b,e_3,a,e_2,d,e_6,b,e_8,h,e_{10},$ $e,e_9,c,e_7,a,e,e_{11},f,e_{12},g,e_{15},e]$ and the *stack* has $[a,d,c,b,a,d,b,h,e,c,a,e,$ $f,g,e]$. The second circuit starting and ending with the vertex e has been found. Since all the unused edges neighboring e have been exhausted, we backtrack to g which is found to have unused edges.

Third circuit: The vertex g becomes the starting vertex of a new circuit and we arrive at $[g,e_{14},h,e_{16},f,e_{13},g]$. We find that all unused edges neighboring g have been traveled and $E(G)$ has no more edges. The algorithm terminates at this point and the *path* is found to be $[a,e_1,d,e_5,c,e_4,b,e_3,a,e_2,d,e_6,b,e_8,h,e_{10},e,e_9,c,e_7,$ $a,e,e_{11},f,e_{12},g,e_{15},e,g,e_{14},h,e_{16},f,e_{13},g]$

It is obvious that the trail given by *path* does not quite seem like the Eulerian trail we are looking for, since it contains three different circuits. However, a little re-arrangement of the vertices and circuits leads to one complete Eulerian circuit, consisting of all the edges of G. The second circuit must be appended to the *path* structure at the point where we backtracked and reached the vertex e which is the start of the second circuit. Similarly, the third circuit must be appended to *path* at the point where we backtracked and reached g. The refined Eulerian trail now looks like $[a,e_1,d,e_5,c,e_4,$ $b,e_3,a,e_2,d,e_6,b,e_8,h,e_{10},e,e_{11},f,e_{12},g,e_{14},h,e_{16},f,e_{13},g,e_{15},e,e_9,c,e_7,a]$.
Hierholzer's algorithm works in a way that we create sub-tours or sub-circuits which are later combined to form one large circuit that traverses all the edges and vertices of the graph.

The complexity of the odd-degree check takes $O(n+m)$ while the main loop consisting of stack operations and edge traversal is $O(m)$. Combining all the steps, the overall complexity of Hierholzer's algorithm is $O(n+m)$ since each vertex and each edge is processed once, making it an efficient algorithm for finding Eulerian paths or circuits.

6.5 Chinese postman problem

In this section, we discuss an application of the Eulerian graph, called the Chinese postman problem (CPP). A postman picks all letters from the post office and travels along each street to deliver letters and comes back to the post office to return all undelivered letters. He wishes to travel all the streets (at least) once to deliver all the letters, but covering the least possible distance. In 1962, a Chinese mathematician *Meigu Guan* was interested in granting the postman's wish and found the shortest route for him. The problem was thus named as the Chinese postman problem in his honour. CPP helps one to find a closed path starting at a vertex v(post office) and traversing each edge (street) at least once and then coming back to the same vertex v such that the total length of the path is minimum.

We can rephrase the CPP as follows: Form a weighted graph G in which each edge represents a street, each vertex represents a junction of streets and the weight assigned to each edge represents the length of the street between junctions. Denote the weight of edge e_i as $w(e_i)$ and define the weight of a tour $T : v_0, e_1, v_1, e_2, \ldots, e_n, v_0$ as $w(T) = w(e_1) + w(e_2) + \ldots + w(e_n)$. Now the solution to the CPP is to find a tour of minimum weight in a weighted connected graph. If the constructed graph G is Euler, then any Euler circuit of G is the required closed path that contains each edge exactly once and thus the weight of the Euler circuit is minimum. In that case, we can use Fleury's algorithm to obtain an Euler circuit. If G is not Euler, then there exist vertices of odd degree, which may be either odd or even in number. An edge e is said to be *duplicated* when its ends are joined by a new edge with the same weight $w(e)$ as of e. We duplicate suitable edges in the tour to make odd degree vertices into even degree vertices. The graph G, together with the shortest path between odd degree vertices through duplicated edges forms Euler supergraph G^*. Then any Euler circuit in G^* is the solution for CPP.

We summarize the procedure to find the solution to CPP:

(a) If the weighted graph G is Eulerian, then any Eulerian circuit is the solution. Use Fleury's algorithm to find one such circuit.

(b) If G is not Eulerian, then G has odd degree vertices. Suppose G has exactly two vertices of odd degree u and v. Use Dijkstra's algorithm to find a shortest path P between u and v. Then duplicate all the edges involved in the path P. The resulting graph G^* is Eulerian and we are done.

(c) If G has 4 odd vertices a, b, c and d. Then there are three possible pairings: (i) ab and cd, (ii) ac and bd, or (iii) ad and bc. Again use Dijkstra's algorithm to find the shortest paths for the two pairs. Select the pairing whose total value is less than the other two pairings, say for example, the sum of weights of shortest paths for *ab* and *cd* is minimum, then duplicate all the edges involved in these two paths.

(d) The same procedure will be applied if *G* has more odd degree vertices, but the process becomes more and more tedious if the number of odd vertices increases.

We illustrate the above procedure with the following example.

EXAMPLE 6.5.1 Solve the CPP for the graph shown in Figure 6.9.

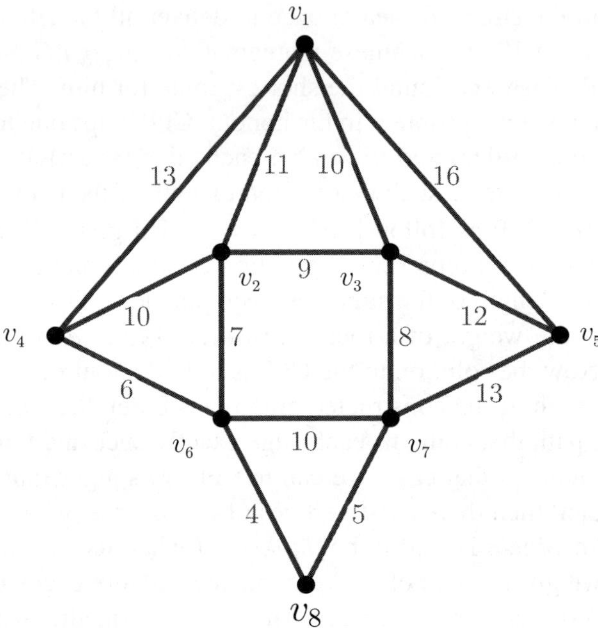

Figure 6.9   Graph *G* with two odd vertices

The graph *G* is not Eulerian as it contains two odd vertices v_4 and v_5. We need to connect these two vertices by a shortest path. By Dijkstra's algorithm, we get a shortest path $P : v_4, v_6, v_8, v_7, v_5$. So we duplicate the edges v_4v_6, v_6v_8, v_8v_7, and v_7v_5. The resulting supergraph G^* is Eulerian. Duplicated edges are shown in dotted lines in Figure 6.10. As G^* is Eulerian, we apply Fleury's algorithm to get an Euler circuit $v_1, v_4, v_6, v_8, v_7, v_5, v_1, v_2, v_4, v_6, v_8, v_7, v_5, v_3, v_2, v_6, v_7, v_3, v_1$.

In the next example, we solve a more complicated CPP having four odd vertices.

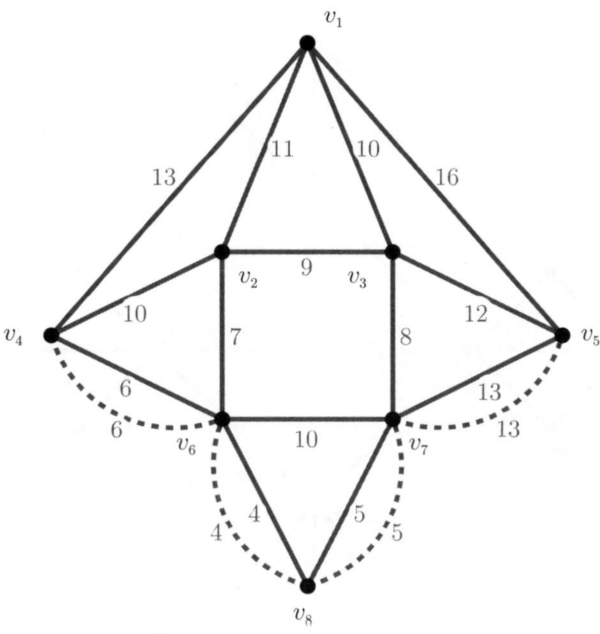

Figure 6.10   Supergraph G^* of G with duplicated edges

EXAMPLE 6.5.2 Solve the CPP for the graph shown in Figure 6.11.

The graph G is not Eulerian as it has four odd vertices a, n, k and g. Then, the following are the three possible paths:

(i). $a - g$ and $k - n$,

(ii). $a - k$ and $g - n$, and

(iii). $a - n$ and $g - k$.

We apply Dijkstra's algorithm to get a shortest path for the above pairs with their weights as shown in Table 6.2. From Table 6.2, we can see that the total weight of the shortest

Table 6.2   Shortest paths for the pairs

Shortest path	Weight	Shortest path	Weight	Total weight
a, d, e, f, g	9	k, j, i, q, n	14	23
a, d, r, i, j, k	18	g, f, e, r, q, n	13	31
a, d, r, q, n	12	g, f, e, i, j, k	16	28

paths for the pair $a - g$ and $k - n$ is minimum. Hence we duplicate all the edges belonging to these two paths. These duplicate paths are shown in dotted lines in Figure 6.12. Now the

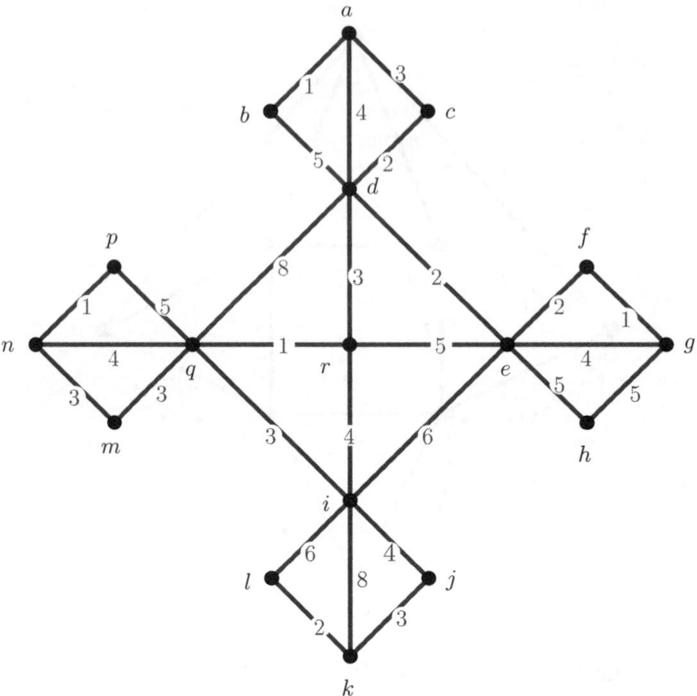

Figure 6.11   Graph G with four odd vertices

supergraph G^* is Eulerian, we can get the Eulerian circuit:
$a,d,b,a,c,d,e,f,g,f,e,g,h,e,r,d,e,i,j,k,j,i,l,k,i,q,r,i,q,n,q,m,n,p,q,d,a.$

To write the pseudocode for an algorithm that determines the solution of a CPP, one must keep in mind, that when the odd degree vertices are two in number, the solution is much simpler than when the odd degree vertices are four in number. The solution becomes progressively more tedious to obtain when the number of odd degree vertices increases. If the graph has an Euler circuit, it naturally becomes the solution to the CPP as well. Such an Euler circuit can be constructed using Fleury's algorithm mentioned in Algorithm 1 or Hierholzer's algorithm in Algorithm 2, when the graph is Eulerian. The pseudocode for the CPP can be found in Algorithm 3.

The algorithm initializes an *oddvertices* array with vertices of odd degree. The characterization of Eulerian graphs does not allow for more than two vertices of odd degrees. But the solution to the CPP does not make that restriction. We can have four or other multiples of two as the number of odd degree vertices. In the *oddvertices*[ ] array, when there are more than two vertices, *start* vertex is any one of the vertices from the *oddvertices*[ ] array, while the variable *shortest path* and *vChosen* are initialized as empty sets or null. For each pair of vertices taken from *oddvertices*[ ], the function

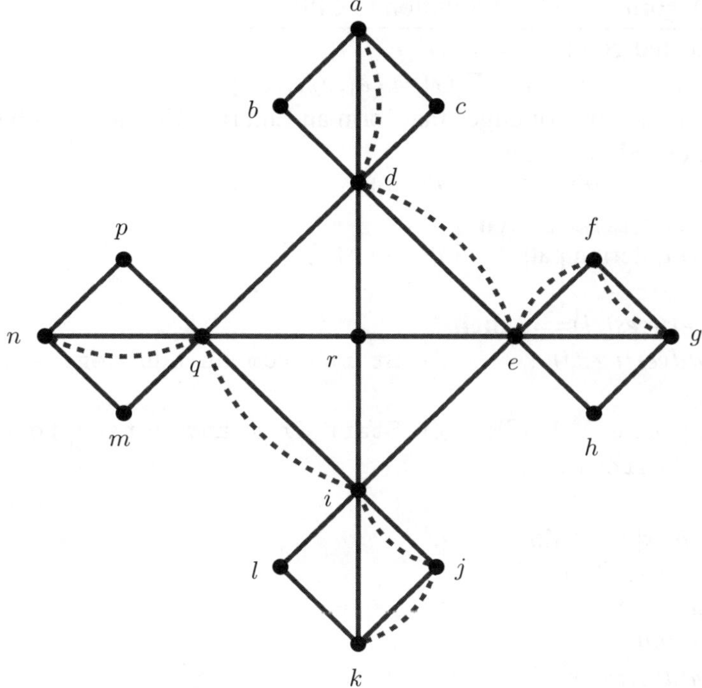

Figure 6.12   Supergraph G^* with duplicate edges

Findshortestpath(u,v) is executed, which is basically any shortest path algorithm, and the length of the shortest path is assigned to *pathlength* variable and the edges of the path are assigned to *pathEdges* array.

In order to find *vChosen*, the variable *shortestpath* gets updated with the *pathlength* each time *Findshortestpath*(u,v) is executed and is found to be larger than the current *pathlength*. After the *Findshortestpath*(u,v) between u and v is executed for every pair of vertices in *oddvertices* array, based on the length of the shortest path, *vChosen* is assigned the vertex v. The *shortestpathEdges* also receives a list of edges from *pathEdges* corresponding to vertex v. These edges are then added to the path between u and v, thus making the odd degree vertices even. Once this is done, the graph is now Eulerian, with only vertices of even degree. At this point, we may resort to Fleury's or Hierholzer's algorithm to find the Eulerian circuit, thus leading us to the solution of the CPP.

In the inner loop, *FindShortestPath* is invoked $O(k^2)$ times. Each *FindShortestPath* call with Dijkstra's algorithm has a complexity of $O(m+nlogn)$. The overall complexity of the pairing step is $O(k^2(m+nlogn))$ where $k \leq n$. Hierholzer's algorithm has a complexity of $O(m+n)$. Hence the overall complexity of the CPP is $O(k^2(m+nlogn))+O(m+n)$. Since $k \leq n$ this can be simplified to $O(n^2(m+nlogn))$.

Algorithm 3: Algorithm to Find Solution to CPP

Data: A connected graph $G = (V(G), E(G))$
with $V(G) = \{v_1, v_2 \ldots v_n\}$ and $E(G) = \{e_1, e_2 \ldots e_m\}$
Result: *path*, a sequence of edges that form an Eulerian Circuit or Path of G
Initialization of OddVertices
$oddvertices[i] \leftarrow$ list of vertices in G with odd degree
if $length(oddvertices) > 2$ **then**
| Return "No Eulerian path or circuit exists"
end
if $length(oddvertices[i])==2$ **then**
| $start \leftarrow oddvertices[0]$;   /* Start from an odd degree vertex */
else
| $start \leftarrow$ any vertex $\in V(G)$;   /* Start from any vertex to construct
| Eulerian circuit */
end
while $oddvertices[\ ] \neq \emptyset$ **do**
| $u \leftarrow oddvertices[i].pop()$
| $shortestpath \leftarrow \infty$
| $vChosen \leftarrow null$
| **for** ($v \in oddvertices[\ ]$) {
| $pathlength, pathEdges \leftarrow Findshortestpath(u, v)$; /* Use Dijkstra to
| find the shortest path between u and v */
| **if** $pathlength < shortestpath$ **then**
| $shortestpath \leftarrow pathlength$
| $vChosen \leftarrow v$
| $shortestpathEdges \leftarrow pathEdges$
| **end**
| }
| Append $shortestpathEdges$ to $E(G)$; /* Adding edges to make degrees
| even */
| $oddvertices.remove(vChosen)$
end
Return $Hierholzer(V(G), E(G)$; /* Perform Hierholzer's algorithm, to
find the Eulerian circuit since all vertices of G have even
degree */

Function $Findshortestpath(u, v)$;   /* Implement Dijkstra's or any
shortest path algorithm */
; /* Return pathlength and list of edges in the shortest path */

6.6 Hamiltonian graphs

The *Icosian game* is a puzzle that Sir William Hamilton created using the dodecahedron. Every one of the puzzle's twenty vertices was labeled with the name of a global capital city. The goal of the puzzle was to build a circle with all of the cities passing through the edges so that no city was visited more than once. Subsequently, the existence of such a cycle in graphs came to be called a Hamiltonian cycle. A graph that contains a Hamiltonian cycle is referred to as a Hamiltonian graph.

DEFINITION 6.6.1 A path in a graph G is called a *Hamiltonian path* if it contains every vertex of G. A cycle in a graph G is called a *Hamiltonian cycle* if it contains every vertex of G.

DEFINITION 6.6.2 A graph G is called a *Hamiltonian graph* if it has a Hamiltonian cycle. A *non-Hamiltonian graph* is a graph that is not Hamiltonian.

EXAMPLE 6.6.3 The dodecahedron graph contains a Hamiltonian cycle as shown in Figure 6.13, the edges of the cycle are represented as bold edges.

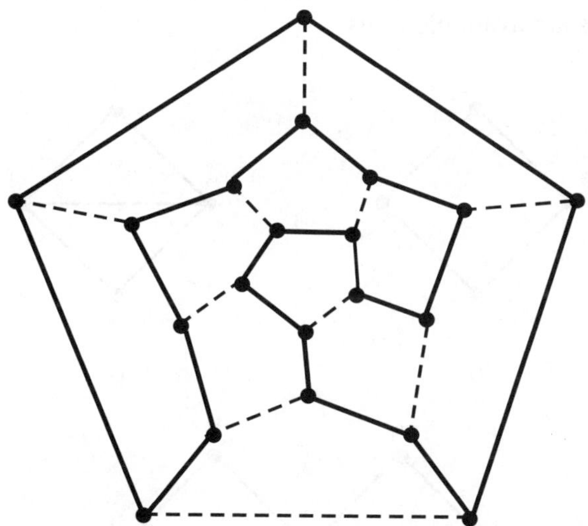

Figure 6.13   Bold edges form a Hamiltonian cycle in dodecahedron graph

EXAMPLE 6.6.4 Consider the graphs G_1, G_2 and G_3 given in Figure 6.14. Clearly G_1 has no Hamiltonian path, G_2 has a Hamiltonian path but not a Hamiltonian cycle, and G_3 has a Hamiltonian cycle. Therefore G_3 is a Hamiltonian graph whereas G_1 and G_2 are non-Hamiltonian graphs.

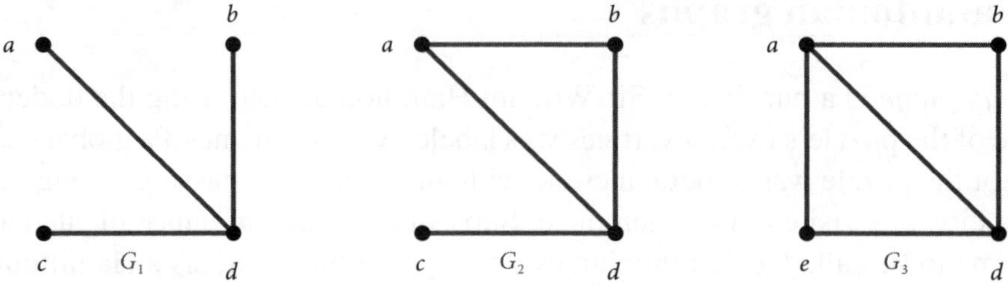

Figure 6.14   Examples of Hamiltonian and non-Hamiltonian graphs

EXAMPLE 6.6.5 The n-cycle C_n and the complete graph K_n on n vertices are Hamiltonian.

REMARK 6.6.6 Let's remember that a Hamiltonian graph has a cycle that includes all of the graph's vertices, but an Eulerian graph has a circuit that includes all of the graph's edges. Keep in mind that every cycle is a circuit by nature. Thus, it raises an intriguing question: Is every Hamiltonian graph an Eulerian graph, or vice versa? Examine the G_1, G_2, and G_3 graphs found in Figure 6.15. It is obvious that G_1 is both Hamiltonian and Eulerian, G_2 is Hamiltonian but not Eulerian, and G_3 is Eulerian but not Hamiltonian. Thus, all three options are available to us.

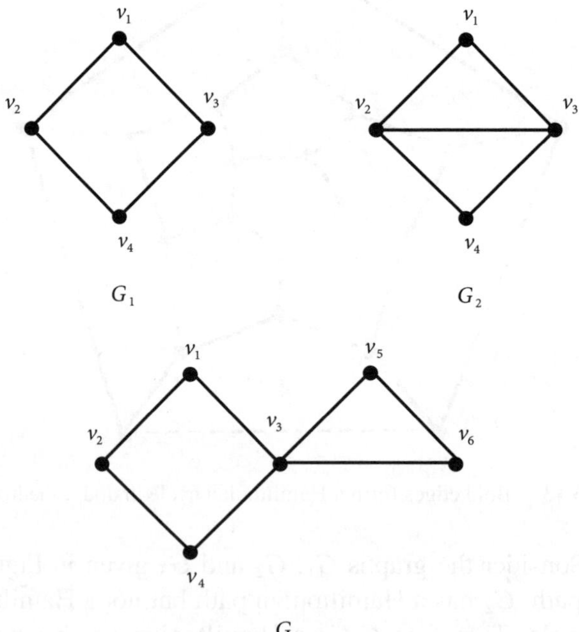

Figure 6.15   Graphs that illustrate the Eulerian and Hamiltonian characteristics

6.7 Criteria for Hamiltonian graphs

In this section, we will discuss some necessary conditions and some sufficient conditions for a graph to be Hamiltonian. Unlike Eulerian graphs, we don't have specific characterization for Hamiltonian graphs.

In the following result, we provide a sufficient condition for a graph to be Hamiltonian.

THEOREM 6.7.1 (Dirac's Theorem) *Let G be a graph with n vertices, where $n \geq 3$. If $deg(v) \geq \frac{n}{2}$ for every vertex v of G, then G is Hamiltonian.*

Proof. Let G be a graph with n vertices, where $n \geq 3$. If $n = 3$, then $deg(v) = 2$ for each $v \in V(G)$ and thus $G \cong K_3$ which is Hamiltonian. So assume that $n \geq 4$. Let P denotes a path having the maximum number of vertices in G. Denote the path P by $P : v_1, e_1, v_2, e_2, v_3, \ldots, v_{k-1}, e_{k-1}, v_k$ where $k \in \{2, 3, \ldots, n\}$, refer Figure 6.16. Since there is no path in G having more vertices than P, each vertex adjacent to v_1 in G lies on P. Similarly, each vertex adjacent to v_k in G also lies on P. Since v_1 is adjacent to at least $\frac{n}{2}$ vertices, which all lie on P, we get that P has at least $1 + \frac{n}{2}$ vertices. That is $k \geq 1 + \frac{n}{2}$. We now claim that there exists a vertex v_i on P, where $2 \leq i \leq k - 1$, such that v_1 is adjacent to v_i and v_k is adjacent to v_{i-1}. If this is not true, then for each vertex v_i adjacent to v_1, the vertex v_{i-1} is not adjacent to v_k in G. Since at least $\frac{n}{2}$ vertices are adjacent with v_1, we have at least $\frac{n}{2}$ vertices that are not adjacent with v_k in G. So $deg(v_k) \leq (n-1) - \frac{n}{2} < \frac{n}{2}$, which is a contradiction to the assumption. Hence there exists a vertex v_i on P such that v_1 is adjacent to v_i and v_k is adjacent to v_{i-1} as shown in Figure 6.17.

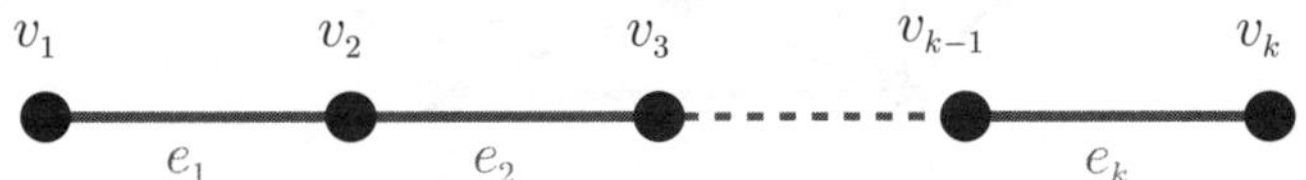

Figure 6.16   The path P with k vertices

As in Figure 6.17, we can see that the vertices of the path P form a cycle $C : v_1, v_i, v_{i+1}, v_{i+2}, \ldots, v_{k-1}, v_k, v_{i-1}, v_{i-2}, \ldots, v_2, v_1$ in G. Next, we claim that C contains all the vertices of G. Suppose there exists a vertex u of G not in C. Since C contains at least $1 + \frac{n}{2}$ vertices, the number of vertices outside C is strictly less than $\frac{n}{2}$. As $deg(u) \geq \frac{n}{2}$, the vertex u is adjacent with v_j in C for some $j \in \{1, 2, \ldots, k\}$. Then the edge uv_j together with the path from v_j to v_{j-1} via the cycle C has one more vertex than P, which is a contradiction. Hence the cycle C contains all the vertices of G implying that C is a Hamiltonian cycle. Hence G is Hamiltonian.   □

EXAMPLE 6.7.2 Consider the graph G in Figure 6.18. The number of vertices in G is $n = 6$. Also, $deg(u_1) = deg(u_6) = 4$ and $deg(u_2) = deg(u_3) = deg(u_4) = deg(u_5) = 5$ which implies $deg(v) > \frac{n}{2}$ for all $v \in V(G)$. Thus, by Theorem 6.7.1, G is Hamiltonian.

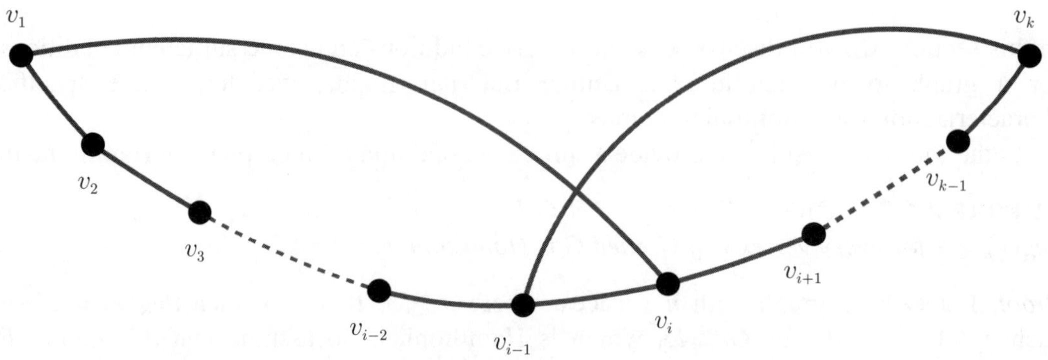

Figure 6.17   Vertices of P forming a cycle

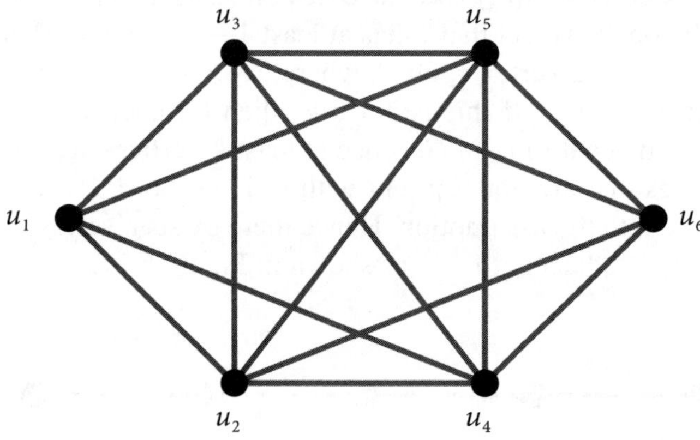

Figure 6.18   A Hamiltonian graph G

The condition for the graph to be Hamiltonian given in Theorem 6.7.1 is not a necessary condition. For example, the graph C_6 (cycle on 6 vertices) is Hamiltonian but the degree of each vertex is 2.

We will now give a necessary condition for the graph to be Hamiltonian.

THEOREM 6.7.3 *If G is a Hamiltonian graph, then $c(G-S) \leq |S|$ for every non empty proper subset S of $V(G)$, where $c(G-S)$ denotes the number of components of the graph $G-S$.*

Proof. Since $S \neq \emptyset$, let $w \in S$. Let C be a Hamiltonian cycle in G. Let $k = c(G-S)$. Suppose $k = 1$. In this case, $c(G-S) = 1 \leq |S|$. So assume that $k \geq 2$. Let $G_1, G_2, \ldots, G_k$ be the components of $G-S$. Since C contains all the vertices of G, let u_i be the last vertex of

C that belongs to G_i for all $1 \le i \le k$. Also let v_i be the next vertex in C just after u_i for $1 \le i \le k$. So $v_i \notin G_i$. Note that u_i and v_i are adjacent vertices in C. If $v_i \in G_j$ for some $1 \le j \ne i \le k$, then G_i and G_j have a common edge $u_i v_j$ in $G - S$, a contradiction to G_i and G_j are distinct components of G. Therefore $v_i \notin G_j$ for any $1 \le j \ne i \le k$. Thus, for any $1 \le i \le k$, we have $v_i \notin G_j$ for all $j \in \{1, 2, \ldots, k\}$. That is, $v_i \in S$ for all $i \in \{1, 2, \ldots, k\}$. Hence $|S| \ge k = c(G - S)$. $\square$

Suppose v is a cut-vertex of a graph G. If we take $S = \{v\}$, then $c(G - S) \ge 2 > 1 = |S|$. So, by Theorem 6.7.3, G is not Hamiltonian. This leads to the following result.

COROLLARY 6.7.4 *If a graph G contains a cut-vertex, then G is not Hamiltonian.*

EXAMPLE 6.7.4 The graph shown in Figure 6.19 is not Hamiltonian as the vertex v is the cut-vertex. Also, the Graph G shown in Figure 6.20 is not Hamiltonian because $c(G - \{c, g, j\}) = 4 > 3 = |S|$.

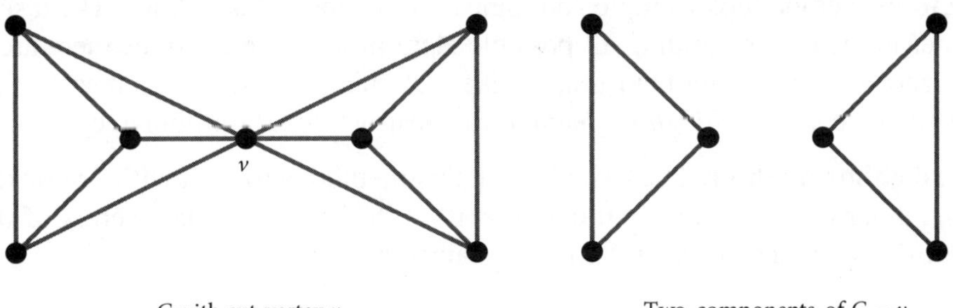

Figure 6.19   A non-Hamiltonian graph

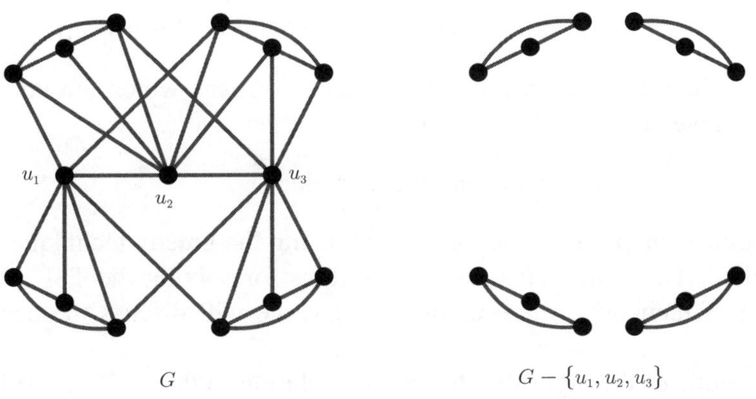

Figure 6.20   A non-Hamiltonian graph

6.8 Traveling salesman problem

In this section, we discuss an application of Hamiltonian graphs, called the traveling salesman problem (TSP). A salesman has customers in some cities and he wants to visit each of those cities with the least total cost. Is it possible to determine the cheapest route possible for the salesman? More precisely, is it possible to construct a tour for the salesman that begins from his home town, going through each city exactly once and help him return to his place with the shortest distance traveled? This problem of finding the shortest route, while touching all the cities is called the TSP. Graph theoretically, TSP is about constructing a Hamiltonian cycle of minimum weight for the weighted graph, in which the vertices are cities and the weighted edge is the distance between the two cities.

To find the Hamiltonian cycle of the least weight, one obvious way is to find all possible Hamiltonian cycles in a graph and then choose the one with minimum weight. Such a method of listing out all possibilities is called the *brute force method*. Though in theory, the brute force method looks simple and logical, in practice, it is tedious. The task is made even harder in the case of finding all possible Hamiltonian cycles, since there are no easy characterizations for Hamiltonian graphs like Eulerian graphs. So we resort to a simple method called the *nearest neighbor method* to construct our Hamiltonian cycle.

Nearest neighbor method: The basic idea in this method is to start with a vertex, find the nearest neighbor of this vertex, and continue it all the way to the last vertex of the graph while choosing one edge at every step as explained below.

Step 1: Start with any vertex v.

Step 2: Among all edges incident to v, choose one edge with the smallest weight. If more than one edge is suitable to our needs, pick one randomly.

Step 3: From vertex v move to the other end of the chosen edge, to vertex say u.

Step 4: Repeat Step 2 and Step 3 for the edges of vertices which are not chosen in the process till we produce a Hamiltonian path with end vertex w.

Step 5: Connect w and v to get a Hamiltonian cycle.

We will discuss the pseudocode for the TSP using the greedy technique of the nearest neighbor method. There are many more algorithms for solving the TSP. The reader can seek them out of their own interest. But for now, we will discuss the nearest neighbor method alone.

The pseudocode of the algorithm to find the solution to the TSP begins by initializing the *start*, *current* and *visited* variables along with the array *path*. The algorithm executes until all the vertices are visited which therefore translates to a while-loop that continues until the *path* array has all the vertices of the graph. The variable *minDistance* is

---

Algorithm 4: Algorithm to Find Solution to TSP

Data: A connected graph $G = (V(G), E(G))$
with $V(G) = \{v_1, v_2 \ldots v_n\}$ and edges $E(G) = \{e_1, e_2 \ldots e_m\}$
Result: *path*, a sequence of edges that form a Hamiltonian cycle or Path of G
$n \leftarrow |V(G)|$
if $n < 3$ **then**
$\quad$| **return** No Hamiltonian cycle exists
end
$start \leftarrow$ any vertex$\in V(G)$; /* We initialize the path and the visited
$\quad$ list with the current vertex */
$current \leftarrow start$
$path \leftarrow [current]$; $\quad$ /* Initialize a path list and a set of visited
$\quad$ vertices */
$visited \leftarrow \{current\}$
while $length(path) < n$ **do**
$\quad$| $nearest \leftarrow \emptyset$; $\quad$ /* Visit the nearest neighbor until all vertices
$\quad$| $\quad$ are visited */
$\quad$| $minDistance \leftarrow \infty$
$\quad$| **for** (*each vertex* $v \in V(G)$) {
$\quad$| $\quad$| **if** $v \notin visited$ *and* $d(current, v) < minDistance$ **then**
$\quad$| $\quad$| $\quad$ $nearest \leftarrow v$
$\quad$| $\quad$| $\quad$ $minDistance \leftarrow d(current, v)$; $\quad$ /* The variable minDistance
$\quad$| $\quad$| $\quad$ $\quad$ keeps adding the distances of the nearest vertex from
$\quad$| $\quad$| $\quad$ $current$ */
$\quad$| $\quad$| **end**
$\quad$| }
$\quad$| **if** $nearest = \emptyset$ **then**
$\quad$| $\quad$| **return** No Hamiltonian cycle exists ; $\quad$ /* If no unvisited neighbor
$\quad$| $\quad$| $\quad$ is found, then cycle is not possible */
$\quad$| **end**
$\quad$| $current \leftarrow nearest$
$\quad$| $path.append(current)$
$\quad$| $visited.add(current)$
end
; /* We check if there is an edge from the last vertex to the
$\quad$ start vertex $\qquad\qquad\qquad\qquad\qquad\qquad\qquad\qquad\qquad\qquad\qquad$ */
if $(current, start) \in E(G)$ **then**
$\quad$| $path.append(start)$ **return** $path$; $\quad$ /* Hamiltonian cycle found */
else
$\quad$| **return** No Hamiltonian cycle exists
end

---

initialized with ∞ but updates to the distance between *current* and *v* if it is lesser than the value assigned to *minDistance*. The *nearest* variable is also assigned the vertex *v* if the *minDistance* value exceeds the distance between *current* and *v*. Effectively, this *for*-loop helps to add only those vertices, which are nearest to the *current* vertex and have the smallest distance $d(current, v)$. Although the pseudocode of the algorithm, works in a general manner, considering distances between vertices in the conventional sense, it will work just as efficiently in the case of a weighted graph, with $d(current, v)$ being equated to the weight of the edge. If these conditions are satisfied, *current* vertex is updated to *nearest* while the original *current* vertex is appended to the *path* array and the *visited* list. The process continues until all the vertices are added to *path*.

The last few lines of the pseudocode helps to check if there is an edge connecting *current* and the *start* thus helping to complete the cycle. If it is not available, *path* merely provides the Hamiltonian path, instead of the cycle.

We will now calculate the complexity of the algorithm. The outer while loop runs until all the n vertices are added to the path, which requires n iterations (i.e) 1 per vertex. In each iteration of the while loop, it runs a for loop over all n vertices to find the nearest unvisited neighbor. This requires $O(n)$ operations in each iteration as it calculates the distance from the current vertex to each unvisited vertex to identify the nearest one. Since there are n iterations of the while loop and each iteration involves a $O(n)$ search for the nearest neighbor, the total complexity of the main loop is $O(n^2)$. Thus the time complexity of the TSP, nearest neighbor heuristic algorithm is $O(n^2)$. The quadratic complexity makes it efficient for small to moderately sized graphs, though it may not yield the optimal solution for TSP.

NOTE 6.8.1 An important note about the TSP is that in general, this problem falls under the category of intractable problems which do not have efficient algorithms. Since we have used a heuristic rule (nearest neighbor) for our algorithm, we have been able to obtain an algorithm of polynomial time complexity. The general TSP can only be solved using algorithms of exponential time complexity. For more information on the general TSP and theorems that prove the intractability of TSP we direct the reader to [9, Theorem 35.2, Theorem 35.3]. For further information on the relationship between Job sequencing problem and the TSP, we direct the reader to [5, Section 8.3].

EXAMPLE 6.8.2 Solve the TSP for the graph shown in Figure 6.21.

(a) Let's start with vertex a.

(b) The edges incident to a are ab, ac and ad. Among these edges, ac is of minimum weight, so we choose the edge ac.

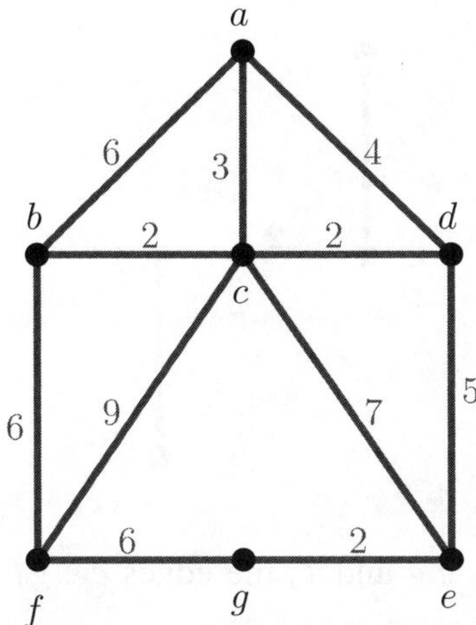

Figure 6.21   Illustration of traveling salesman problem

(c) Now for the vertex c with degree 5, there are only two edges incident on c, which have the same minimum weight. Of these two, we may select any one edge, so we select the edge cd.

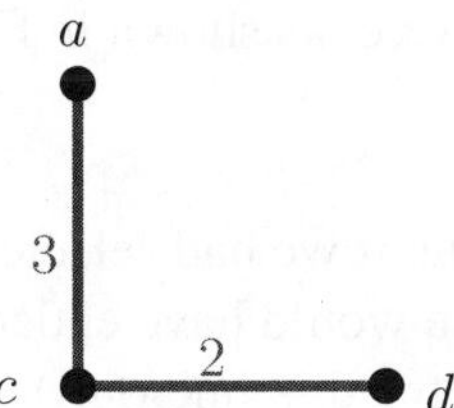

(d) For the vertex d with degree 3, there is only one choice de since the other edges, despite having lesser weight, lead to vertices that have already been visited. We take care not to repeat vertices, since the objective is to visit each vertex, exactly once.

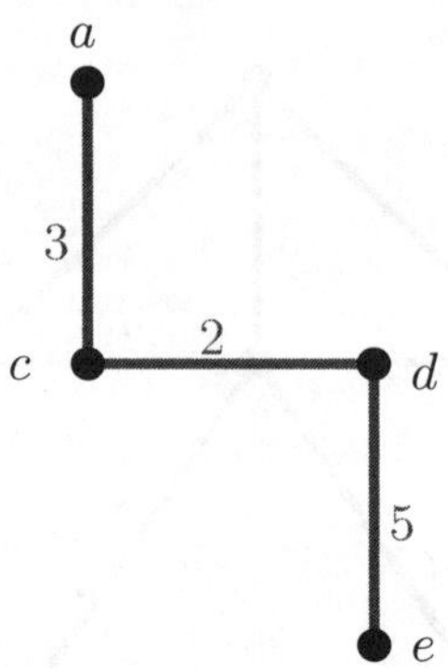

(e) Similarly, for the vertices e, g and f, the edges eg, gf and fb are the only suitable edges.

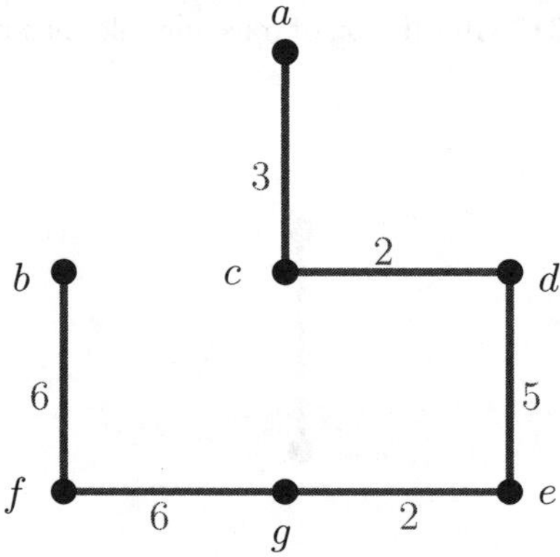

(f) The path we have thus constructed is a Hamiltonian path with end vertex b.

(g) Since we need a Hamiltonian cycle, we connect the vertices b and a using edge ba to get the required Hamiltonian cycle, as shown in Figure 6.22. The total weight of this Hamiltonian cycle is 30.

REMARK 6.8.3 In the above example, if we had selected the edge bc at Step 2, as it has the same weight as dc and proceeded, we would have ended up with a Hamiltonian cycle with a total weight of 28 (the reader can try this one out!). This is the one flaw of the nearest neighbor method, where the road not taken might have produced a favorable result. On the upside, this algorithm produces an optimal result which comes close to the best possible result. Even though a bit of randomness in choosing the edges might skew the final result, it does not skew them afar.

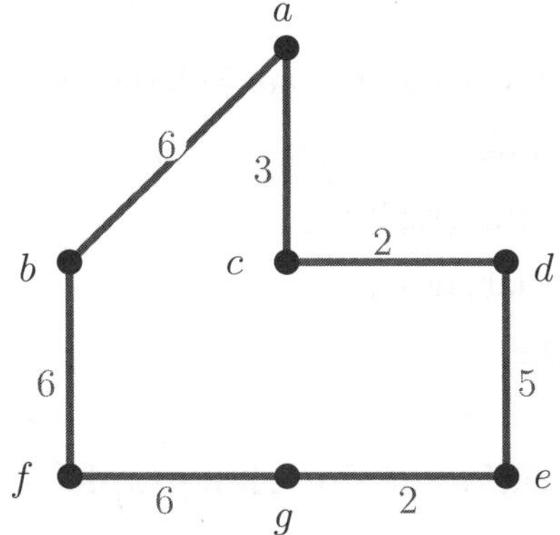

Figure 6.22   Hamiltonian cycle found by TSP

Summary

In this chapter, we have introduced the reader to an efficient form of traversal that ensures that either all edges are included or all vertices are included while avoiding repetition in either case. The former case is expanded upon as the Eulerian trail, Eulerian circuit and Eulerian path while the latter is the Hamiltonian path or the Hamiltonian cycle. The necessary and sufficient conditions for the existence of these desirable characteristics are elaborated through theorems. Additionally, we have explained the operational details of Fleury's algorithm and Hierholzer's algorithm to enable a reader to determine an Eulerian circuit or Eulerian path. The practical applications of Eulerian circuits and Hamiltonian cycles are explored through the CPP and the TSP. We have analyzed these problems comprehensively, through the use of examples and algorithms.

6.9   Exercises

Section 6.1: Eulerian graph

1. Draw a tour for the Königsberg bridge problem so that the number of bridges crossed more than once is as minimum as possible.

2. A bridge builder has arrived in Königsberg with the intention of constructing more bridges so that one could cross each bridge exactly once and arrive at the same place. What number of bridges need to be constructed?

3. Is the union of two Eulerian graphs with the same vertex set also Eulerian?

4. Determine which of the following graphs are Eulerian or semi-Eulerian:

    (i). The complete graph K_7.

    (ii). The complete bipartite graph $K_{3,5}$.

    (iii). The graph of an octahedron.

    (iv). The graph of a cube.

Section 6.2: Characterization of Eulerian graphs

5. Suppose that G is a connected graph in which there are exactly four odd vertices. Show that there are two trails in G such that each edge of G belongs to exactly one of these trails.

6. Suppose G is a connected graph in which there are exactly $2k$ odd vertices, where $k \geq 1$. Prove or disprove that there are k-open trails in G such that each edge of G belongs to exactly one of these trails.

7. Prove or disprove:

    (i). All Euler bipartite graphs have an even number of edges.

    (ii). All simple Euler graphs with an even number of vertices have an even number of edges.

    (iii). Suppose G be an Euler graph with edges c and d such that they share a common vertex, then G has an Euler circuit in which c and d appear consecutively.

8. Prove that the Petersen graph is Eulerian.

9. Determine whether the following statements are true or false. Justify your answer with a proof or counterexample:

    (i). Every Eulerian graph is connected.

    (ii). Every connected graph with all vertices of even degree is Eulerian.

    (iii). The complement of an Eulerian graph is also Eulerian.

Section 6.3: Fleury's algorithm

10. Find which among the graphs in Figure 6.23 have (a) Euler trail and (b) Euler tour. For those that have one, using Fleury's algorithm produce such a trail or tour.

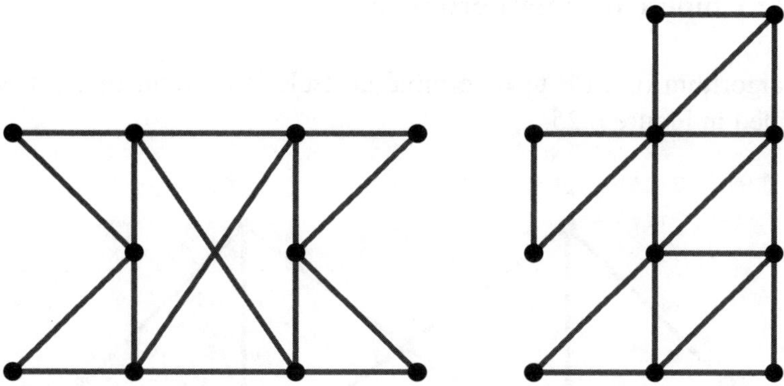

Figure 6.23   Graphs for verifying existence of Euler tour and Euler trail

Section 6.4: Hierholzer's algorithm

11. Use (i) Fleury's algorithm and (ii) Hierholzer's algorithm to construct an Euler path for the graph in Figure 6.28.

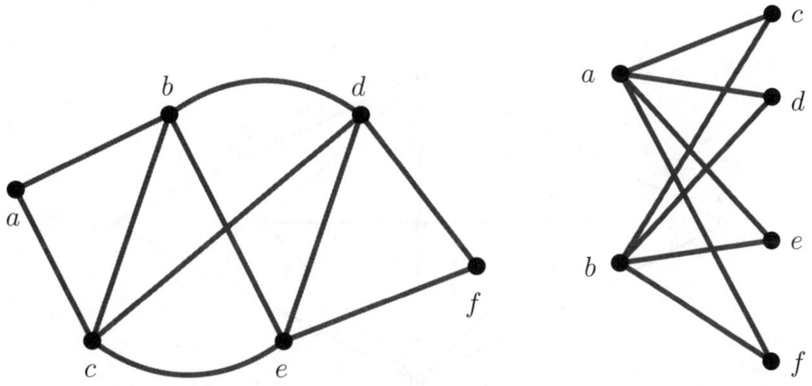

Figure 6.24   Graph for Fleury's and Hierholzer's algorithm

12. Rewrite the pseudocode of Hierholzer's algorithm to give an output as the Eulerian trail with the individual circuits properly integrated into the trail *path* at the suitable positions, to provide the complete trail, for Figure 6.6.

Section 6.5: Chinese postman problem

13. Apply the algorithm of CPP to determine an Eulerian circuit of least weight for the graph provided in Figure 6.25.

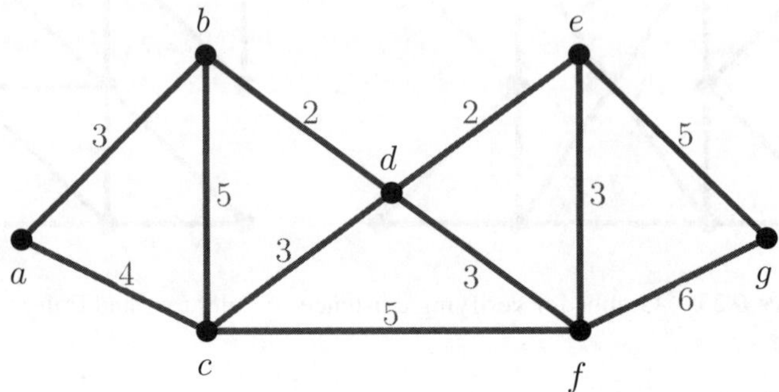

Figure 6.25   Graph for Chinese postman problem

14. Use the CPP algorithm to determine an Euler tour of least weight for the weighted graph of Fig 6.26

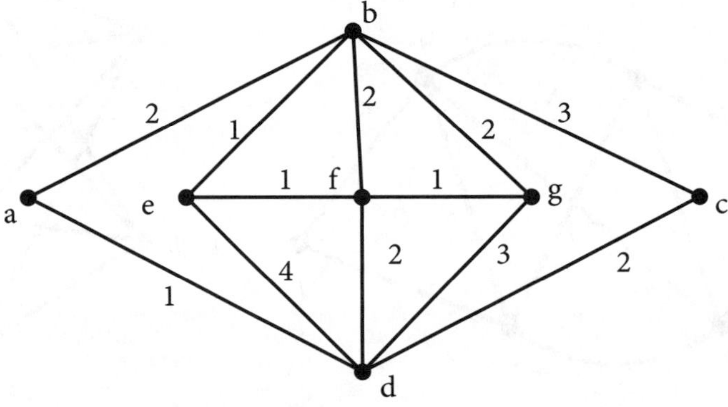

Figure 6.26   Chinese postman problem

15. Find the shortest path from vertex a to d in the weighted graph G 6.27, using CPP.

Section 6.6: Hamiltonian graphs

16. Provide an example of a graph that has an Euler circuit and a Hamiltonian circuit where the two circuits are not the same.

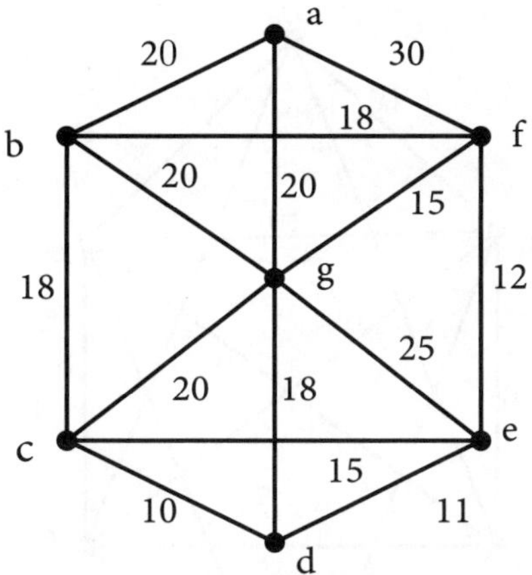

Figure 6.27   Shortest path from vertex *a* to *d*

17. Prove that in a graph, there exists no path longer than a Hamiltonian path (if it exists).

18. Check whether Grotzsch graph in Figure 6.28 is Hamiltonian.

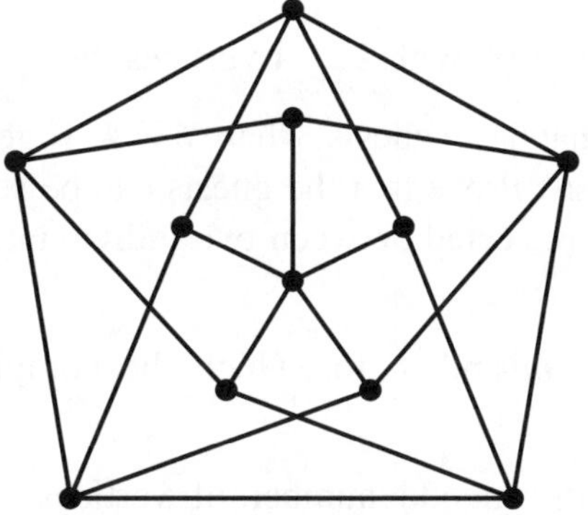

Figure 6.28   Grotzsch graph

19. Prove that if n is odd, it is impossible for a knight to visit all the squares of an $n \times n$ chessboard exactly once using knight's moves and return to the starting point.

20. Find the Hamiltonian cycles of highest and lowest weight in the graph of Figure 6.29.

Section 6.7: Criteria for Hamiltonian graphs

21. Suppose G is a Hamiltonian graph. Prove that G does not have a cut-vertex.

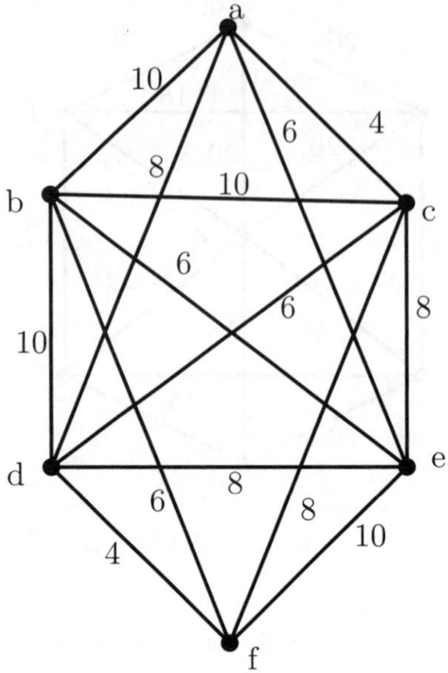

Figure 6.29   Hamiltonian cycles

22. If G is a simple k-regular graph with $2k - 1$ vertices then prove that G is Hamiltonian.

23. Consider a dinner party that has n guests, where $n \geq 4$. Together, any two of these guests know of other $n - 2$ guests. Prove that the guests can be seated around a circular table in such a way that they are seated between two individuals who are acquainted (know each other).

24. Determine the values of m and n, for which the complete bipartite graph $K_{m,n}$ is Hamiltonian.

25. Let G be a bipartite with an odd number of vertices. Then prove that G is non-Hamiltonian.

26. Prove that if a graph is Hamiltonian, then it must be connected. Is converse true? Justify your answer.

27. Prove or disprove: If a graph contains a Hamiltonian path, it must also contain a Hamiltonian cycle.

28. Show that for a graph to be Hamiltonian, there is no vertex with degree less than $\frac{n}{2}$ where n is the number of vertices in the graph.

29. Let G be a Hamiltonian graph, and let S be any set of k vertices in G. Prove that the graph $G - S$ has no more than k components.

Section 6.8: Traveling salesman problem

30. Apply the heuristic algorithm of the TSP for the graph provided in Figure 6.30.

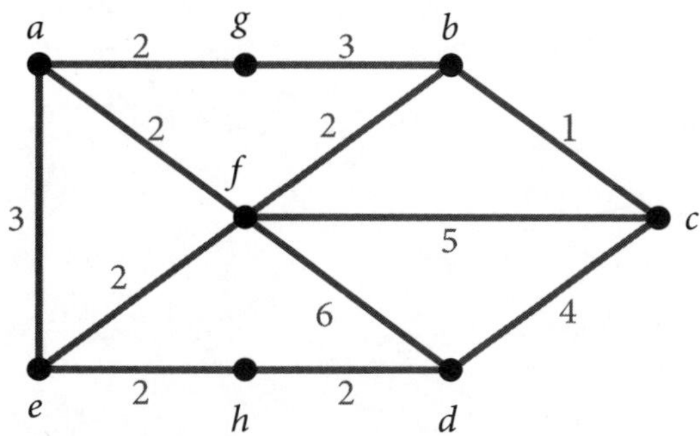

Figure 6.30 Graph for traveling salesman problem

31. Solve the TSP for the weighted graph of Figure 6.31

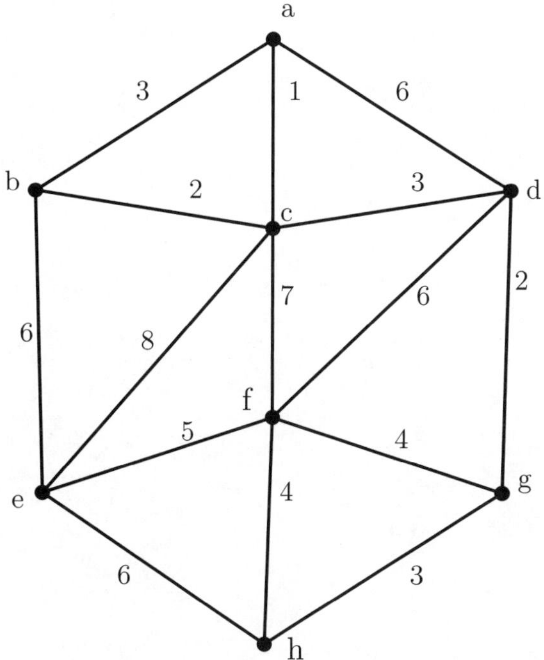

Figure 6.31 Traveling salesman problem

7
Matchings

How do you pair individuals or objects in an optimal manner? This question was first framed in the context of marriage and other pairings, but spurred a great deal of research that had combinatorial optimization as its goal. Matchings are a cornerstone concept in graph theory, offering powerful solutions to optimization problems in network design, resource allocaton and scheduling. The concept of matchings play a significant role in finding perfect matches in partnership scenarios, solving complex assignment problems thus providing the framework for making optimal pairings. This chapter also deals with coverings that provide insights into efficient resource allocation. In addition to these concepts, we will explore an algorithm designed to find the maximum matching in bipartite graphs which culminates in the powerful Kuhn–Munkres algorithm for solving the personnel assignment and optimal assignment problems.

Let G be a graph with vertex set $V = V(G)$ and edge set $E = E(G)$. We will assume throughout this chapter that G has no loops.

DEFINITION 7.0.1 A *matching* in a graph G is a subset $M = \{e_1, e_2, \ldots, e_k\}$ of edges $E(G)$ such that no two of the edges in M are incident. Equivalently, M is said to be a matching if every edge in M shares no vertex with any other edge. That is, each vertex in matching M has degree one.

EXAMPLE 7.0.2 Consider the graph G in Figure 7.1. Then $M_1 = \{e_1, e_7\}$, $M_2 = \{e_5, e_8\}$, $M_3 = \{e_1, e_7, e_4\}$ and $M_3 = \{e_1, e_3, e_6\}$ are some of the matchings of G, whereas $\{e_1, e_8\}$ is not a matching of G as the vertex a is incident with both the edges e_1 and e_8. The matching $M_3 = \{e_1, e_7, e_4\}$ is represented in dashed lines in Figure 7.1.

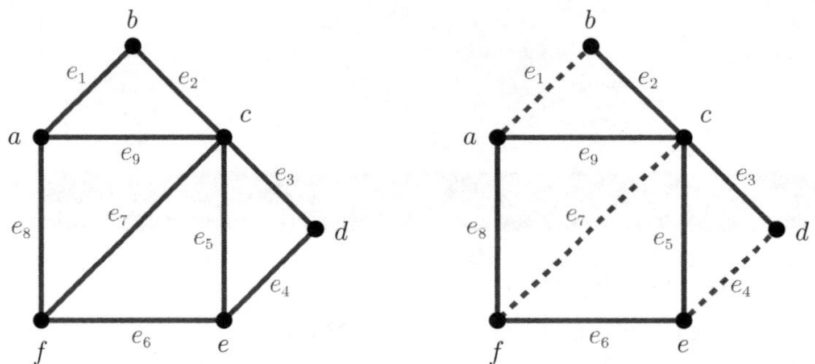

Figure 7.1   A graph G and one of its matchings

DEFINITION 7.0.3 A matching M is said to be a *maximal matching* if for each edge $e \notin M$, $M \cup \{e\}$ is not a matching of G.

DEFINITION 7.0.4 The size of matching M is the number of edges in M. A matching M is said to be a *maximum matching* in a graph G, if there does not exist a matching of size bigger than the size of M. That is, if M' is any matching of G, then $|M'| \leq |M|$. The *matching number* of a graph G is the size of a maximum matching in G and it is denoted by $\alpha'(G)$.

Clearly, a maximum matching is a maximal matching, but not the converse. For example, the graph G in Figure 7.2, the matching $M = \{e_2, e_4\}$ is maximal but not maximum as a bigger matching $M' = \{e_1, e_3, e_5\}$ exists in G.

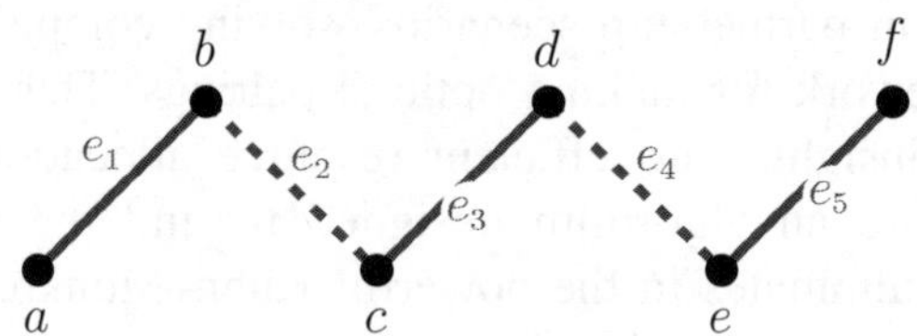

Figure 7.2   A maximal matching but not maximum

We shall compute the matching number of some standard graphs.

The path P_n: Consider the $P_n : v_1, e_1, v_2, e_2, \ldots, e_{n-1}, v_n$. Let either $k = n - 1$ if n is even or $k = n - 2$ if n is odd. Then $\{e_1, e_3, e_5, \ldots, e_k\}$ is a maximum matching of P_n and so $\alpha'(P_n) = \lfloor \frac{n}{2} \rfloor$.

The cycle C_n: For the n-cycle $C_n : v_1, e_1, v_2, e_2, \ldots, e_{n-1}, v_n, e_n, v_1$ with $n \geq 3$. Let either $k = n - 1$ if n is even or $k = n - 2$ if n is odd. Then $\{e_1, e_3, e_5, \ldots, e_k\}$ is a maximum matching of C_n. Therefore the matching number $\alpha'(C_n) = \lfloor \frac{n}{2} \rfloor$.

The complete graph K_n: For the complete graph K_n, let $V(K_n) = \{v_1, v_2, \ldots, v_n\}$. Let either $k = n$ if n is even or $k = n - 1$ if n is odd. Then the edge subset $\{v_1 v_2, v_3 v_4, v_5 v_6, \ldots, v_{k-1} v_k\}$ is a maximum matching of K_n. Therefore $\alpha'(K_n) = \lfloor \frac{n}{2} \rfloor$.

The complete bipartite graph $K_{m,n}$: In the case of the complete bipartite graph $K_{m,n}$, consider a bipartition X, Y of $K_{m,n}$. Let $X = \{x_1, x_2, \ldots, x_m\}$ and $Y = \{y_1, y_2, \ldots, y_n\}$. Without loss of generality, assume that $m \leq n$. Then the edge subset $\{x_1 y_1, x_2 y_2, \ldots, x_m y_m\}$ is a maximum matching of $K_{m,n}$. Thus $\alpha'(K_{m,n}) = \min\{m, n\}$.

7.1 Alternating and augmenting paths

An alternating path is a sequence of edges in a graph where the edges alternate between those that are part of a matching and those that are not. This idea is essential for algorithms that seek to find maximum matchings in bipartite graphs.

DEFINITION 7.1.1 Let M be a matching in a graph G. A vertex v is said to be *M-saturated* if v is incident with some edge in the matching M. Otherwise, v is said to be *M-unsaturated*. Further, if v is M-saturated, then we say that M *saturates* v.

EXAMPLE 7.1.2 Consider the graph given in Figure 7.1. For the matching $M_1 = \{e_1, e_7\}$, the vertices a, b, f and c are M_1-saturated, whereas the vertices d and e are M_1-unsaturated. On the other hand, for the matching $M_3 = \{e_1, e_7, e_4\}$, all the vertices of G are M_3-saturated.

DEFINITION 7.1.3 A matching M in a graph G is said to be *perfect matching* if each vertex of G is M-saturated. That is $\alpha'(G) = \frac{|V(G)|}{2}$.

Note that the even path $P_{2\ell}$, the even cycle $C_{2\ell}$, the even complete graph $K_{2\ell}$ and the complete bipartite graph $K_{n,n}$ contains a perfect matching where $\ell, n \in \mathbb{N}$.

EXAMPLE 7.1.4 The matchings $M_3 = \{e_1, e_7, e_4\}$, $M_4 = \{e_2, e_8, e_4\}$ and $M_5 = \{e_1, e_6, e_3\}$ are the perfect matchings of the graph given in Figure 7.1.

Clearly, any perfect matching is a maximum matching, but not the converse. For example, the matching $M = \{e_6, e_9, e_{10}\}$ of the graph in Figure 7.3, shown by the dashed edges, is a maximum matching in G but not a perfect matching in G, as the vertex a is M-unsaturated.

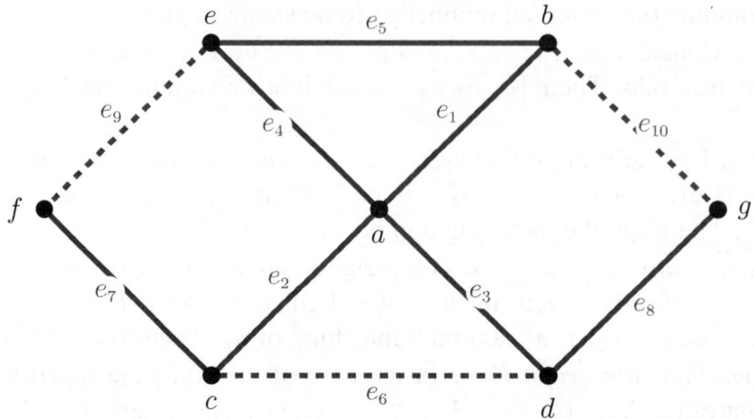

Figure 7.3   A maximum matching but not perfect

DEFINITION 7.1.5 Let M be a matching in a graph G. An *M-alternating path* in G is a path whose edges are alternately in M and $E(G) \setminus M$. That is, a path $P : v_1, e_1, v_2, e_2, \ldots, e_k, v_{k+1}$ is an M-alternating path if $e_i \in M$, then $e_{i+1} \in E(G) \setminus M$ for all $i \in \{1, 2, \ldots, k-1\}$.

EXAMPLE 7.1.6 Consider the graph G with a matching $M_1 = \{e_1, e_3, e_5\}$ as shown in Figure 7.4. The path from a to f is an M_1-alternating path because $e_1, e_3, e_5 \in M$ and $e_2, e_4 \in E(G) \setminus M$.

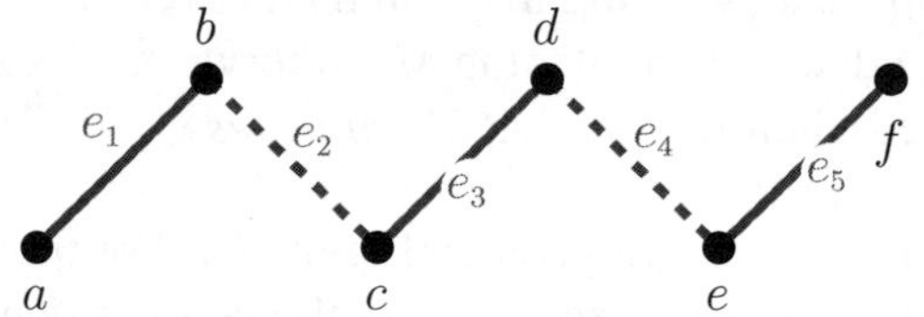

Figure 7.4   An M-alternating path $P : a - b - c - d - e - f$

DEFINITION 7.1.7 An M-alternating path whose origin and terminus are M-unsaturated is called an *M-augmenting path*.

EXAMPLE 7.1.8 Consider the graph G which is shown in Figure 7.4. Consider the matching $M_1 = \{e_1, e_3, e_5\}$ and $M_2 = \{e_2, e_4\}$. Then the path $P : a, e_1, b, e_2, c, e_3, d, e_4, e, e_5, f$ is an M_2-augmenting path as the origin a and the terminus f are M_2-unsaturated. Also the path P is not an M_1-augmenting path.

We now give the characterization of maximum matchings in terms of augmenting paths.

THEOREM 7.1.9 (Berge, 1957) *A matching M in a graph G is a maximum matching if and only if G contains no M-augmenting path.*

Proof. Let M be a maximum matching in G. Suppose there exists an M-augmenting path P in G with edges $e_1, e_2, \cdots, e_k$. Since P is M-augmenting path, the edges $e_1, e_3, \ldots, e_k \in E(G) \setminus M$ and $e_2, e_4, \ldots, e_{k-1} \in M$. This implies that k is an odd integer. Now let $M' = (M \cup \{e_1, e_3, \ldots, e_k\}) \setminus \{e_2, e_4, \ldots, e_{k-1}\}$. Then M' is a matching of G. Since k is odd, we get $|M'| > |M|$, a contradiction to M being a maximum matching. Thus G has no M-augmenting path.

Conversely, suppose that M is a matching in G having no M-augmenting path in G. Let M' be any maximum matching of G. Then from the above argument, G has no M'-augmenting path. Let H be the subgraph of G with edge set

$$E(H) = M \triangle M' = (M \setminus M') \cup (M' \setminus M)$$
$$= \{e \in M \text{ and } e \notin M'\} \cup \{e \in M' \text{ and } e \notin M\}.$$

We discard the isolated vertices of H as their components are trivial. Let v be a non-isolated vertex of H. If v is incident with an edge of $M \setminus M'$ as well as an edge of $M' \setminus M$ in G, then $deg_H(v) = 2$. In this case, since both M and M' are matchings in G, exactly one edge from each M and M' is incident with v. If v is incident with an edge of $M \setminus M'$ or $M' \setminus M$ but not both in G, then $deg_H(v) = 1$. Thus if v is a non-isolated vertex of H, then the degree of v in H is either 1 or 2.

If a component of H has all the vertices of degree 2, then it is an even cycle with edges from M and M' alternatively. If a component of H contains a vertex of degree 1, then it is a path with edges from M and M' alternatively. Since M and M' do not have augmenting paths, such a component should start with an edge from one matching and end with an edge with another matching. Therefore, in either case, any non-trivial component of H has exactly half of its edges from M and exactly half of its edges from M'. Thus, we have $|M \setminus M'| = |M' \setminus M|$. Then $|M| = |M \cap M'| + |M \setminus M'| = |M' \cap M| + |M' \setminus M| = |M'|$. Since M' is a maximum matching in G and $|M| = |M'|$, we conclude that M is also a maximum matching in G. $\qquad\square$

7.2 Matching in bipartite graphs

We recall that a graph G is bipartite if the vertex set $V(G)$ can be partitioned into two sets V_1 and V_2 such that no edge in $E(G)$ has both of its endpoints neither in V_1 nor in V_2.

In this section, we focus on the question: When does a bipartite graph have a matching of V_1? To give context to the question, let us suppose that the vertices in V_1 stand for the students of the class, and the vertices in V_2 stand for seminar topics. In the event that student u would like to present on the topic v, we add an edge from a vertex $u \in V_1$ to a vertex $v \in V_2$. Naturally, a student's vertex would have a degree greater than one if they wanted to present on multiple topics. The instructor wishes to assign each student a distinct topic. Therefore, when we are looking for a matching of V_1, we select a subset of the edges such that no topic is matched to more than one student and every student is matched with exactly one topic.

The necessary condition for a bipartite graph to have a perfect matching is given in the following lemma. Recall that a matching M in G is said to be perfect if it saturates each vertex of G. Therefore the number of edges in a perfect matching M is $\frac{|V(G)|}{2}$.

LEMMA 7.2.1 *Let G be a bipartite graph with vertex partition V_1 and V_2. If G has a perfect matching, then $|V_1| = |V_2|$.*

Proof. Let $V_1 = \{v_1, v_2, \ldots, v_k\}$ and $V_2 = \{u_1, u_2, \ldots, u_\ell\}$ be the vertex partition of G. Suppose there exists a perfect matching M in G. Since M is a perfect matching in G, M covers every v_i for $1 \leq i \leq k$. Since M is a matching in G, there exists a unique $j \in \{1, 2, \ldots, \ell\}$ such that $u_j v_i \in E(G)$. Thus $k = \ell$. $\qquad\square$

Note that if G is a bipartite graph with vertex partition $V(G) = V_1 \cup V_2$, then a perfect matching of G is a matching with edges from every vertex in V_1 to a vertex in V_2.

EXAMPLE 7.2.2 For the complete bipartite graph $K_{3,3}$, two perfect matchings are shown in Figure 7.5 as dashed edges.

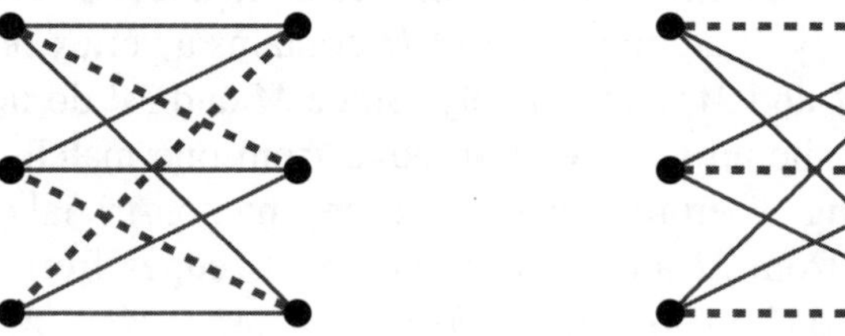

Figure 7.5   Two perfect matchings in $K_{3,3}$

In Example 7.2.2, we have given two perfect matchings in $K_{3,3}$. But are there any other perfect matchings that exist in $K_{3,3}$? More generally, we can ask how many perfect matchings exist in $K_{n,n}$? We answer this question in the following lemma.

LEMMA 7.2.3 *The number of perfect matchings in the complete bipartite graph $K_{n,n}$ is $n!$.*

Proof. Let $V_1 = \{v_1, v_2, \ldots, v_n\}$ and $V_2 = \{u_1, u_2, \ldots, u_n\}$ be the vertex partition of $K_{n,n}$. The vertex v_1 can be matched in n ways by selecting any one of the vertex in V_2. Select $v_1 u_i$, $1 \leq i \leq n$, as the first edge in a matching M of G. Then the vertex v_2 can be matched with $n-1$ vertices in $V_2 \setminus \{u_i\}$ and so on, and the last vertex v_n can be matched in only one way with a vertex of V_2 in M. Therefore the number of distinct perfect matchings in G is $n \cdot (n-1) \cdot (n-2) \cdots 2 \cdot 1 = n!$. $\qquad\square$

7.2.1   Marriage theorem

Let's say that a number of women have a list of men they would be willing to marry. Any woman who would be happy to marry him would also make any man happy. One may now wonder if it is possible to find suitable partners for each woman to marry so that they can all live happy lives. We refer to this as a marriage problem.

It is possible to formulate the problem in terms of graph theory. Create a bipartite graph $G = (V, E)$ with a bipartition $V = V_1 \cup V_2$, where the k men are represented by $Y = \{m_1, \ldots, m_k\}$, and the ℓ women are represented by $V = \{w_1, \ldots, w_\ell\}$. If the women w_i, $1 \le i \le \ell$, are content to marry the men m_j, $1 \le j \le k$, then there is an edge between w_i and m_j.

EXAMPLE 7.2.4 Assume that a bipartite graph representing the relationships between the four women, w_1, w_2, w_3, w_4, and the five men, m_1, m_2, m_3, m_4, m_5, is as displayed in Figure 7.6. Here, we have to choose one of the men that each woman would like to marry. One such matching is represented by the dashed edges in Figure 7.6 and is as follows: w_3 to marry m_4, w_4 to marry m_1, w_1 to marry m_3, and w_2 to marry m_2.

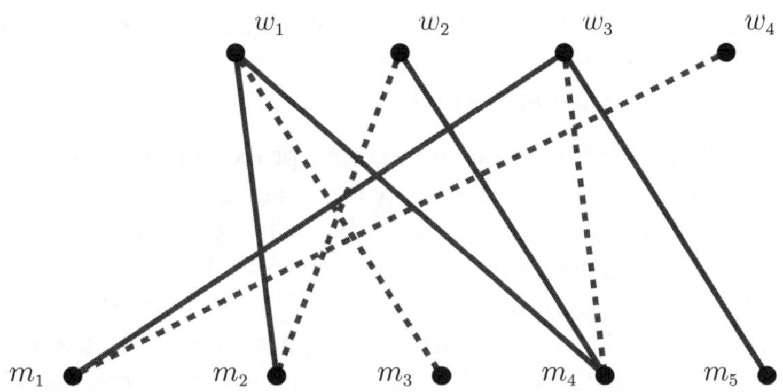

Figure 7.6   The bipartite graph showing the relationship between women and men

The following can be applied to Example 7.2.4: Let G be a bipartite graph, and let $V(G) = V_1 \cup V_2$ be its bipartition. Is it possible to locate a matching in G that saturates each vertex in V_1? A solution for this problem was provided by Philip Hall, which we will demonstrate in this section.

Consider the bipartite graph discussed in Example 7.2.4. Consider the sets $S_1 = \{w_1, w_4\}$, $S_2 = \{w_2, w_3\}$ and $S_3 = \{w_4\}$. Then the neighborhood sets are $N(S_1) = \{m_1, m_2, m_3, m_4\}$, $N(S_2) = \{m_1, m_2, m_4, m_5\}$ and $N(S_3) = \{m_1\}$. We can observe that $|N(S_i)| \ge S_i$ for all i. This is precisely Hall's marriage theorem.

Let G be a graph and S be a subset of $V(G)$. In what follows, the $N(S)$, the neighborhood of S, is the set of all vertices adjacent to some vertex of S. That is, $N(S) = \{v \in V(G):$

v is adjacent to some vertex $u \in S$}. Note that if G is a bipartite graph with bipartition X and Y and $S \subset X$, then $N(S) \subset Y$.

THEOREM 7.2.5 *[Hall's Marriage Theorem] Let G be a bipartite graph with bipartition $V(G) = X \cup Y$. Then G contains a perfect matching if and only if $|N(S)| \geq |S|$ for every subset S of X. Here the condition $|N(S)| \geq |S|$ for every subset S of X is called Hall's condition.*

Proof. Let G be a bipartite graph with partition X and Y. Suppose M is a perfect matching in G. Then for each vertex $x \in X$, there exists a unique vertex $y \in Y$ such that y is adjacent to x. This gives us an injective mapping $f : X \to Y$ given by $f(x) = y$. Thus, for any subset S of X, we have $|S| = |f(S)|$. Let $x \in S$. Since y is adjacent to x, $f(x) = y \in N(S)$ implies $f(S) \subseteq N(S)$. Thus $|N(S)| \geq |S|$.

Conversely, suppose that $|N(S)| \geq |S|$ for every subset S of X. We prove that there exists a perfect matching of G using induction on $|X|$.

Suppose $|X| = 1$. Let $X = \{x\}$. Now consider $S = \{x\}$. Since $|N(S)| \geq |S| = 1$, we get that there exists $y \in Y$ such that $y \in N(S)$. Then $e = xy$ is an edge and the matching $M = \{e\}$ is a perfect matching of X into Y.

Now assume that every bipartite graph with bipartition X' and Y' with $|X'| < |X|$ satisfying the condition $|N(S)| \geq |S|$ for every subset S of X' has a perfect matching of X' into Y'. We split the given assumption into two cases.

Case 1: Suppose $|N(S)| > |S|$ for every non-empty proper subset S of X. Let $x \in X$ and $S = \{x\}$. Then there exists $y \in N(S)$, that is there exists an edge $e = xy$. Let H be the induced subgraph of G induced by $G \setminus \{x,y\}$. That is, $H = G[V(G) \setminus \{x,y\}]$. Then clearly H is bipartite. Now to show H satisfies Hall's condition, let $S' \subset X \setminus \{x\}$. Then $N_H(S') = N_G(S')$ or $N_H(S') = N_G(S') \setminus \{y\}$. Therefore $N_H(S') \geq N_G(S') - 1$. Since $|N_G(S')| > |S'|$, we get $N_H(S') \geq N_G(S') - 1 \geq |S'|$. Thus H is a bipartite graph with bipartition $X \setminus \{x\}$ and $Y \setminus \{y\}$ satisfies Hall's condition. Since $|X \setminus \{x\}| < |X|$, by induction hypothesis, H has a perfect matching M' of $X \setminus \{x\}$ into $Y \setminus \{y\}$. Since e is an edge from x to y, we get a perfect matching $M = M' \cup \{e\}$ of X into Y in G.

Case 2: Suppose $|N(S)| = |S|$ for some non-empty proper subset S of X. Let $X' = S$, $Y' = N(S)$, $X'' = X \setminus S$ and $Y'' = Y \setminus N(S)$. Consider the subgraphs $H' = G[X' \cup Y']$ and $H'' = G[X'' \cup Y'']$. Then clearly H' and H'' are bipartite with $|X'| < |X|$ and $|X''| < |X|$. To show H' satisfies Hall's condition, let $S' \subset X'$. Then $N_{H'}(S')$ is the set of vertices of H' adjacent to S', which is nothing but $N_G(S')$. Thus $|N_{H'}(S')| = |N_G(S')| \geq |S'|$. Now to show H'' satisfies Hall's condition, let $S'' \subset X'' = X \setminus S$. Then by the same arguments as above, $N_G(S'') = N_{H''}(S'')$. Hence

$$N_G(S'' \cup S) = N_G(S'') \cup N_G(S)$$
$$= N_{H''}(S'') \cup N_G(S).$$

Since $S'' \subset X \setminus S$, we get that $N_{H''}(S'')$ and $N_G(S)$ are disjoint. Therefore, $|N_{H''}(S'') \cup N_G(S)| = |N_{H''}(S'')| + |N_G(S)|$. Also by Hall's condition, we have $|N_G(S'' \cup S)| \geq |S'' \cup S|$. Since $|N_G(S)| = |S|$, we get

$$\begin{aligned}
|N_{H''}(S'')| &= |N_G(S'' \cup S)| - |N_G(S)| \\
&\geq |S'' \cup S| - |S| \\
&= |S''| + |S| - |S| = |S''|
\end{aligned}$$

As both H' and H'' satisfy Hall's condition, by induction hypothesis, there exists a perfect matching M' of $X' = S$ into $Y' = N(S)$ and a perfect matching M'' of $X'' = X \setminus S$ into $Y'' = Y \setminus N(S)$. Then $M' \cup M''$ is a perfect matching from X into Y in G. $\quad\square$

COROLLARY 7.2.6 *Every regular bipartite graph has a perfect matching.*

Proof. Let G be a r-regular bipartite graph with bipartition X and Y. Let $|X| = k$ and $|Y| = \ell$. Since the degree of each vertex is r, there are rk edges from X to Y. On the other hand, similarly there are $r\ell$ edges from Y to X. Therefore $rk = r\ell$. Since $r > 0$, we have $k = \ell$. Now to show G satisfies Hall's condition, let S be a subset of X. Let E be the set of edges incident with some vertices of S. As G is r-regular, we get $|E| = r|S|$. Similarly, let E' be the set of edges incident with vertices of $N(S)$. Then $|E'| = r|N(S)|$. Let $e = xy \in E$ for some $x \in S$. Then $y \in N(S)$ and so the edge $e \in E'$, which implies $E \subset E'$. Therefore $|E| \leq |E'|$. Thus $r|N(S)| = |E'| \geq |E| = r|S|$ implies $|N(S)| \geq |S|$. Hence by Hall's marriage theorem; refer to Theorem 7.2.5, G contains a perfect matching M of X into Y. Since $|X| = |Y|$, the edges in the matching M saturates every vertex in Y. Thus M is a perfect matching in G. $\quad\square$

7.3   Covering

We search for maximum matching in a situation in which a graph G does not have a perfect matching. According to Theorem 7.1.9, which we have already shown, a graph has a maximum matching M if and only if it lacks an M-augmenting path. However, there will be a lot of matchings in a graph, and it will take time to show that none of the matchings have augmenting paths. Rather, we establish a graph's minimum covering and relate it to the maximum matching.

DEFINITION 7.3.1 Let G be a graph. A subset W of $V(G)$ is called a *vertex covering* of G if given any edge $e = uv$ in $E(G)$, then either $u \in W$ or $v \in W$. That is, the vertex cover of G is a subset of $V(G)$ containing at least one endpoint of every edge of $E(G)$.

DEFINITION 7.3.2 A vertex covering W of G is said to be a *minimal vertex covering* of G if $W \setminus \{u\}$ is not a covering for each vertex $u \in W$. A vertex covering W of G is said to be a *minimum vertex covering* of G if $|W| \leq |W'|$ for all vertex coverings W' of G. That is, a minimum vertex covering is the smallest possible vertex covering. The *vertex covering number* is the size of the minimum vertex cover of G and is denoted by $\beta(G)$.

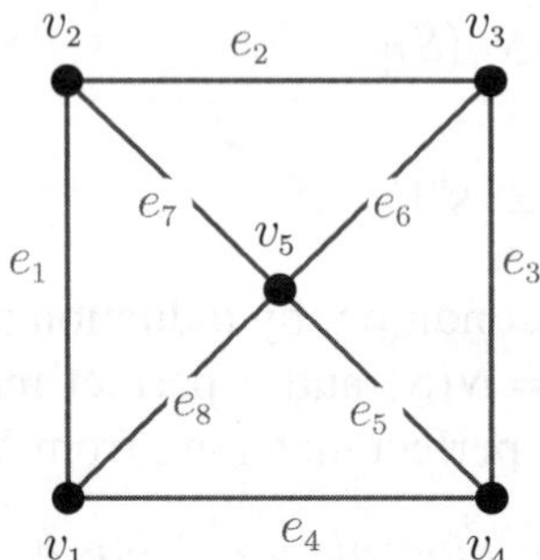

Figure 7.7   Graph G considered in Example 7.3.3

EXAMPLE 7.3.3 Consider the graph given in Figure 7.3.3. Consider the sets $W_1 = \{v_1, v_3, v_5\}$, $W_2 = \{v_1, v_2, v_3, v_5\}$, $W_3 = \{v_1, v_2, v_3, v_4\}$ and $W_4 = \{v_1, v_3, v_4\}$. Then

- W_1, W_2 and W_3 are vertex covering sets of G.

- W_4 is not a vertex covering of G as no endpoint of the edge e_7 belongs to W_4.

- W_1 and W_3 are minimal vertex covering of G but W_2 is not minimal as $W_2 \setminus \{v_2\}$ is still a covering of G.

- W_1 is a minimum vertex covering of G.

We shall compute the vertex covering number of some graphs that are familiar to us.

The path P_n: Consider the $P_n : v_1, e_1, v_2, e_2, \ldots, e_{n-1}, v_n$. Assume that, either $k = n$ if n is even or $k = n - 1$ if n is odd. Then the set $W = \{v_2, v_4, \ldots, v_k\}$ is a minimum vertex cover of P_n and so $\beta(P_n) = \left\lfloor \frac{n}{2} \right\rfloor$.

The cycle C_n: For the n-cycle C_n with $n \geq 3$, the vertex covering number $\beta(C_n) = \left\lceil \frac{n}{2} \right\rceil$, since each vertex in C_n covers at most two edges in C_n.

The complete graph K_n: For the complete graph K_n, any set of $n - 1$ vertices covers all the edges in K_n. Let $W \subset V(K_n)$ with $|W| \leq n - 2$. Choose $u, v \in V(K_n) \setminus W$. Clearly uv is an edge of K_n which is not covered by any vertex in W and so W is not a vertex cover of K_n. Therefore $\beta(K_n) = n - 1$.

The complete bipartite graph $K_{m,n}$: In the case of the complete bipartite graph $K_{m,n}$, consider a bipartition X, Y of $K_{m,n}$. Clearly X (or Y) is a vertex cover of $K_{m,n}$. Choose $x \in X$ and $y \in Y$. Then $X \setminus \{x\}$ is not a vertex cover of $K_{m,n}$ because the edge xy in $K_{m,n}$ is not covered by $X \setminus \{x\}$. Similarly, $Y \setminus \{y\}$ is not a vertex cover of $K_{m,n}$. In addition, the set $(X \setminus \{x\}) \cup (Y \setminus \{y\})$ is not a vertex cover of $K_{m,n}$. Therefore either X or Y is a minimum vertex cover of $K_{m,n}$. Thus $\beta(K_{m,n}) = \min\{m, n\}$.

THEOREM 7.3.4 *Let G be a bipartite graph. Then the maximum size of a matching is equal to the minimum size of a vertex cover of G. That is $\alpha'(G) = \beta(G)$.*

Proof. Let G be a bipartite with vertex partition X and Y. Let W be a vertex cover in G and M be a matching in G. Since no two edges in M share a common vertex, different vertices are needed to cover each distinct edge of M. Therefore $\min\{|W| : W \text{ is a vertex cover in } G\} \geq \max\{|M| : M \text{ is a matching in } G\}$. Thus $\beta(G) \geq \alpha'(G)$.

Let W' be a minimum covering in G. We will construct a matching of size $|W'|$ in G. Clearly $W' = W' \cap V = W' \cap (X \cup Y) = (W' \cap X) \cup (W' \cap Y)$. Let $X' = W' \cap X$ and $Y' = W' \cap Y$. Then $W' = X' \cup Y'$. First we claim that there is no edge from $X \setminus X'$ to $Y \setminus Y'$ in G. If there exists some $x \in X \setminus X'$ is adjacent to $y \in Y \setminus Y'$ in G, then $x, y \notin X' \cup Y' = W'$, a contradiction to W' is a vertex cover of G. Therefore the claim holds true.

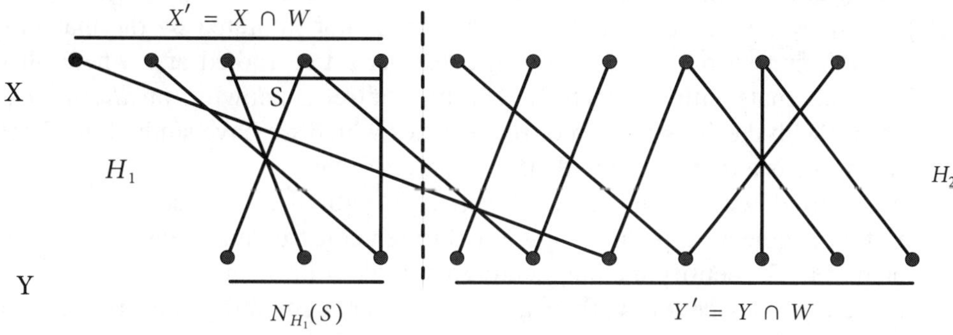

Figure 7.8   A structural representation for the proof of Theorem 7.3.4

Consider the induced subgraphs $H_1 = G[X' \cup (Y \setminus Y')]$ and $H_2 = G[Y' \cup (X \setminus X')]$, in Figure 7.8. Next, we claim that H_1 and H_2 satisfy Hall's condition. In order to prove the claim, let $S \subseteq X'$. Then $N_{H_1}(S) \subseteq Y \setminus Y'$. Suppose $|N_{H_1}(S)| < |S|$. Since $N_{H_1}(S)$ covers all the edges which are incident to S which are not covered by Y', we get $W'' = (X' \setminus S) \cup N_{H_1}(S) \cup Y'$ is a vertex covering of G. Since $|N_{H_1}(S)| < |S|$, we have $|W''| < |X' \cup Y'| = |W'|$, a contradiction to W' is a minimum covering of G. Thus $|N_{H_1}(S)| \geq |S|$. Hence H_1 satisfies Hall's condition. Similar arguments leads to the satisfaction of Hall's condition for H_2. Thus, by Theorem 7.2.5, H_1 and H_2 contain a perfect matching, say M_1 and M_2 respectively. In essence, the matching M_1 saturates X' and the matching M_2 saturates Y' in G. Since H_1 and H_2 are disjoint graphs, we get that the matching $M_1 \cup M_2$ saturates $W' = X' \cup Y'$ in G. Therefore $M_1 \cup M_2$ is a matching of size $|W'|$ in G. Since W' is a minimum covering of G and $\alpha'(G)$ is the size of the maximum matching of G, we have $\beta(G) \leq \alpha'(G)$. Hence $\beta(G) = \alpha'(G)$. $\qquad\square$

7.4 Algorithm to find the maximum matching in a bipartite graph

We have established the need for finding a maximum matching in a graph, in the previous section. An algorithm to find the maximum matching in any graph, is certainly more desirable than an algorithm that would determine the maximum matching, specifically for a bipartite graph. But bipartite graphs form a huge subclass of graphs that arise frequently in real-life applications and therefore deserve an algorithm that would deal with them efficiently.

In this section, we will study the technique to build rooted trees with alternating paths, that eventually lead to the maximum matching of a bipartite graph. The maximum matching algorithm operates on a bipartite graph G, with an initial matching M and vertex set $V(G) = \{v_1, v_2 \ldots v_n\}$. Let v be a matching that is not saturated by the matching M. Using a breadth-first search (BFS), we can construct a tree rooted at v where all paths starting from v alternate with respect to M. This type of tree is known as an *alternating tree*. We are familiar with the construction of rooted trees which we have studied in Chapter 3. An alternating tree is a rooted tree with M-alternating paths.

We have started with an unsaturated vertex v, with the purpose of revealing the formation of an augmenting path, ending at v through constructing the M-alternating tree that is rooted at v. To identify the augmenting path, we follow this process: if there is an unsaturated vertex u adjacent to v, the path vu forms an augmenting path. By augmenting M along this path, we obtain a matching of size $|M| + 1$. If every vertex adjacent to v is matched, we continue building the alternating tree. We start by placing v at level 0 and its adjacent vertices $u_1, u_2 \ldots u_k$ at level 1. We connect v to each u_i for $1 \leq i \leq k$. Let $u_i v_i \in M$ for $1 \leq i \leq k$ and place the vertices $v_1, v_2 \ldots v_k$ at level 2, joining each u_i to v_i. This process constructs the second level of the alternating tree. Suppose the tree is built up to level m, where m is even and no augmenting path has been found. We proceed by examining each vertex x at level m and checking its neighbors. If a neighbor y is already in the tree, we skip it. If y is not part of the tree, it is either a saturated vertex or an unsaturated one. If y is saturated , say $yz \in M$, and z is not found in the alternating tree, then we add both y and z to levels $m + 1$ and $m + 2$ of the tree respectively, connecting x to y and y to z.

If y is an unsaturated vertex then an augmenting path from v to y has been found. This augmenting path, consists of the alternating path from v to x in the tree followed by the edge xy. The construction of the alternating tree is complete, when either an augmenting path is detected or it becomes impossible to add more level based on established rules. In the case, where an augmenting path is identified, we increase the matching M by augmenting along that path resulting in a new matching of size $|M| + 1$. If there is still an unsaturated vertex, with respect to this new matching that hasn't yet been the root of an alternating tree, we proceed by constructing another alternating tree rooted at that vertex. However if no unsaturated vertex remains, we have determined that the current matching is the maximum matching.

In the second case, where no augmenting path is found ending at v, the alternating tree construction is halted. If there is still an unsaturated vertex say r in G, that has not been considered as the root of the alternating tree, we use r to start a new alternating tree. If no unsaturated vertices remain, it confirms that M is a maximum matching for the graph.

The pseudocode for this algorithm can be found in Algorithm 1. The reader may try this out with any simple bipartite graph, and arrive at expected results.

The outer loop iterates up to $O(|V(G)|)$ times, while constructing the alternating tree in each iteration involves up to $O(|E(G))$ operations. Finding and augmenting paths within the tree takes $O(|V(G)|^2)$ in the worst case due to path augmentations. Thus the total time complexity is $O(|V(G)| \times |E(G)|) + O(|V(G)|^2)$ which is approximately $O(|V(G)| \times |E(G)|)$. Hence the time complexity of this algorithm is $O(|V(G)| \times |E(G)|)$. This complexity is suitable for bipartite graphs with relatively sparse edges but it may become costly for dense graphs.

EXAMPLE 7.4.1 For the bipartite graph G found in Figure 7.9 determine the maximum matching.

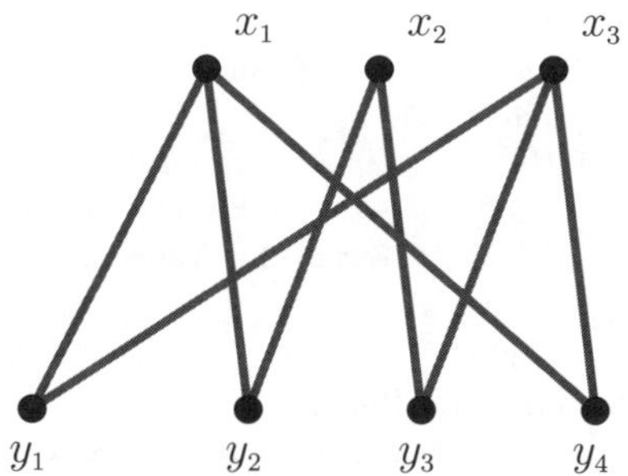

Figure 7.9   A bipartite graph G

The bipartite graph G consists of two sets of vertices, $X = \{x_1, x_2, x_3\}$, $Y = \{y_1, y_2, y_3, y_4\}$. We need to find a maximum matching that connects X to Y.

Step 1: We start with an empty matching $M = \emptyset$, since no vertices are matched yet.

Step 2: To find augmenting paths, we construct alternating trees. These trees grow from unsaturated vertices, alternating between saturated edges and unsaturated edges. Let us start with an unsaturated vertex in X. We choose x_1 as the root of the alternating tree and expand the tree. x_1 is adjacent to y_1, y_2 and y_4. Since $M = \emptyset$, y_1, y_2 and y_4 are unsaturated. Let us choose y_1, the matching M is now $\{(x_1, y_1)\}$.

Algorithm 1: Maximum Matching Algorithm for Bipartite Graphs

Data: A bipartite graph $G = (V(G), E(G))$
with $V(G) = \{v_1, v_2 \dots v_p\}$ and an initial matching M_1
Result: A maximum matching M

$i \leftarrow 1$;   /* Index to iterate */
$M \leftarrow M_1$;   /* M is initialized with the current matching */
$Q \leftarrow \emptyset$;   /* An empty queue is initialized */
Iterate through all vertices in the graph
while $i < |V(G)|$ **do**
  if $v_i \in M$ **then**
    $i \leftarrow i+1$;   /* Skip this vertex if it is already saturated */
  else
    $v \leftarrow v_i$;   /* Choose an unsaturated vertex v */
    $Q \leftarrow Q \cup \{v\}$; /* Add v to the queue to start building the alternative tree
    */
  end
  Alternating tree construction
  while $Q \neq \emptyset$ **do**
    $x \leftarrow RemoveFirst(Q)$;   /* Remove the first vertex from the queue */
    $N(x) \leftarrow \{y_1, y_2 \dots y_k\}$;   /* Neighbors of vertex x */
    for ($j = 1$ *to* k) {
      $y \leftarrow y_j$;   /* Explore the neighbors y of x */
      if y *is unsaturated* **then**
        $P \leftarrow AlternatingPath(x, y)$; /* If an augmenting path is found, get the
        alternating path P */
        $M \leftarrow Augment(M, P)$;   /* Augment the matching along path P */
        $Q \leftarrow \emptyset$;   /* Empty the queue after augmenting */
        break; ;   /* Exit the loop after augmenting the path */
      else
        $z \leftarrow MatchedVertex(y)$;   /* Get the vertex matched with y */
        if z *is not in alternating tree* **then**
          $Q \leftarrow Q \cup z$;   /* Add z to the queue for further exploration */
          Mark y and z as part of the alternating tree
          $Parent(y) \leftarrow x$;   /* Set x as the parent of y */
          $Parent(z) \leftarrow y$;   /* Set y as the parent of z */
        end
      end
    }
    After the alternating tree construction
    if *An augmenting path is found* **then**
      $M \leftarrow Augment(M, P)$
    else
      $i \leftarrow i+1$;   /* Move to the next unsaturated vertex to build the next
      alternating tree */
    end
  end
  Repeat until no more augmenting paths can be found
end
return M ;   /* The maximum matching is found */

Step 3: We will now look for augmenting paths, which can help to increase the size of the matching. Let us start with another unsaturated vertex , say x_2. We now expand the tree from x_2. We can see that x_2 is adjacent to y_2 and y_3. We will add y_2 to the matching. Hence the matching $M = \{(x_1, y_1), (x_2, y_2)\}$.

Step 4: We will continue to build alternating trees. Let us start with the last unmatched vertex x_3. Let us explore the edges incident on x_3, it is connected to y_1, y_3 and y_4. Both y_3 and y_4 are unmatched, so we can choose either vertex. We select y_4 and add (x_3, y_4) to the matching.

Step 5: After completing the exploration, the final matching is $M = \{(x_1, y_1), (x_2, y_2), (x_3, y_4)\}$. This is the maximum matching for the bipartite graph. All vertices from set X are matched with vertices from set Y and no augmenting paths remain.

7.5 Optimal assignment problem

Suppose there are n teachers available for n classes, each teacher being qualified to teach one or more of these classes. Can all the teachers be assigned as per their qualifications such that there is one teacher per class? This problem is called the *personnel assignment problem.*

We can rewrite the problem in terms of a bipartite graph G as follows: Form the vertex set $V = X \cup Y$ where

$$X = \{x_1, \ldots, x_n\} \text{ is the set of teachers}$$

and

$$Y = \{y_1, \ldots, y_n\} \text{ is the set of classes}$$

There is an edge from teacher x_i to class y_j if and only if x_i is qualified to teach class y_j. Then the problem is to find a matching in G, which saturates Y.

One more example of the assignment problem is a problem where a company wants to assign n different employees to n different jobs, where each employee is qualified to one or more jobs and each employee is assigned to one of the jobs.

We either find a matching in G which saturates every vertex of Y or exhibits a subset S of X with $|N(S)| < |S|$.

In the case of the personnel assignment problem, we may also think of a weighted bipartite graph G in which there is a certain cost to the assignment of a certain job X. For example worker A may demand more wages for doing job X when compared to worker B. Likewise worker C may demand even lesser wage than B, for doing the same job X. But the same worker C may demand a higher wage than A or B when doing a job Y. Representing the wages as the weight of the edges connecting jobs to the workers, we are able to construct a weighted bipartite graph. The problem is to find the most suitable assignment of jobs to workers, while keeping the total cost minimum. This is the *optimal assignment problem* where we aim to find a perfect matching in a weighted bipartite graph.

7.6 Kuhn–Munkres algorithm

To solve the optimal assignment problem, we need to apply the *Hungarian algorithm* or the *Kuhn–Munkres algorithm*. The algorithm is attributed to various origins and independent discoveries. But if we were to go by recent history, it was published in 1955 by *Harold Kuhn*, who named it the *Hungarian method* after the two Hungarian mathematicians *Dénes König* and *Jenő Egerváry*, whose work he relied on to create the algorithm. It was reviewed by *James Munkres* in 1957, therefore making it the *Kuhn–Munkres algorithm*. The Kuhn–Munkres algorithm is a polynomial time algorithm that helps to solve the optimal assignment problem.

Let G be a weighted bipartite graph with partite sets V_1 and V_2. We begin by constructing a complete weighted bipartite graph G' that contains G as a subgraph. The partite sets of G' denoted by U_1 and U_2 satisfy $|U_1| = |U_2| = max\{|V_1|, |V_2|\}$ with $V_1 \subseteq U_1$ and $V_2 \subseteq U_2$. If $x \in U_1$ and $y \in U_2$, then the weight of the edge xy in G' denoted by $w_{G'}(xy)$ equals $w_G(xy)$ if xy is an edge in G and $w_{G'}(xy) = 0$ otherwise.

If M is a maximum weight matching of G', it can be assumed to be a perfect matching of G'(where some edges of M may have weight of 0). As a result, $M \cup E(G)$ is a maximum weight matching in G. Thus it suffices to focus on finding a maximum weight matching in weighted complete bipartite graphs. Let G now represent a weighted complete bipartite graph with partite sets $V_1 = \{v_1, v_2 \ldots v_p\}$ and $V_2 = \{u_1, u_2 \ldots u_p\}$. One approach, to solve the optimal assignment problem, would be to identify perfect matchings of G and select one with the maximum weight. However, this method is not efficient. Instead the *Kuhn–Munkres algorithm* is employed to efficiently solve the optimal assignment problem.

DEFINITION 7.6.1 A *feasible vertex labeling* is defined as a real valued function ℓ on the vertex set $V_1 \cup V_2$ such that for all $v \in V_1$ and $u \in V_2$

$$\ell(v) + \ell(u) \geq w(uv)$$

Here $\ell(x)$ represents the label of vertex x. We can initiate a feasible vertex labeling ℓ by setting

$$\ell(v) = max\{w(uv) | u \in V_2\} \text{ for all } v \in V_1$$

and

$$\ell(u) = 0 \text{ for all } u \in V_2$$

This ensures that every weighted complete bipartite graph has a feasible vertex labeling. Now define the edge set as

$$E_\ell = \{vu \in E(G)|v \in V_1 \text{ and } \ell(v) + \ell(u) = w(vu)\}$$

The Kuhn–Munkres algorithm works as follows to find a maximum weight matching in a weighted complete bipartite graph.

Step 1: Initialize a feasible vertex labeling for the graph with the following rules:
For each $v \in V_1$, set $\ell(v) = max\{w(uv)|u \in V_2\}$
For each $u \in V_2$, set $\ell(u) = 0$.
Let H_ℓ be the spanning subgraph of G with edge set E_ℓ and let G_ℓ be the underlying graph of H_ℓ.

Step 2: Apply the maximum matching algorithm to determine a maximum matching M in G_ℓ.

Step 3: If every vertex of V_1 is saturated by M, output M and stop. Else proceed to Step 4.

Step 4: Let x be the first vertex in V_1 that is unsaturated by M. Build an alternating tree, rooted at x with respect to M. If an augmenting path P is discovered, then augment M along P and return to Step 3. Otherwise, construct the alternating tree T and if no further expansion is possible in G_ℓ continue to Step 5.

Step 5: Calculate m_ℓ where
$m_\ell = min\{\ell(v) + \ell(u) - w(vu)|v \in V_1 \cap V(T) \text{ and } u \in V_2 - V(T)\}$. Update the labels as follows:

$$\ell(v) = \begin{cases} \ell(v) - m_\ell \text{ for } v \in V_1 \cap V(T) \\ \ell(v) + m_\ell \text{ for } v \in V_2 \cap V(T) \\ \ell(v) \text{ otherwise} \end{cases}$$

Step 6: Replace ℓ with the new labeling ℓ', reconstruct G_ℓ and return to Step 4. The iterative process continues until the algorithm outputs a perfect matching of maximum weight.

The pseudocode of Kuhn–Munkres algorithm can be found in Algorithm 2. The input is a complete bipartite graph with two partite sets U and V, its edges in set E and the weight function W mapping edges to weights. The initial labels $L_U[u]$ for vertices are set to the maximum weight of the edges connected to u while $L_V[v]$ is initialized to 0, for vertices in V. The sets S and T track the visited vertices in U and V respectively, to avoid revisiting nodes unnecessarily, during the search for augmenting paths.

We define a "tight edge condition" which checks whether an edge is "tight", i.e., the sum of the labels equals the weight of the edge. The qualifying condition is $L_U[u'] +$

$L_V[v'] = W[u',v']$. Only tight edges are explored. If an unsaturated vertex v is found, an augmenting path is constructed and the matching is updated. Otherwise the exploration continues with the vertex u'' matched to v'.

When no augmenting path is found, the labels are adjusted by computing the minimum slack values α for unsaturated vertices. This ensures that the graph is updated for the next round of augmenting path searches while maintaining the feasibility of the matching. The algorithm repeats this process, until a maximum weight matching is found for all vertices in U.

The correctness of the Kuhn–Munkres algorithm or Hungarian algorithm stated as: "The Hungarian algorithm finds a maximum weight matching and a minimum cost cover"can be found in [13, Theorem 3.2.11].

The Hungarian Algorithm finds a maximum weight matching in a bipartite graph and has a time complexity of $O(n^3)$, where n is the number of vertices in one partite set (either U or V) of the bipartite graph.

If an augmenting path is found, the algorithm updates the matching M. This update, and storing the augmenting paths, can take up to $O(|V|)$ in each iteration.

In the worst case, finding an augmenting path and updating the matching can take $O(|U| \times |V|)$. If no augmenting path is found, the algorithm computes the minimum slack value α, which takes $O(|U| \times |V|)$ to check all relevant vertices. After finding α, labels in U and V are adjusted, taking $O(|U|+|V|)$ time in each iteration.The algorithm iterates until it finds an augmenting path for each vertex in U. Since this can require up to $|U|$ augmenting path searches, and each search involves operations bounded by $O(|U| \times |V|)$, the total complexity can be approximated as:

$$O(|U|^2 \times |V|) + O(|U| \times |V|^2) = O(n^3)$$

where $n = |U| = |V|$ for simplicity, assuming the graph is balanced. Hence the total time complexity of this algorithm is $O(n^3)$ where n is the number of vertices in either partite set (assuming a balanced bipartite graph). This is typical of the Hungarian algorithm used for finding maximum weight matchings in bipartite graphs.

EXAMPLE 7.6.2   Determine an optimal assignment for the bipartite graph G in Figure 7.10 with edge weights given in adjacency matrix .

$$\mathbf{A} = \begin{matrix} & \begin{matrix} x_1 & x_2 & x_3 & x_4 & x_5 \end{matrix} \\ \begin{matrix} y_1 \\ y_2 \\ y_3 \\ y_4 \\ y_5 \end{matrix} & \begin{pmatrix} 6 & 5 & 1 & 1 & 2 \\ 3 & 4 & 6 & 1 & 1 \\ 4 & 3 & 0 & 5 & 2 \\ 3 & 4 & 3 & 0 & 5 \\ 6 & 0 & 2 & 1 & 4 \end{pmatrix} \end{matrix}$$

Here $U = \{y_1,y_2,y_3,y_4,y_5\}$ while $V = \{x_1,x_2,x_3,x_4,x_5\}$.

Algorithm 2: Kuhn–Munkres Algorithm

Data: $G = (U, V, E, W)$
; /* A complete bipartite graph where U and V are the two partite sets, E is the set of edges and W is the weight function mapping edges to weights */
Result: M, a maximum matching of G
$L_U[u] \leftarrow max(W[u,v])$ for all $u \in U$; /* Label each vertex u in U by the maximum weight of the edges incident on u */
$L_V[v] \leftarrow 0$ for all $v \in V$; /* Label each vertex v in V by 0 */
$M \leftarrow \emptyset$; /* Initialize an empty matching */
Iterate through each vertex $u \in U$
for (*each* $u \in U$) {
    $S \leftarrow \emptyset$; /* Set of visited vertices in U */
    $T \leftarrow \emptyset$; /* Set of visited vertices in V */
    $prev[v] \leftarrow NULL$ for all $v \in V$; /* An array to store augmenting paths */
    Find an augmenting path
    while *no augmenting path is found for u* **do**
        Choose $u' \in U$ such that $u' \notin S$
        $S \leftarrow S \cup \{u'\}$; /* Mark u' as visited */
        for (*each* $v' \in V$) {
            if $v' \notin T$ *and* $L_U[u'] + L_V[v'] = W[u',v']$ **then**
                $T \leftarrow T \cup \{v'\}$; /* Mark v' as visited (tight edge condition) */
                if v' *is unsaturated* **then**
                    Augment the path from u to v'
                    Update the matching M
                    break;
                else
                    Let u'' be the vertex matched to v'
                    $S \leftarrow S \cup u''$; /* For further exploration of u'' */
                    $prev[v'] \leftarrow u'$; /* Store the augmenting paths */
                end
        end
        if *No augmenting paths are found* **then**
            $\alpha \leftarrow Min(L_U[u'] + L_V[v'] - W[u',v'])$ for all $u' \in S, v' \notin T$; /* Calculate the minimum slack for unvisited vertices */
            $L_U[u'] \leftarrow L_U[u'] - \alpha$ for all $u' \in S$; /* Decrease labels for vertices in U */
            $L_V[v'] \leftarrow L_V[v'] + \alpha$ for all $v' \in T$; /* Increase labels for vertices in V */
        end
    }
    end
    Repeat until an augmenting path is found for each u in U.
}
return M

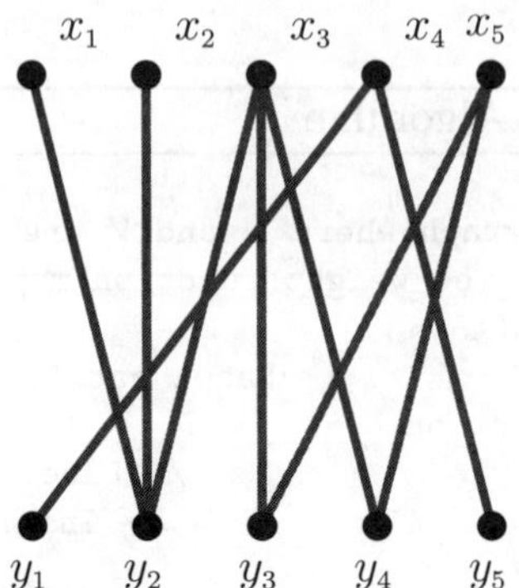

Figure 7.10   Bipartite graph G

We will first initialize all the labels for y_i to the maximum edge weight while for $x_i \in V$, we set the label to 0. It follows that

$$L_U(y_1) = 6 \quad L_U(y_2) = 6 \quad L_U(y_3) = 5 \quad L_U(y_4) = 5 \quad L_U(y_5) = 6$$

and $L_V(x_j) = 0$ for all j. We will find an initial feasible matching. We look for edges for which $L_U(y_i) + L_V(x_i) = w(x_j, y_i)$, as these are the "tight" edges. We arrive at the matching $\{(x_1, y_1), (x_3, y_2), (x_4, y_3), (x_5, y_4)\}$.

$$
\mathbf{A} =
\begin{bmatrix}
 & x_1 & x_2 & x_3 & x_4 & x_5 & L_U(y_i) \\
y_1 & 6 & 5 & 1 & 1 & 2 & 6 \\
y_2 & 3 & 4 & 6 & 1 & 1 & 6 \\
y_3 & 4 & 3 & 0 & 5 & 2 & 5 \\
y_4 & 3 & 4 & 3 & 0 & 5 & 5 \\
y_5 & 6 & 0 & 2 & 1 & 4 & 6 \\
L_V(x_i) & 0 & 0 & 0 & 0 & 0 & -
\end{bmatrix}
$$

Clearly, the vertex x_2 is unsaturated by this matching. Let us adjust the values by calculating the slack values, so that the new edges become tighter. Using $\alpha = min(L_U[u] + L_V[v] - W(u,v))$ for all vertices $y_i \in S$ and for all $x_i \notin T$, we do the following calculations. $\alpha = min\{L_V(x_2) + L_U(y_1) - w(x_2, y_1), L_V(x_2) + L_U(y_2) - w(x_2, y_2), L_V(x_2) + L_U(y_3) - w(x_2, y_3), L_V(x_2) + L_U(y_4) - w(x_2, y_4), L_V(x_2) + L_U(y_5) - w(x_2, y_5)\}$ which gives us $\alpha = min\{1, 2, 2, 1, 6\}$ and therefore $\alpha = 1$. Let us update the labels using the condition $L_U[y_i] \leftarrow L_U[y_i] - \alpha$ for all $y_i \in U \cap V(X)$ and $L_U[y_i] \leftarrow L_U[y_i] + \alpha$ for all $x_i \in V \cup V(X)$. Here X represents the alternating tree. We will arrive at another matrix A' consisting of the modified values.

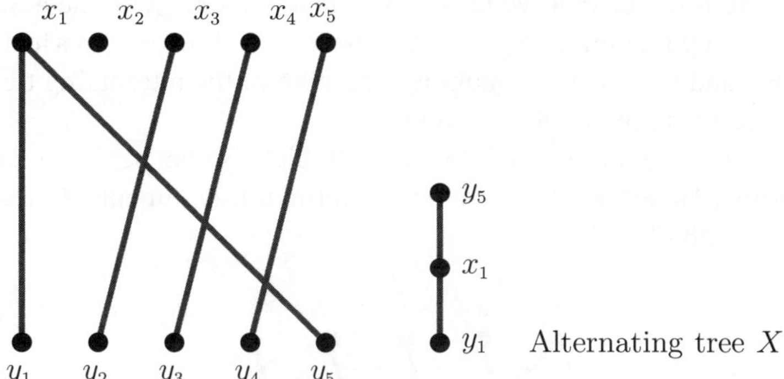

Figure 7.11   Initial matching M for bipartite graph G

$$\mathbf{A}' = \begin{bmatrix} & x_1 & x_2 & x_3 & x_4 & x_5 & L_U(y_i) \\ y_1 & 6 & 5 & 1 & 1 & 2 & 5 \\ y_2 & 3 & 4 & 6 & 1 & 1 & 6 \\ y_3 & 4 & 3 & 0 & 5 & 2 & 5 \\ y_4 & 3 & 4 & 3 & 0 & 5 & 5 \\ y_5 & 6 & 0 & 2 & 1 & 4 & 5 \\ L_V(x_i) & 1 & 0 & 0 & 0 & 0 & - \end{bmatrix}$$

By checking the tight edge condition again, we obtain another matching $M' = \{(x_2, y_1), (x_3, y_2), (x_4, y_3), (x_5, y_4)\}$.

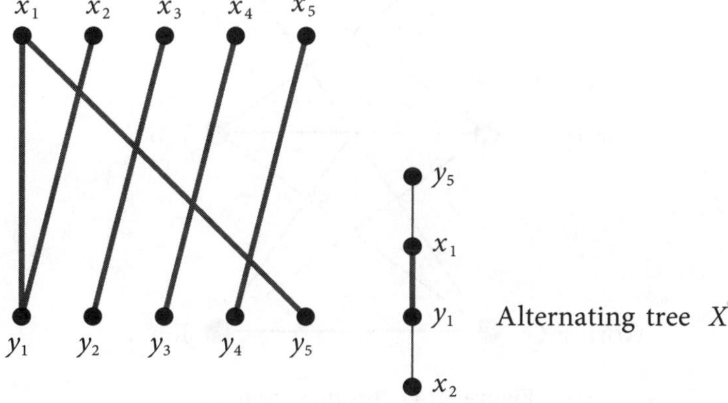

Figure 7.12   Matching M' for bipartite graph G

In Figure 7.12 we now see that we have a new matching M' with the augmenting paths found. x_2 is saturated by this matching, but it is clear that M' is not a perfect matching. By considering (x_1, y_5) and (y_1, x_2) in the augmenting path of the alternating tree X (shown in lighter edges), we arrive at the final matching
$M'' = \{(x_1, y_5), (x_2, y_1), (x_3, y_2), (x_4, y_3), (x_5, y_4)\}$, which is the perfect matching for the given bipartite graph. Hence we have found an optimal assignment M'' that saturates all the vertices of the graph G.

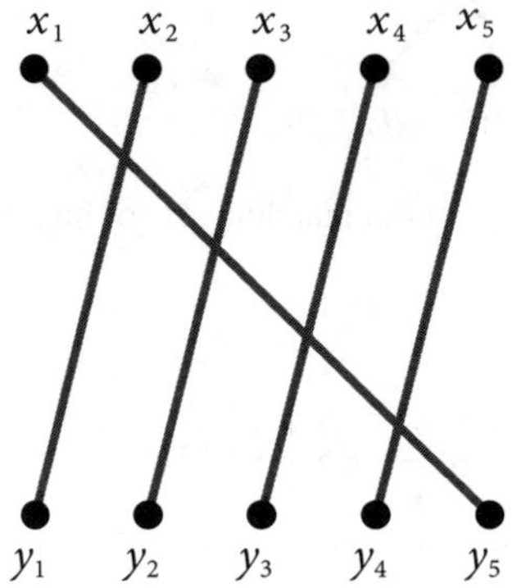

Figure 7.13   Matching M'' for bipartite graph G

EXAMPLE 7.6.3   Determine an optimal assignment for the bipartite graph G in Figure 7.14, with edge-weights given in Table 7.1.

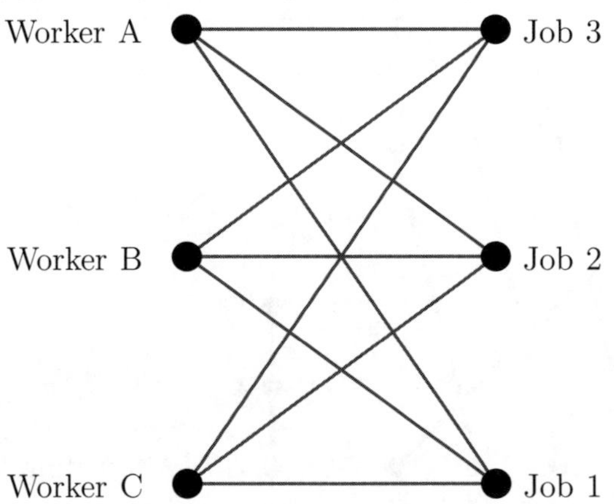

Figure 7.14   Bipartite graph G

Here U is the set of workers while V is the set of jobs.

Step 1: $L_U[A] = max(20, 60, 45) = 60$

Table 7.1 Assignment problem

Edge weight	Job 1	Job 2	Job 3
Worker A	20	60	45
Worker B	15	25	70
Worker C	45	60	20

$L_U[B] = max(15, 25, 70) = 70$
$L_U[C] = max(45, 60, 20) = 60$
For each vertex $v \in V$, $L_V[v] = 0$. Hence the initial labels for V are $L_V[A] = 0, L_V[B] = 0, L_V[C] = 0$

Step 2: Find an initial matching. We will try to find an augmenting path by inspecting the edges that satisfy the equality condition. $L_U[u] + L_V[v] = W[u, v]$. For Worker A,
$L_U[A] + L_V[1] = 60 + 0 = 60 \neq W(A, 1) = 20$ (not tight)
$L_U[A] + L_V[2] = 60 + 0 = 60 = W(A, 2)$ (tight)
$L_U[A] + L_V[3] = 60 + 0 = 60 \neq W(A, 3) = 45$ (not tight)
Hence the edge connecting Worker A to Job 2 is tight. Add edge (Worker A, Job 2) to the matching.
For Worker B,
$L_U[B] + L_V[1] = 70 + 0 = 70 \neq W(B, 1) = 15$ (not tight)
$L_U[B] + L_V[2] = 70 + 0 = 70 \neq W(B, 3) = 25$ (not tight)
$L_U[B] + L_V[3] = 70 + 0 = 70 = W(B, 3)$(tight) Add edge (Worker B, Job 3) to the matching.
For Worker C,
$L_U[C] + L_V[1] = 60 + 0 = 60 \neq W(C, 1) = 45$ (not tight)
$L_U[C] + L_V[2] = 60 + 0 = 60 = W(C, 2) = 60$ (tight)
$L_U[C] + L_V[3] = 60 + 0 = 60 \neq W(C, 3) = 20$ (not tight)
But Job 2 is already matched to Worker A.
At this point the matching is $M = \{(A, 2), (B, 3)\}$.

Step 3: We will now update the labels. Since Worker C is not matched with a job yet, we need to update the labels to create a new tight edge for Worker C. We calculate the minimum slack α as $\alpha = min(L_U[u] + L_V[v] - W(u, v))$ for $u \in S$ and $v \notin T$ where T represents the workers saturated by M and S consists of jobs that are saturated by M. For the vertex C,
$\alpha = min(60 + 0 - 45, 60 + 0 - 60, 60 + 0 - 20)$
Hence, $\alpha = min(15, 0, 40) = 0$. Since $\alpha = 0$ no label updates are needed.

Step 4: We will determine the augmenting path for Worker C. Since Job 2 is already taken by Worker A, we cannot proceed. But Job 3 is taken by B, which leaves us with Job 1 for C. By using alternating edges, we augment the matching by swapping $A \rightarrow 2$

with $C \to 1$ and again with B, thus arriving at the optimal assignment: Worker A$\to$ Job 1, Worker B $\to$ Job 2, Worker C$\to$ Job 3, which equals a total assignment cost of $20 + 25 + 20 = 65$. The matching $M\{(A,1),(B,2),(C,3)\}$ is the optimal matching.

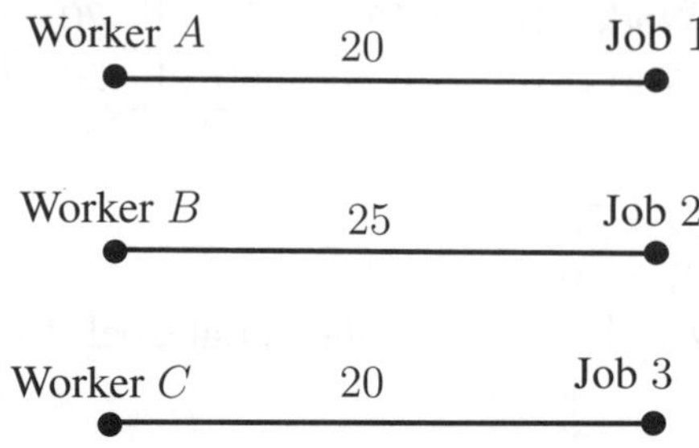

Figure 7.15   Bipartite graph G with assignments

Summary

In this chapter, we have examined in depth, the concept of matching, that helps one to make optimal pairings in a graph. We have differentiated between a perfect matching and a maximum matching in a graph. In the process, we have introduced the reader to the concept of alternating and augmenting paths with respect to a certain matching M. We have given special attention to bipartite graphs, whose matchings have many interesting properties and outcomes. Hall's marriage theorem is discussed and its applications to optimal pairings are explained in detail. The concept of covering is introduced and its relation to maximum matching is illustrated with the help of theorems and examples. The optimal assignment problem whose solution is given in the form of the maximum matching algorithm for bipartite graphs and Kuhn–Munkres algorithm is examined comprehensively through examples and pseudocodes.

7.7   Exercises

Section 7.1: Alternating and augmenting paths

1. For the graphs of Figures 7.16, 7.17, 7.18 and 7.19 with matchings M shown by the dashed edges, find

   (i) an M-alternating path which is not M-augmenting,

   (ii) an M-augmenting path if one exists, and, if so, use it to obtain a bigger matching.

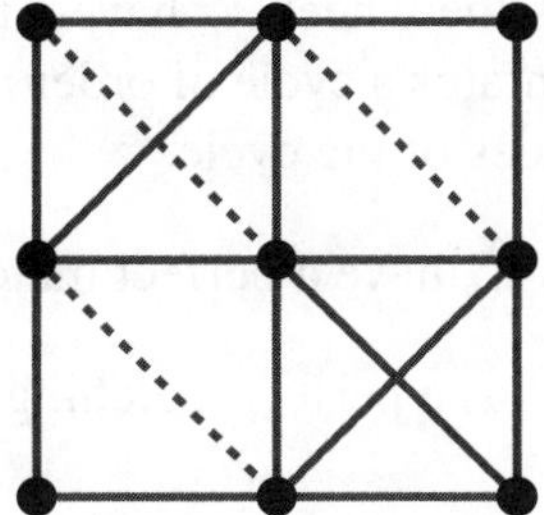

Figure 7.16   Graph 1

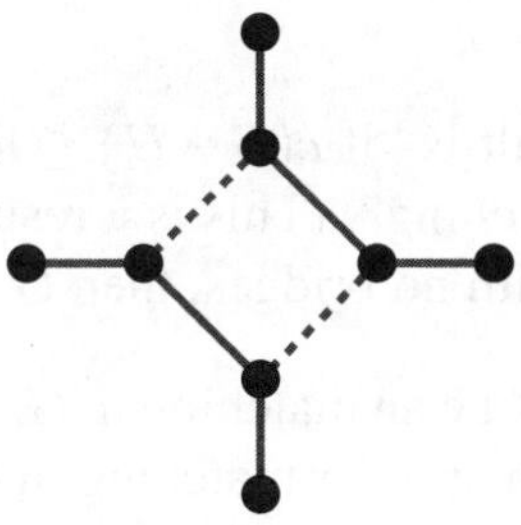

Figure 7.17   Graph 2

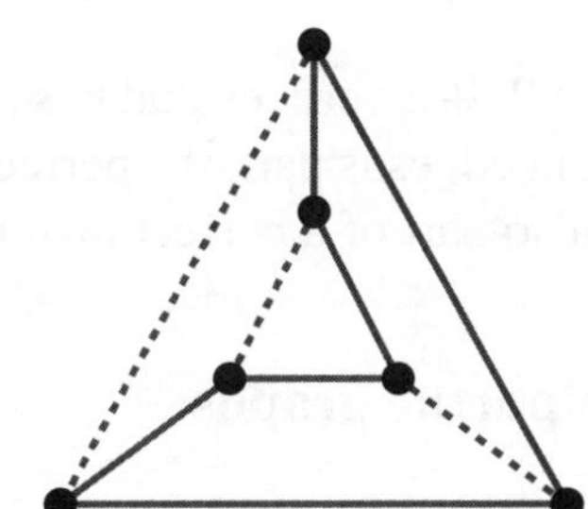

Figure 7.18   Graph 3

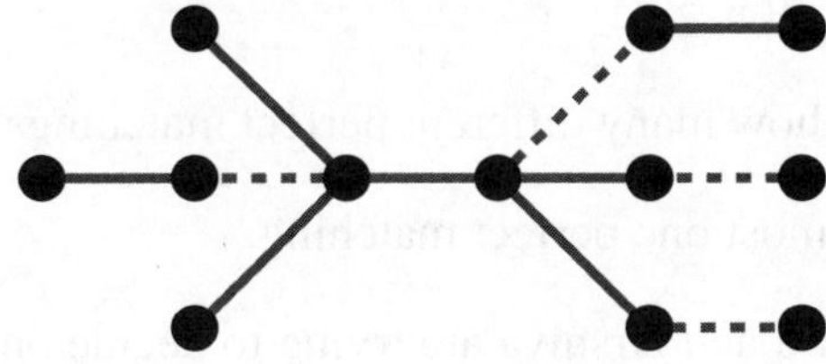

Figure 7.19   Graph 4

2. For which values of $n \geq 4$ does the wheel W_n have a perfect matching? A *wheel graph* W_n of order n is a graph that contains a cycle of order $n-1$ together with a vertex which is adjacent to all the $n-1$ vertices of the cycle.

3. For which values of $n \geq 2$ does K_n have a perfect matching?

4. Prove that a 2-regular graph G has a perfect matching if and only if each component of G is an even cycle.

5. For a graph G, recall that $c(G)$ denotes the number of connected components of G having an odd number of vertices. Let $U \subseteq V(G)$. If G has a perfect matching, then prove that $c(G-U) \leq |U|$.

6. The converse of the above result is "if $c(G-U) \leq |U|$ for all the proper subsets U of $V(G)$, then G has a perfect matching". (This is a result due to Tutte.) Using this, prove that if G is a 3-regular graph with no bridges, then G has a perfect matching.

7. Let G be a bipartite graph and M be an matching in G. Then prove that if M is suboptimal i.e., it contains fewer edges than another matching in G, then G contains an augmenting path with respect to M. Does this property extend to matchings in non-bipartite graphs?

8. Does every r-regular graph have a perfect matching? Justify your answer.

9. Consider a graph G with at least $2r+2$ vertices that has a perfect matching. Prove that if every collection of r independent edges is part of a perfect matching, then any collection of $r-1$ independent edges is also part of a perfect matching.

Section 7.2: Matching in bipartite graphs

10. Which complete tripartite graph of the form $K_{n,n,n}$ possesses a perfect matching?

11. Prove that the *n-cube* Q_n has a perfect matching for each $n \geq 2$. Note that $Q_3 = C_4 \square K_2$ and $Q_n = Q_{n-1} \square K_2$ for all $n \geq 4$.

12. For each integer $n \geq 2$, how many different perfect matchings are there for K_{2n}?

13. Prove that a tree has at most one perfect matching.

14. Mano, Subathra, Vardhini and Arshiya are trying to decide on who to take to the Study High School. Table 7.2 shows which boys they would not mind taking. Is it possible for all four girls to choose suitable partners? Answer this by first setting up a bipartite graph and checking in detail to see if Hall's condition is satisfied by the set X consisting of the four girls.

Table 7.2   Suitable partners

Mano	Matt, Pete
Subathra	Nick
Vardhini	Pete, Rod, Steve
Arshiya	Rod, Steve

15. Prove that a non-empty bipartite graph G has a perfect matching if and only if $|S| \leq |N(S)|$ for all subsets S of $V(G)$. Show, by example, that the bipartite condition cannot be omitted.

16. Prove that if $G = G(V_1, V_2)$ is a bipartite graph where the degree of every vertex in V_1 is at least as large as the degree of every vertex in V_2, then G contains a complete matching.

17. Show that the marriage condition guarantees a solution to the marriage problem if each boy has $n(n > 1)$ choices among the girls, and each girl has n choices among the boys.

18. Prove that a bipartite graph G that is n-regular has a complete matching. Furthermore, deduce that the chromatic index of G is n. [Hint: Exercise 3.]

19. Show that a bipartite graph with a minimum degree of at least r and containing a perfect matching has at least $r!$ perfect matchings. Is this the best possible outcome?

20. In a school with 30 professors, 12 committees are formed, each consisting of 8 professors. Each professor is a member of exactly 4 committees. Show that it is possible to choose a distinct representative from each committee.

21. Let G be a bipartite graph with vertex set $V = V_1 \cup V_2$ and each vertex of V_1 is of odd degree. Suppose that any two vertices of V_1 have an even number of common neighbors. Show that G has a matching saturating V_1.

22. Prove that a bipartite graph G has a matching of size at least $|E(G)|/\Delta(G)$.

Section 7.3: Covering

23. Determine $\alpha(G), \alpha'(G), \beta(G)$ and $\beta'(G)$, for the graph G in Figure 7.20.

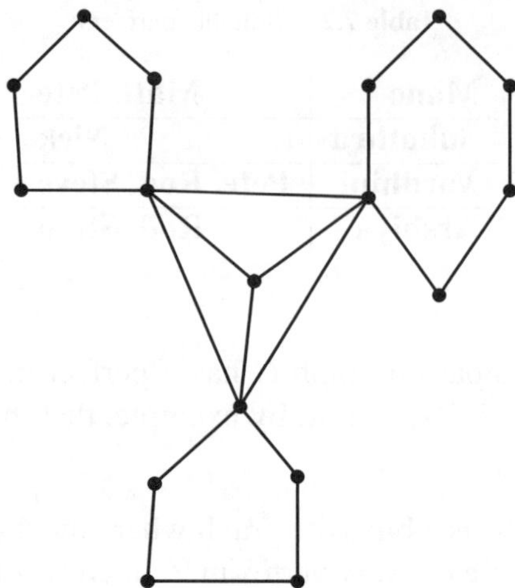

Figure 7.20   Graph 5

Section 7.4: Algorithm to find the maximum matching in a bipartite graph

24. Apply the Maximum matching algorithm to the bipartite graphs shown in Figures 7.21, 7.22 and 7.23 starting with a matching consisting of a single edge to find out if the graphs have matchings that saturate every vertex of the bipartition subset X. (Apply the algorithm methodically by taking into account the vertices in the subsets X and Y in the specified order, that is $x_1, x_2, \ldots$ and $y_1, y_2, \ldots$).

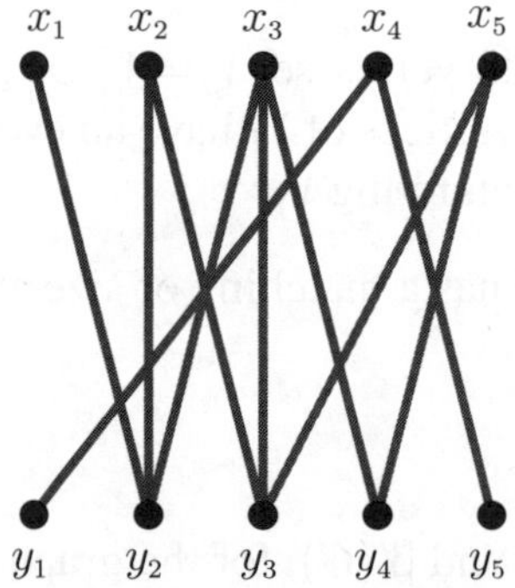

Figure 7.21   Graph 6

25. For the graph in Figure 7.24, use maximum matching algorithm for a bipartite graph to determine a maximum matching. Start with the initial matching $M = \{(x_1, y_1), (x_2, y_3), (x_6, y_4)\}$.

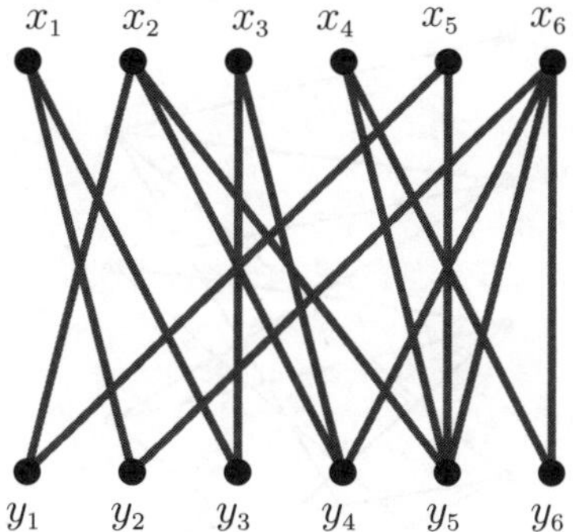

Figure 7.22   Graph 7

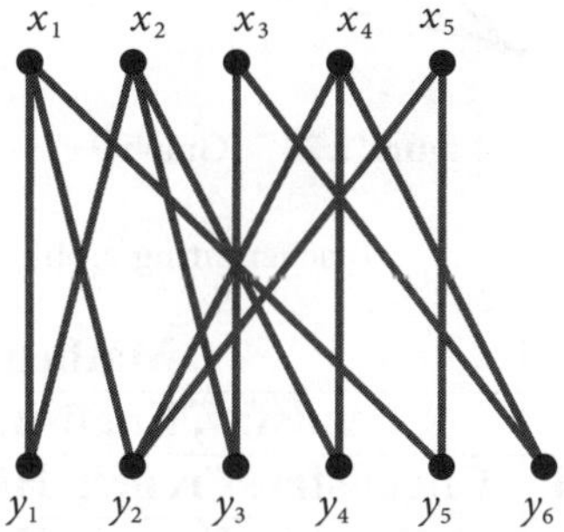

Figure 7.23   Graph 8

26. Two players take turns playing a game on a graph. Starting from any vertex, each player selects a vertex adjacent to the one chosen in the previous move. The game ends when a player can no longer make a valid move, and the last player to make a move wins. Show that the first player has winning strategy if and only if the graph has no perfect matching.

Section 7.5: Optimal assignment problem

27. There are seven teaching positions available at Hope School, one for each of the following: Mathematics, Physics, Chemistry, Geography, History, French and English. Even so, candidates who possess the qualifications to teach multiple subjects have applied for the available positions. Table 7.3 lists the applicants along with their subjects. To illustrate this scenario, create a bipartite graph and figure out the highest number of (suitably qualified) teachers that Hope School can hire.

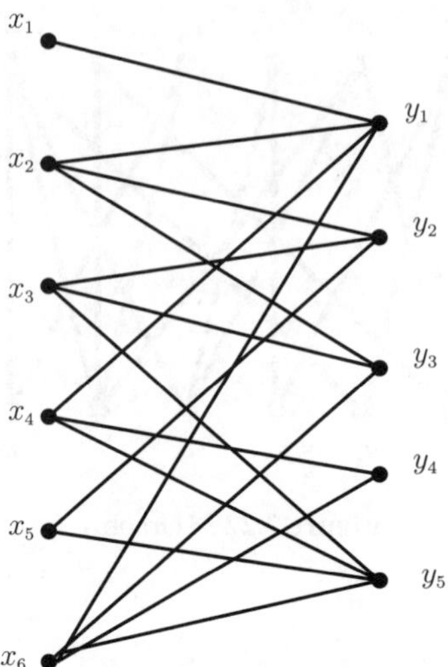

Figure 7.24   Graph 9

Table 7.3   Teacher hiring problem

Ms Rabikka	**Mathematics, Physics**
Ms Ashitha	**Chemistry, English, Mathematics**
Mr Cheri Paul	**Chemistry, French, History, Physics**
Mr Akhil	**English, French, History, Physics**
Ms Sanusha	**Chemistry, Mathematics**
Ms Esha	**Mathematics, Physics**
Ms Ankitha	**English, Geography, History**

28. A building contractor is hiring for the following positions: bricklayer, carpenter, plumber, and toolmaker. The contractor receives applications from five candidates with the following qualifications:

One applicant is qualified for the position of bricklayer.

One applicant is qualified for the position of carpenter.

One applicant is qualified for both bricklayer and plumber.

Two applicants are qualified for both plumber and toolmaker.

(i) Draw the bipartite graph representing the applicants and the jobs they are qualified for.

(ii) Verify whether the marriage condition is satisfied for this problem.

(iii) Determine whether it is possible to fill all the positions with the given applicants.

Section 7.6: Kuhn–Munkres algorithm

29. Apply the Kuhn–Munkres algorithm to the bipartite graphs with vertices $x_1, \ldots, x_6$ in one partition X while $y_1, \ldots, y_6$ form the other partition Y. Their edge weights are given in Table 7.4. Find the assignment with the least total cost.

Table 7.4 Assignment problem

Edge weight	x_1	x_2	x_3	x_4	x_5	x_6
y_1	30	25	50	25	20	30
y_2	20	60	15	40	30	25
y_3	30	15	35	60	45	30
y_4	40	20	45	20	25	30
y_5	25	35	20	40	30	45
y_6	15	50	45	50	15	60

30. Determine an optimal assignment for a complete bipartite graph $K_{6,5}$ with the edge weights given in Table 7.5.

Table 7.5 Assignment problem

Edge weight	Job 1	Job 2	Job 3	Job 4	Job 5	Job 6
Worker A	30	60	45	50	20	80
Worker B	25	70	55	40	90	60
Worker C	80	20	50	40	60	30
Worker D	70	85	25	90	45	55
Worker E	65	55	80	35	70	25

8

Planar Graphs

- Define a planar graph and recognize its visual characteristics.
- Differentiate between planar and non-planar graphs.
- State and apply Euler's formula for connected planar graphs.
- State Kuratowski's theorem to identify planar graphs.
- State Wagner's theorem to identify planar graphs.
- Discuss characteristics of planar and non-planar graphs.
- Construct the dual of a planar graph.

In graph theory, planar graphs introduce a fascinating area of study that intersects geometry, topology and network analysis. A graph is called planar if it can be drawn on a plane without any edges crossing each other. This property of planarity has significant implications in fields like circuit design, geographic mapping and urban planning where minimizing crossings leads to more efficient and visually accessible structures. In this chapter we will be covering important topics like the Euler's formula which provides a basis for understanding the structural constraints of a planar graph. We will also be discussing Kuratowski's and Wagner's theorems that provide a criteria for non-planarity. As we will see later in Chapter 9, one of the intriguing mathematical characteristics of planar graphs is that their vertices can be colored by at most four colors.

DEFINITION 8.0.1 A graph G is *planar* if it can be drawn in the plane such that none of its edges cross. That is, G can be drawn on the plane (such as on a piece of paper) in such a way that its edges intersect only at the endpoints. If such a drawing is not possible, then the corresponding graph is called a *non-planar* graph.

Some pictures of a planar graph might have crossing edges, but it's possible to redraw the picture to eliminate the crossings.

For instance, in Figure 8.1, the complete graph on 4 vertices G_1 and a graph on 7 vertices G_3 are planar graphs. In fact the graphs G_2 and G_4 are redrawings of G_1 and G_3 respectively.

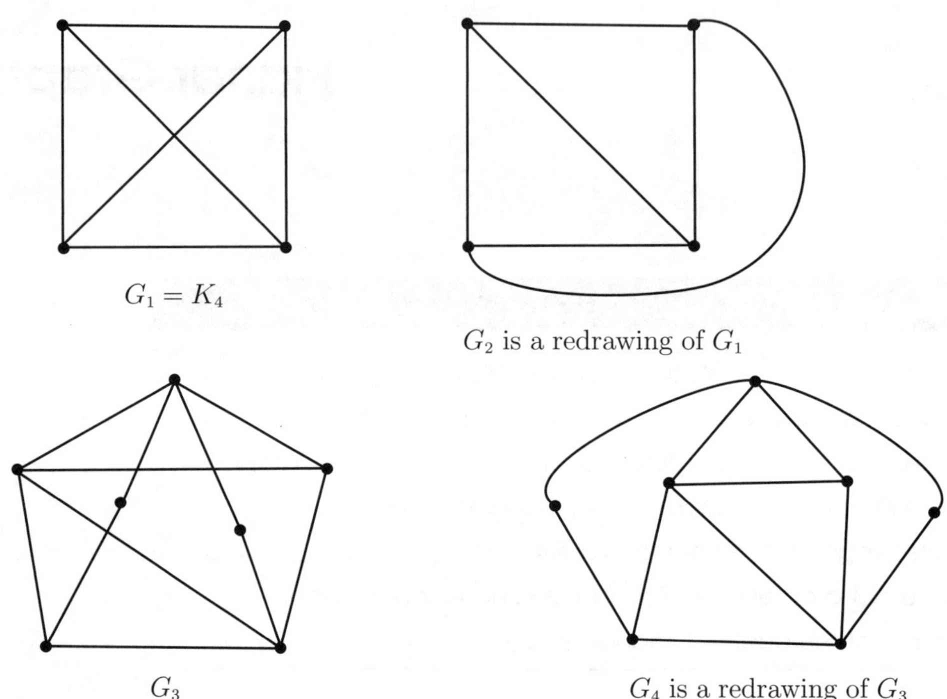

Figure 8.1   Examples of planar graphs

Is it possible to redraw any connected graph so that no two edges cross over? Alternatively stated, is every graph planar? The reader may attempt to draw the complete graph K_6 in the plane so that no two edges of K_6 cross each other, but the answer is no.

Is an infinite planar graph possible? To our infinite surprise, the answer is yes. Look no further than the infinite path P_∞ or the infinite star graph $K_{1,\infty}$.

Further, the pattern we notice is that all the subgraphs of a planar graph are planar and if a subgraph of a graph G is non-planar, then G is a non-planar graph.

8.1   Euler's formula

When a planar graph is drawn on the plane without any edges crossing, it divides the plane into regions called faces. Let us see the formal definition of faces in planar graphs.

DEFINITION 8.1.1 The edges of any planar graph split the planes into various regions. The regions enclosed by the edges of the planar graph are called the *interior faces* of the graph. The region surrounding the planar graph is called the *exterior face* of the graph. The interior and exterior faces of the graph are included in the term *faces* of the graph. The number of faces of a planar graph G is denoted by $f(G)$ or simply by f.

EXAMPLE 8.1.2 Consider the planar graph G as shown in Figure 8.2. Then G has four interior faces f_1, f_2, f_3 and f_4 and one external face f_5. Thus $f(G) = 5$.

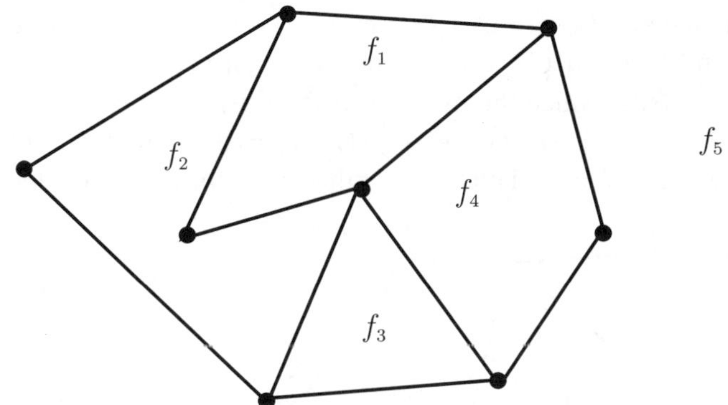

Figure 8.2   A planar graph G with five faces

Is there a relationship between the number of vertices n, the number of edges m and the number of faces f in any connected planar graph? Let's look into this. We can associate a triple (n, m, f) containing the number of vertices, edges, and faces with each planar graph. For instance, $(8, 11, 5)$ is the triple for the graph G shown in Figure 8.2. Furthermore, $(4, 6, 4)$ and $(7, 11, 6)$ are the triples for the graphs G_2 and G_4 depicted in Figure 8.1.

If we were to create numerous triples (n, m, f), containing different combinations of integers, which triples can be realized into a planar graph?

This interesting question helps one to understand the relationship between a planar graph's vertex, edge, and face counts and enables one to determine which triples can and cannot be realized into a graph. In general, Euler's formula establishes a relationship between a planar graph's vertices, edges, and faces.

THEOREM 8.1.3 (Euler's formula) *Let G be a connected planar graph and let n, m, and f denote the number of vertices, edges and faces of G, respectively. Then $n - m + f = 2$.*

Proof. Let G be a connected planar graph with n number of vertices, m number of edges and f number of faces. We prove the result by induction on f, the number of faces of G. If $f = 1$, then G has only one face, the exterior face. If G contains a cycle C, then C forms a face which is bounded, this is not possible as G has only the exterior face. Therefore G has no cycles, which implies G is a tree. Then, by Theorem 3.1.4, the number of edges m of G is $n-1$. Thus $n-m+f = n-(n-1)+1 = 2$. Hence the result is true for the case $f = 1$.

Now assume that $f \geq 2$ and the result is true for all connected planar graphs having less than f faces. Since $f \geq 2$ and any planar graph has exactly one exterior face, we have G with at least one internal face. Therefore G contains a cycle. Thus G is not a tree. Then, by Corollary 4.1.4, G has an edge e which is not a bridge. So the subgraph $G-e$ is a connected subgraph of G. As G is a planar graph, its subgraph $G-e$ is also a planar graph. Since the edge e is not a bridge, by Theorem 4.1.3, it must be a part of some cycle in G. So removing the edge e from G reduces two faces into a single face, as shown in Figure 8.3. Hence the number of faces in $G-e$ is $f-1$. Let $n(G-e), m(G-e)$ and $f(G-e)$ denote the number of vertices, edges and faces of $G-e$ respectively. Then $n(G-e) = n$, $m(G-e) = m-1$ and $f(G-e) = f-1$. Now, since the number of faces in $G-e$ is less than f, by induction hypothesis, we have $n(G-e) - m(G-e) + f(G-e) = 2$. That is $n - (m-1) + f - 1 = 2$, which implies that $n - e + f = 2$. Thus the result is true for all connected planar graphs.

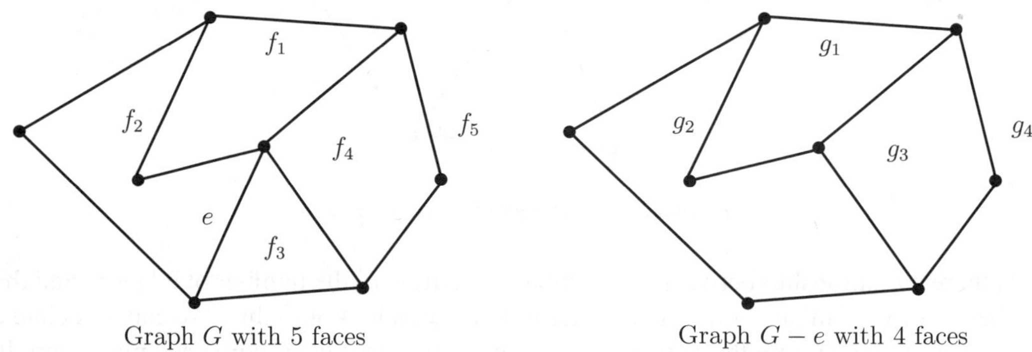

Graph G with 5 faces   Graph $G-e$ with 4 faces

Figure 8.3   Number of faces in G and $G-e$ where e is not a bridge

$\square$

COROLLARY 8.1.4 *If G is a planar graph with n vertices, m edges, f faces and k connected components, then $n - f + m = k + 1$.*

Proof. Let G be a planar graph with n vertices, m edges, f faces and k connected components. The result is true for $k = 1$, as it is Euler's formula. Let $G_1, G_2, \ldots, G_k$ be the connected components (blocks) with $n(G_i) = n_i, m(G_i) = m_i$ and $f(G_i) = f_i$ for all $i \in \{1, 2, \ldots, k\}$. By Euler's formula, refer Theorem 8.1.3, we have $n_i - m_i + f_i = 2$ for $1 \leq i \leq k$. Clearly $n = \sum_{i=1}^{k} n_i$ and $m = \sum_{i=1}^{k} m_i$. Since the counting of f_i includes the one external face of G_i for each $i \in \{1, 2, \ldots, k\}$ but the counting of f includes only one external face of G, we have

$$f = \sum_{i=1}^{k} f_i - (k-1).$$

Therefore,

$$
n - m + f = \left(\sum_{i=1}^{k} n_i - \sum_{i=1}^{k} m_i + \sum_{i=1}^{k} f_i \right) - (k-1)
$$

$$
= \sum_{i=1}^{k} (n_i - m_i + f_i) - (k-1)
$$

$$
= \sum_{i=1}^{k} 2 - (k-1) = 2k - k + 1 = k + 1.
$$

$\square$

COROLLARY 8.1.5 *Let G_1 and G_2 be two distinct redrawings of a graph G. Then G_1 and G_2 have the same number of faces.*

Proof. Let $n(G_1)$ and $n(G_2)$ be the number of vertices in G_1 and G_2 and, let $e(G_1)$ and $e(G_2)$ be the number of edges in G_1 and G_2, respectively. Since G_1 and G_2 are isomorphic to G, we have $n(G_1) = n(G_2)$ and $m(G_1) = m(G_2)$. Using Euler's formula, we get $n(G_1) - m(G_1) + f(G_1) = 2 = n(G_2) - m(G_2) + f(G_2)$. Therefore, $f(G_1) = f(G_2)$, as required. $\square$

8.2 Properties of planar graphs

Let us start to examine the properties of planar graphs. We begin with a bound for the number of edges in planar graphs in terms of its vertices.

Let φ be a face of a planar graph G. Then the number of edges on the boundary of φ is denoted $d(\varphi)$. Note that $d(\varphi) \geq 3$ for any interior face φ of a simple planar graph.

THEOREM 8.2.1 *Let G be a simple planar graph with order n and size m where $n \geq 3$. Then $m \leq 3n - 6$.*

Proof. Let G be a simple planar graph with $|V(G)| = n$ and $|E(G)| = m$.

Suppose G is connected. If $n = 3$, then G can have at most three edges, as shown in Figure 8.4. Therefore $3n - 6 = 9 - 6 = 3 \geq m$; thus the result is true for $n = 3$. Now assume that $n \geq 4$. If G is a tree, then $m = n - 1$. As $n \geq 4$, we have $2n \geq 5$. That is $3n - n \geq 6 - 1$,

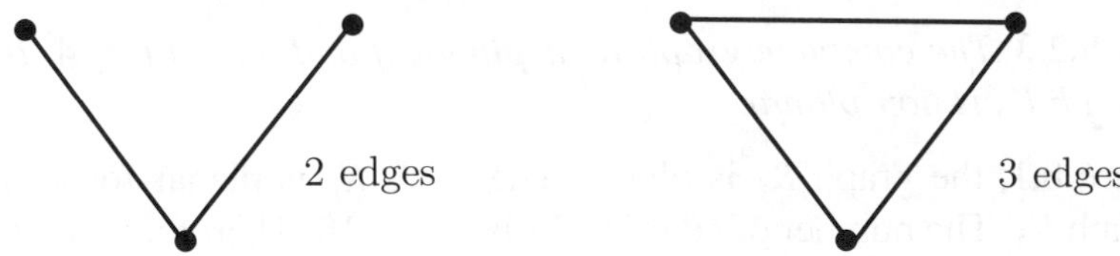

Figure 8.4   Simple connected graphs of order 3

which implies $3n - 6 \geq n - 1 = m$. So assume that G is not a tree. Then G contains a cycle. Then the exterior face of G has at least 3 edges. Therefore, we have $d(\varphi) \geq 3$ for each face φ (interior or exterior) of G. Let $b = \sum_{\varphi \in \Phi} d(\varphi)$, where the summation is done over the set of all faces of G. Since $d(\varphi) \geq 3$, we get $b \geq 3f$, where $f = |\Phi|$ is the number of faces of G. Also note that each edge can be a boundary of at most two faces. Therefore, each edge is counted twice to the maximum when calculating $d(\varphi)$. Thus $b = \sum_{\varphi \in \Phi} d(\varphi) \leq 2m$. By combining the above two inequalities, we get $3f \leq b \leq 2m$ and so $-f \geq \frac{-2m}{3}$. Also by Euler's formula (Theorem 8.1.3), we have $n = m - f + 2 \geq m - \frac{2m}{3} + 2 = \frac{m}{3} + 2$. That is, $3n \geq m + 6$, which implies $m \leq 3n - 6$. Thus, the result is true for connected simple planar graph G.

Now suppose that G is not a connected graph and it has t connected components where $t \geq 2$. Let $G_1, G_2, \ldots, G_t$ be the connected components of G. For each $i \in \{1, 2, \ldots, t\}$, let n_i and m_i denote the number of vertices and edges in G_i respectively. Since each G_i is a simple connected planar graph, we get $m_i \leq 3n_i - 6$ for each i, $1 \leq i \leq t$. Then $\sum_{i=1}^{t} m_i \leq \sum_{i=1}^{t}(3n_i - 6) = 3\sum_{i=1}^{t} n_i - 6t = 3n - 6t \leq 3n - 6$. Thus $m \leq 3n - 6$, as required. $\square$

Is it possible to get a simple planar graph with a minimum degree of 6? The following result provides the question's answer is in the negative.

COROLLARY 8.2.2 *If G is a simple planar graph, then G has a vertex of degree less than 6.*

Proof. Let G be a simple planar graph with $|V(G)| = n$ and $|E(G)| = m$. If $n \leq 6$, then naturally the minimum degree $\delta(G) \leq 5$. So assume that $n \geq 7$. Suppose, in contrary, that $deg(v) \geq 6$ for all $v \in V(G)$.

$$\text{Then} \quad \sum_{v \in V(G)} deg(v) \geq \sum_{v \in V(G)} 6 = 6n$$

$$2m \geq 6n$$

$$m \geq 3n > 3n - 6.$$

This is a contradiction to $m \leq 3n - 6$, refer to Theorem 8.2.1. Thus $\delta(G) \leq 5$. $\square$

Now we can verify the planar nature of the complete graph. To understand what follows, recall that if a graph G is planar and H is a subgraph of G, then H is planar and, if G is a non-planar graph and G is a subgraph of some graph G', then G' is non-planar.

PROPOSITION 8.2.3 *The complete graph K_n is planar if and only if $n \leq 4$. In particular, the complete graph K_5 is non-planar.*

Proof. By Figure 8.1, the graph K_4 is planar. Therefore K_n is planar for all $n \leq 4$. Now consider the graph K_5. The number of edges in K_5 is $\frac{5 \times 4}{2} = 10$. Therefore $3n - 6 = 15 - 6 = 9 < 10$. Thus by Theorem 8.2.1, K_5 is non-planar. Because of the fact that every super graph of a non-planar graph is non-planar, we have K_n is non-planar for all $n \geq 5$. $\square$

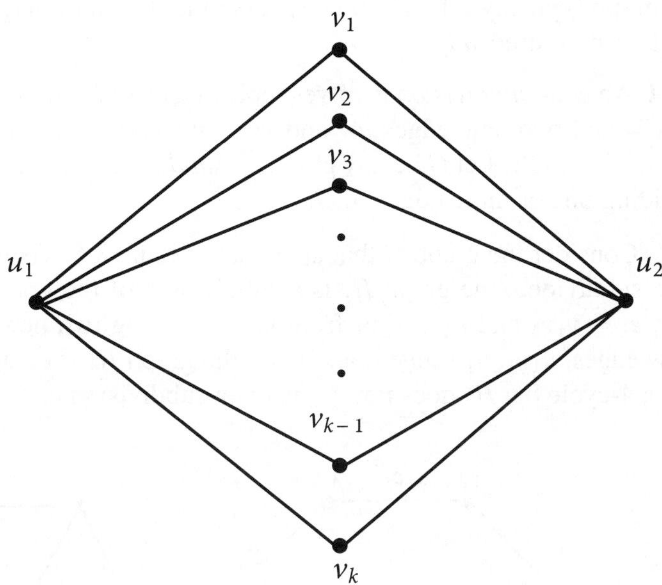

Figure 8.5   Planar representation of $K_{2,k}$ for $k \in \mathbb{N}$

Next let us verify the planarity nature of the complete bipartite graph.

PROPOSITION 8.2.4 *The complete bipartite graph $K_{m,n}$ is planar if and only if either $m \leq 2$ or $n \leq 2$. In particular, the complete bipartite graph $K_{3,3}$ is non-planar.*

Proof. By Figure 8.5, the complete bipartite graph $K_{2,k}$ is planar for any $k \in \mathbb{N}$. Now consider the graph $K_{3,3}$. Suppose, in contrary, that $K_{3,3}$ is planar. Since $K_{3,3}$ is bipartite, by Theorem 1.5.6, it contains no odd cycles. In particular, $K_{3,3}$ does not have a 3 cycle. Note that the length of any even cycle is at least 4. Therefore $d(\varphi) \geq 4$ for all the faces of $K_{3,3}$. Thus $\sum_{\varphi \in \Phi} d(\varphi) \geq 4f$, where $f = |\Phi|$. Since each edge is counted twice to the maximum when calculating $d(\varphi)$, we have $\sum_{\varphi \in \Phi} d(\varphi) \leq 2m$. Thus $4f \leq 2m$ which implies $2f \leq m$. On the other hand, in $K_{3,3}$, the number of edges $m = 9$. Therefore $2f \leq 9$ and so $f \leq 4.5$. Hence $K_{3,3}$ can have at most 4 faces. Now by Euler's formula, $f = 2 - n + m = 2 - 6 + 9 = 5$, which is a contradiction. Thus $K_{3,3}$ is non-planar. $\qquad\square$

8.3 Kuratowski's theorem

We have seen two examples of non-planar graphs, K_5 and $K_{3,3}$, in the previous section. In this section, we establish Kuratowski's theorem, which states that any graph with a subgraph of K_5 or $K_{3,3}$ is non-planar and vice-versa. Let us start with the definition of subdivision of a graph.

Subdividing a graph typically refers to the process of adding new vertices and edges to an existing graph in a structured way.

DEFINITION 8.3.1 An *edge subdivision* involves replacing an existing edge uv in the graph with a new vertex w and two new edges uw and vw. This creates a new vertex w that is adjacent to the vertices u and v. Let G be any graph. A *subdivision of G* is a graph obtained from G by subdividing one or more edges in G.

EXAMPLE 8.3.2 Consider the graph G that appears in Figure 8.6. Then, since the edges e_1 and e_2 in G are subdivided, the graph H_1 is a subdivision of the graph G. By actually deleting the edges $e_1 = v_1v_4$ and $e_2 = v_2v_3$ from G and adding two new vertices, u_1 and u_2, along with new edges, u_1v_1, u_1v_4, u_2v_2, and u_2v_3, the graph H_1 is created. On the other hand, since G has a 4-cycle but H_2 does not, H_2 is not a subdivision of G.

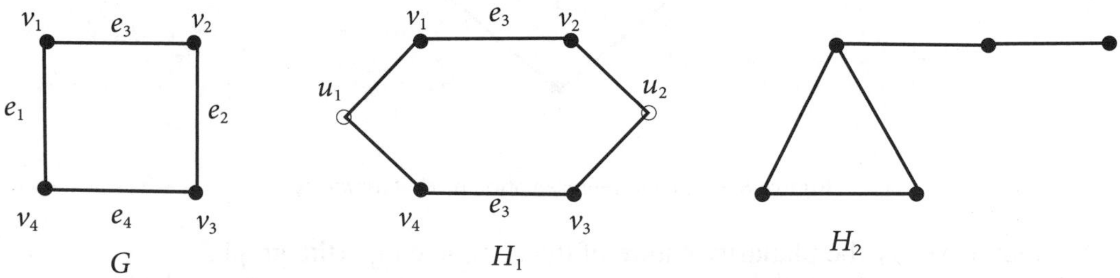

Figure 8.6 Subdivision of a graph

In 1930, Kuratowski gave a necessary and sufficient condition for the planarity of a graph: A graph G is planar if and only if it does not contain K_5 or $K_{3,3}$ as a subdivision. In 1937, Klaus Wagner gave another characterization of planar graphs: A graph G is planar if and only if it does not contain K_5 or $K_{3,3}$ as a minor. This result is, in fact, equivalent to Kuratowski's theorem. The difference between graph minors and subdivisions may appear subtle. However, its consequences are profound: Wagner ventured further to make the enormously bold conjecture, which is false for subdivision, that for any property characterized by graph minors, the set of minors could be considered finite. The proof, completed nearly 70 years later, is the centerpiece of a branch of combinatorics known colloquially as Robertson-Seymour theory. In this section, we prove Kuratowski's theorem in terms of minors instead of subdivisions.

Edge contracting refers to a process where an edge in a graph is merged into a single vertex, thereby reducing the size of the graph. Edge contracting involves selecting an edge uv in a graph and merging its two endpoints u and v into a single vertex. Let us see the formal definition.

DEFINITION 8.3.3 Let G be a simple connected graph and $e = uv$ be an edge of G. Let $N(u) = \{v = u_0, u_1, \ldots, u_{m-1}\}$ and $N(v) = \{u = v_0, v_1, \ldots, v_{n-1}\}$ be the neighborhood set of u and v respectively. A *contraction of the edge e* in G results a new graph, denoted by $G * e$, with the vertex set $V(G * e) = (V(G) \setminus \{u, v\}) \cup \{w\}$, a new vertex w, and the edge set $E(G * e) = (E(G) \setminus \{uv\}) \cup \{wu_1, \ldots, wu_{m-1}, wv_1, \ldots, wv_{n-1}\}$.

DEFINITION 8.3.4 A *minor of a graph G* is a graph obtained from G by contracting edges in G or deleting edges and vertices in G.

REMARK 8.3.5 Contraction of an edge $e = uv$ indicates that all edges that were previously incident with either u or v have now become incident with w, where the vertices u and v fused as one vertex w, and the edge e was eliminated. All that is left on the graph is $G * e$. We provide an example of this in Figure 8.7.

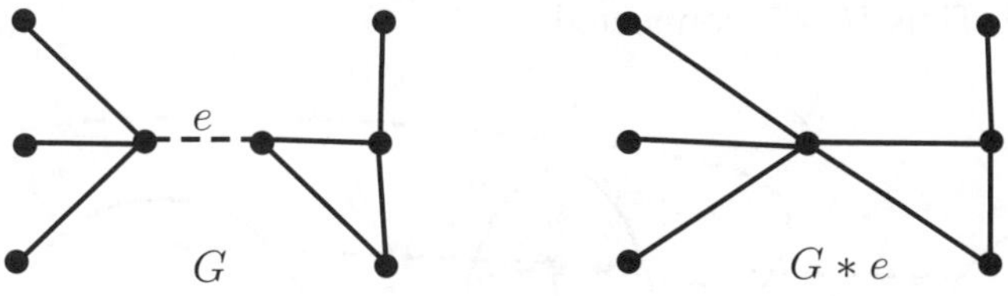

Figure 8.7 Contraction of an edge in a graph

Note that the structural minor of a graph G looks like a subgraph structure of G. Therefore, if a minor of a graph G is non-planar, then G is also non-planar.

The proof of Kuratowski's theorem, our primary finding in this section, requires the understanding of when an edge is contractible, which is connecting the vertex connectivity numbers of a graph G and it's edge contraction graph $G * e$. For any $e \in E(G)$, it is obvious that the vertex connectivity number $\kappa(G * e) \leq \kappa(G)$.

DEFINITION 8.3.6 An edge e of G is said to be *contractible*, if G and $G * e$ have the same vertex connectivity number, that is $\kappa(G) = \kappa(G * e)$.

The question then becomes, which graphs will have contractible edges? And under what conditions? The following result provides an answer that a 3-connected simple graph with at least five vertices has a contractible edge.

THEOREM 8.3.7 *Let G be a simple 3-connected graph with at least five vertices. Then G has a contractible edge.*

Proof. Let G be a simple 3-connected graph with an order of at least five. Suppose G has no contractible edges. Let $e = xy$ be an edge of G and w be the new vertex in $G * e$. Since e is not contractible, $G * e$ is not 3-connected. That is $\kappa(G * e) \leq 2$. So there exists a vertex cut S' of $G * e$ with $|S'| \leq 2$. We claim that $w \in S'$ and $|S'| = 2$. In order to prove the claim, suppose, in contrary, that $w \notin S'$. Then S' is also a vertex cut of G which is a contradiction to $\kappa(G) = 3$. Therefore $w \in S'$. Suppose that $|S'| = 1$. Then $S' = \{w\}$ and so the set $\{x,y\}$ is a vertex cut of G, a contradiction. Thus $w \in S'$ and $|S'| = 2$. Let $S' = \{w,z\}$ for some vertex $z \in V(G)$. Then $S = \{x,y,z\}$ is a vertex cut of G. So we have proved that for each edge $e = xy$, there exists a vertex z such that $S = \{x,y,z\}$ is a vertex cut of G. That is, the subgraph induced by $V(G) \setminus S$ is disconnected.

Now among all the edges and the corresponding vertex cut set S in G, choose an edge $e' = ab$ and the vertex cut $X = \{a,b,c\}$ such that the subgraph induced by $V(G) - X$ has a component C containing maximum vertices. Since G is 3-connected, the vertex c is adjacent to some vertex $u \notin C$, because otherwise $G - \{a,b\}$ itself is disconnected, as shown in Figure 8.8. Now consider the subgraph H of G induced by $C \cup \{a,b\}$. If H is 1-connected with a cut-vertex, say v, then $G - \{v,c\}$ is disconnected, a contradiction to G being 3-connected. Thus H is 2-connected.

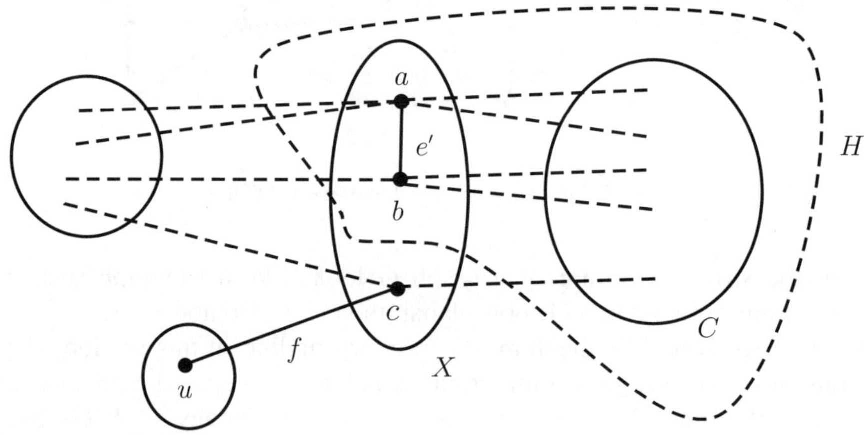

Figure 8.8   Diagrammatic representation for the proof of Theorem 8.3.7

Now consider the edge $f = uc$. As f is not contractible in G, there exists a vertex t in G such that $G - \{t,u,c\}$ is disconnected. Suppose $t \in V(H)$. Since $u,c \notin H$, we have t to be a cut-vertex of H, a contradiction to H is 2-connected. Therefore $t \notin V(H)$. This implies that $H \subseteq G - \{t,u,c\}$. Also H is a connected component of the disconnected graph $G - \{a,b,c\}$. As $V(C) \subsetneq V(H)$, we have a contradiction to the maximality of C. Thus, our initial assumption that G has no contractible edges is wrong and hence G must have a contractible edge. $\square$

DEFINITION 8.3.8 A non-planar graph G is said to be a *minimal non-planar graph* if each subgraph of G is planar.

Next, we will discuss some properties of minimal non-planar graphs.

LEMMA 8.3.9 *Let G be a minimal non-planar graph. Then G is 2-connected.*

Proof. Suppose G is not 2-connected. Then there exists a vertex u in G such that $G - \{u\}$ is disconnected. Without loss of generality, we can assume that $G - \{u\}$ has two components, G_1 and G_2. Since G is minimal non-planar, we have both $G_1 \cup \{u\}$ and $G_2 \cup \{u\}$ are planar subgraphs of G. Note that for every face f in a planar graph, there exists a planar drawing in which f is the exterior face. So consider a planar drawing of $G_1 \cup \{u\}$ and $G_2 \cup \{u\}$ such that u lies on the exterior face. Now as shown in Figure 8.9, combine $G_1 \cup \{u\}$ and $G_2 \cup \{u\}$

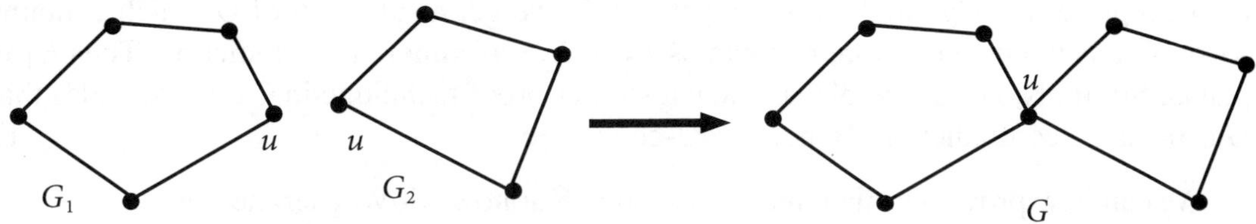

Figure 8.9   Contradicting the planarity of G

by merging the vertex u, we get a planar embedding of G, a contradiction. Therefore, G is 2-connected. $\square$

LEMMA 8.3.10 *Let G be a graph with the fewest edges among all non-planar graphs which do not have K_5 or $K_{3,3}$ as minors. Then G is 3-connected.*

Proof. Assume that G has the fewest edges out of all non-planar graphs without K_5 or $K_{3,3}$ as minors. So G is a minimal non-planar graph. Suppose G is not a 3-connected graph. Then by Lemma 8.3.9, G is 2-connected. Thus there exist two vertices u and v such that $G - \{u, v\}$ is disconnected. Without loss of generality, we can assume that $G - \{u, v\}$ has two components G_1 and G_2. Let G_1' be the subgraph induced by $V(G_1) \cup \{u, v\}$ in G and G_2' be the subgraph induced by $V(G_2) \cup \{u, v\}$ in G. Since G is minimal non-planar, both G_1' and G_2' are planar. We have two cases:

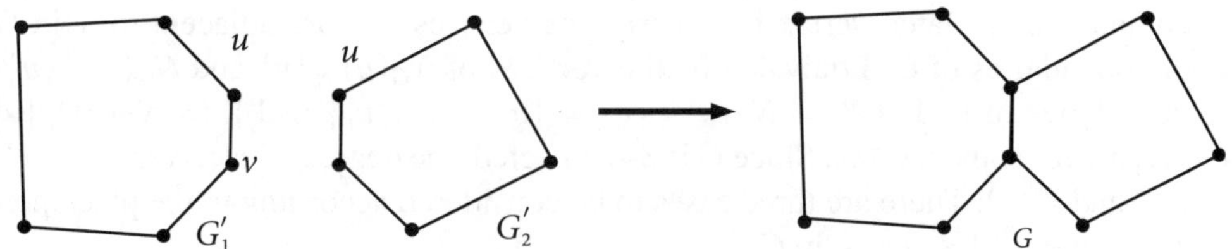

Figure 8.10   Planar embeddings of G_1' and G_2' results in planar embedding of G

Case 1: Suppose u and v are adjacent in G. Since G_1' is planar, we can have a planar drawing of G_1' such that the edge uv belongs to the exterior face of G_1'. Similarly, there is a planar drawing of G_2' with the edge uv that belongs to the exterior face of G_2'. As shown in

Figure 8.10, by combining these two planar embeddings, we get a planar embedding of G, a contradiction to G is non-planar.

Case 2: Suppose u and v are not adjacent in G. Now we consider the graphs $H_1 = G'_1 \cup \{uv\}$ and $H_2 = G'_2 \cup \{uv\}$, by adding the new edge uv. Since $\{u, v\}$ is a minimum vertex cut of G, u and v are adjacent to at least one vertex in $V(G'_2)$. Therefore $|E(G'_2)| \geq 2$, and so H_1 has fewer number of edges than G. Suppose H_1 is non-planar. Clearly H_1 has fewer number of edges than G, by the choice of G. Hence H_1 contains either K_5 or $K_{3,3}$ as minor. Now, in H_1, replace the edge uv by a $u - v$ path of G, we get a subgraph of G which contains K_5 or $K_{3,3}$ as minor. Therefore G contains K_5 or $K_{3,3}$ as minor, a contradiction. Thus H_1 is planar. Similarly, H_2 is also planar. Now a similar proof technique in Case 1 provides that G is planar, a contradiction. Hence G is 3-connected. $\square$

We can now prove our main result, so-called Kuratowski–Wagner theorem.

THEOREM 8.3.11 (Kuratowski–Wagner theorem) *A graph G is planar if and only if it does not have K_5 or $K_{3,3}$ as minors.*

Proof. We prove an equivalent statement: A graph G is non-planar if and only if G contains K_5 or $K_{3,3}$ as minors.

On one side, if G contains K_5 or $K_{3,3}$ as minors, then G is non-planar because K_5 and $K_{3,3}$ are non-planar.

On the other side, we claim that every non-planar contains either K_5 or $K_{3,3}$ as minor. Suppose, in contrary, that there exists a non-planar graph that does not contain K_5 or $K_{3,3}$ as minor. Let G be a non-planar graph with the fewest edges among all the non-planar graphs that does not contain K_5 or $K_{3,3}$ as minor. Then, by Lemma 8.3.10, G is 3-connected. Since every graph with 4 or fewer vertices is planar, we have G has at least five vertices. Then by Theorem 8.3.7, G has a contractible edge $e = uv$ for some $u, v \in V(G)$. Therefore $G * e$ is also 3-connected. Since $G * e$ has fewer number of edges than G and by the choice of G, we get that $G * e$ is planar. Let w be the new vertex in $G * e$ corresponding to the edge e. Consider a plane drawing of $G * e$. Now delete the edges of $G * e$ which are incident with w and the resulting graph is still planar. Note that one of the faces of the resulting graph, say C, contains the isolated vertex w. Clearly, the vertices that are adjacent to w in $G * e$ are on the boundaries of C. Equivalently, the vertices of $N_G(u) \setminus \{v\}$ and $N_G(v) \setminus \{u\}$ are on the boundaries of C. Let $X := N_G(u) \setminus \{v\} = \{u_1, u_2, \ldots, u_r\}$ and $Y := N_G(v) \setminus \{u\} = \{v_1, v_2, \ldots, v_s\}$ for some $r, s \in \mathbb{N}$. Since G is 3-connected, the degree of each vertex is at least 3. So $r \geq 2$ and $s \geq 2$. There are three cases to be considered according to the placement of u_i, $1 \leq i \leq r$, and v_j, $1 \leq j \leq s$, in C.

Case 1: Suppose $X \cap Y = \emptyset$. Then all v_j, $1 \leq j \leq s$, lie between u_i and u_{i+1} for some $i \in \{1, \ldots, r\}$.

In this case, a planar embedding of G can be retrieved from a planar embedding of $G * e$ in the following way:

- Place the vertex v in place of w in C and join all v_j, $1 \leq j \leq s$, with v.

- Then insert the vertex u in C and join u and v.

- Finally, join all u_i, $1 \leq i \leq r$, with u; refer to Figure 8.11.

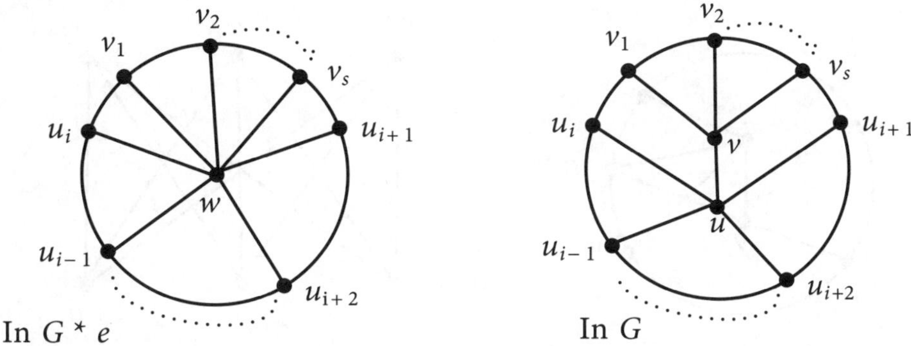

Figure 8.11　　A plane drawing of G from $G * e$

Hence G is planar, a contradiction.

Case 2: Suppose $|X \cap Y| \leq 2$.

Case 2.1: Assume that the vertices of $X \cap Y$ on C are not interlaced. In this case, the possibilities are $u_1 = v_s$ or/and $u_r = v_1$. As shown in Figure 8.12, we get a planar embedding of G, a contradiction for both possibilities.

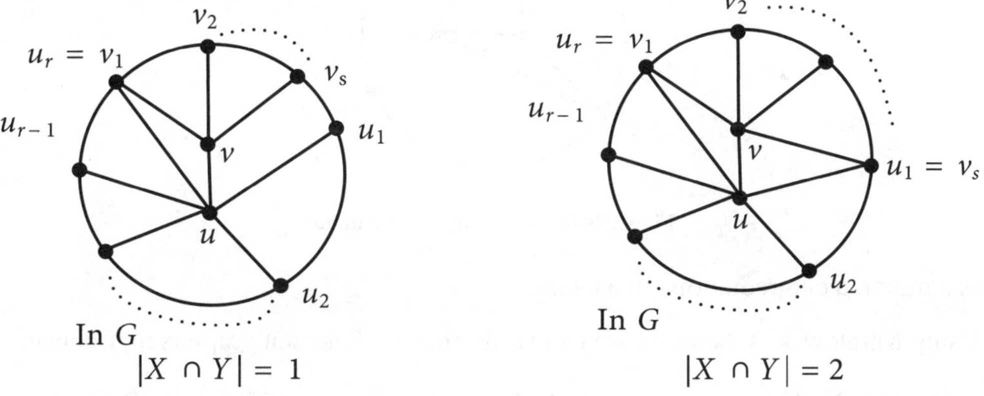

Figure 8.12　　A plane drawing of G

Case 2.2: Assume that the vertices of $X \cap Y$ in C are interlaced. Then $u_i = v_j$ for some $1 \leq i \leq r$ and $1 \leq j \leq s$ with the placement of the vertices $v_{j-1}, u_i = v_j, v_{j+1}$ and u_{i+1} are in either clockwise or anti-clockwise order in C. The existence of at least four vertices is guaranteed, since $deg(u)$ is at least 3. Then, as shown in Figure 8.13, G contains $K_{3,3}$ as a minor, a contradiction to the assumption on G. In Figure 8.13, the notation v_{j+1}^* denotes the contraction edges in C from v_{j+1} to the preceding edge u_{i+1} and u_{i+1}^* denotes the contraction edges in C from u_{i+1} to the preceding edge v_{j+1}.

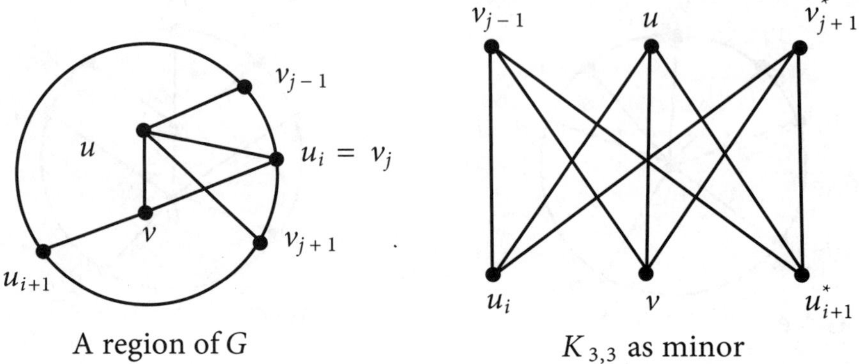

A region of G $K_{3,3}$ as minor

Figure 8.13 G contains $K_{3,3}$ as a minor

Case 3. Suppose $|X \cap Y| \geq 3$, say $x, y, z \in X \cap Y$. Then, as shown in Figure 8.14, G contains

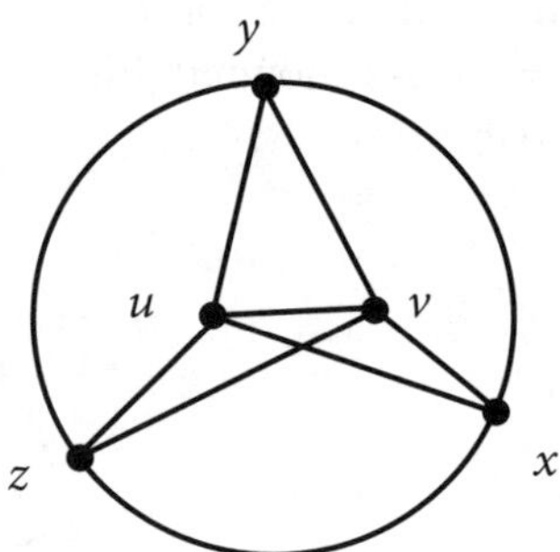

Figure 8.14 G having K_5 as a minor

K_5 as a minor, a contradiction once again. $\square$

Using Kuratowski's theorem, we can verify that the Petersen graph is non-planar.

EXAMPLE 8.3.12 Examine the graph depicted in Figure 8.15 for the Petersen graph. Then, we obtain K_5 as a minor of the Petersen graph by contracting the edges indicated by dashed lines. It is also important to note that $K_{3,3}$ is a minor in the Petersen graph. Referencing Figure 8.16, consider the subgraph H of the Petersen graph after deleting the vertex u_3 and its adjacent edges, namely $u_1 u_3$, $u_3 v_3$, and $u_3 u_5$. Then, we obtain $K_{3,3}$ as a minor of H by redrawing it as given in Figure 8.16. As a result, $K_{3,3}$ is a minor in the Petersen graph. The Petersen graph is therefore not planar.

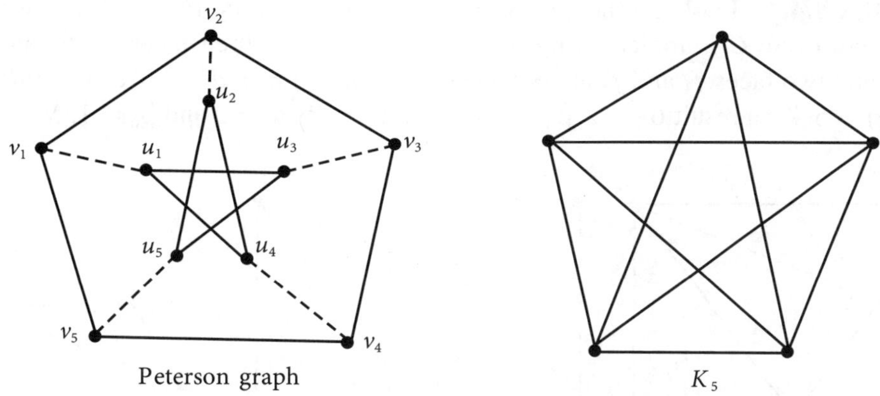

Figure 8.15 K_5 as a minor of the Petersen graph

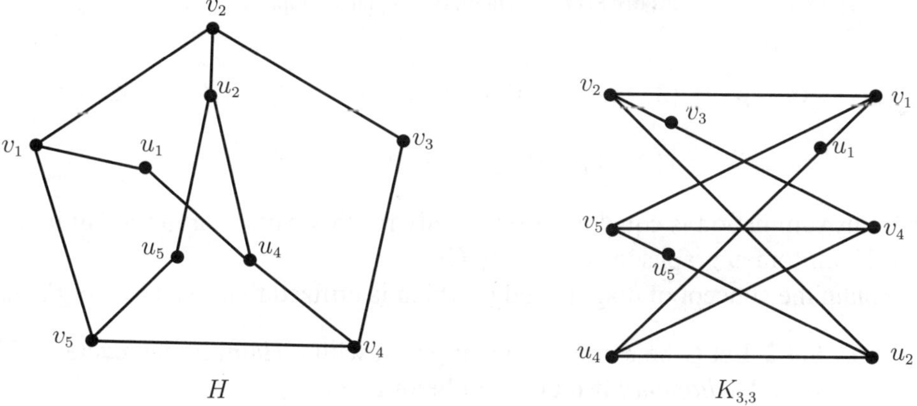

Figure 8.16 $K_{3,3}$ as minor of the Petersen graph

8.4 Non-Hamiltonian planar graphs

In this section, we will introduce the reader to the non-Hamiltonian planar graphs. They form an interesting sub class of planar graphs. We will also discuss the Grinberg's theorem which provides a necessary condition to check whether a planar graph is Hamiltonian.

Denote $d(\varphi)$ as the number of edges on the boundary of a face φ of a planar graph G. Let G be a planar graph with Hamiltonian cycle C. For $i = d(\varphi)$, the number of faces φ in G that lie inside the cycle C is denoted by α_i and the number of faces φ in G that lie outside the cycle C is denoted by β_i. Let's use an example to illustrate the idea in a better manner.

EXAMPLE 8.4.1 Look at the planar graph G depicted in Figure 8.17, where the Hamiltonian cycle C is indicated by dashed edges. Then, three faces f_1, f_2, and f_3 are inside C and two faces f_4 and f_5 are outside C. Also $d(f_1) = 5$, $d(f_2) = d(f_3) = d(f_4) = 3$ and $d(f_5) = 6$. As a result $\alpha_3 = 2, \beta_3 = 1, \alpha_5 = 1, \beta_5 = 0, \alpha_6 = 0$ and $\beta_6 = 1$. Moreover,

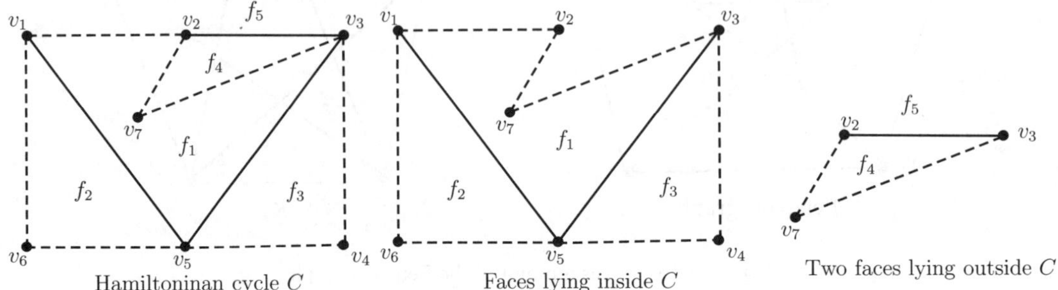

Figure 8.17   A Hamiltonian planar graph G

$$\sum_i (i-2)(\alpha_i - \beta_i) = (3-2)(\alpha_3 - \beta_3) + (5-2)(\alpha_5 - \beta_5) + (6-2)(\alpha_6 - \beta_6)$$

$$= 1(2-1) + 3(1-0) + 4(0-1) = 0.$$

Note that this summation is equal to zero not only for this particular graph but it is true for all planar Hamiltonian graphs, as proved by Grinberg.

We explain the concept of diagonal edge which is utilized to prove Grinberg's theorem.

DEFINITION 8.4.2 Let G be a simple planar graph with a Hamiltonian cycle C. Then an edge e of G is called a *diagonal* if it does not belong to C.

In Example 8.4.1, there are two diagonals interior to the cycle C and one diagonal exterior to C.

THEOREM 8.4.3 (Grinberg's theorem) *Let G be a planar graph without loops. Let G contain a Hamiltonian cycle C, and denote the number of faces φ that lie inside C by α_i and the number of faces φ that lie outside C by β_i where $i = d(\varphi)$. Then*

$$\sum_i (i-2)(\alpha_i - \beta_i) = 0.$$

Proof. Let G be a simple planar graph with a Hamiltonian cycle C. Let $|V(G)| = n$. Suppose that there are s diagonals of G in the interior of the cycle C. Since G is a planar graph, the diagonal edges won't intersect. If we remove all the s diagonal edges from G, then the interior of C is a single face. Now, by adding any one of the diagonal edges into the interior of C, the number of faces will be added by one. So, the drawing of each diagonal

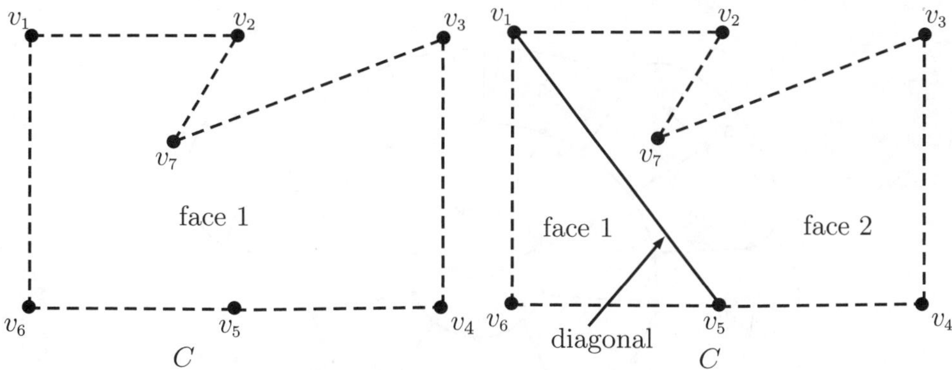

Figure 8.18 Each diagonal contributes to one extra face

will give rise to one extra face in G, as represented in Figure 8.18. Thus after drawing all the s diagonals of G, there are $s+1$ faces in the interior of C. Therefore, we have $\sum_{i\geq 2}\alpha_i = s+1$. Now, let d denote the sum of the $d(\varphi)$ where φ are the faces interior to C. Then $d = \sum_{i\geq 2} i \cdot \alpha_i$. Since each diagonal of G is counted twice in the summation for d, and each of the n edges on the cycle C is counted only once in the summation for d, we have

$$d = 2s + n.$$

Hence

$$\sum_{i\geq 2} i\alpha_i = d = 2\left(\sum_{i\geq 2}\alpha_i - 1\right) + n \tag{8.1}$$

$$\sum_{i\geq 2}(i-2)\alpha_i = n - 2. \tag{8.2}$$

If we apply the same arguments, for the diagonals in the exterior of C, we get

$$\sum_{i\geq 2}(i-2)\beta_i = n - 2. \tag{8.3}$$

Now, by Equation 8.2 and Equation 8.3, we get $\sum_{i\geq 2}(i-2)(\alpha_i - \beta_i) = 0$, as required. $\square$

Grinberg's theorem is mainly used to prove certain planar graphs to be non-Hamiltonian. For instance, consider the graph G shown in Figure 8.19. We claim that G is non-Hamiltonian. By contrary, suppose G has a Hamiltonian cycle C. Then, by Grinberg's theorem (refer Theorem 8.4.3),

$$\sum_i (i-2)(\alpha_i - \beta_i) = 0.$$

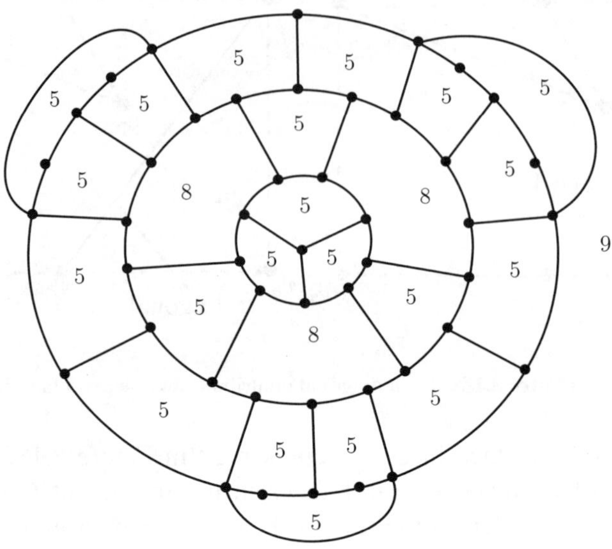

Figure 8.19   A non-Hamiltonian planar graph

Here the values for i are 5, 8 and 9. So the equation becomes

$$3(\alpha_5 - \beta_5) + 6(\alpha_8 - \beta_8) + 7(\alpha_9 - \beta_9) = 0$$
$$3[(\alpha_5 - \beta_5) + 2(\alpha_8 - \beta_8)] = 7(\beta_9 - \alpha_9).$$

This shows that 3 must divide $7(\beta_9 - \alpha_9)$. Since $\gcd(3,7) = 1$, we get $3 \mid (\beta_9 - \alpha_9)$. But there is only one face φ with $d(\varphi) = 9$ in G which implies that $(\beta_9 - \alpha_9) = \pm 1$, which is a contradiction to $3 \nmid \pm 1$. Hence G is a non-Hamiltonian planar graph.

8.5   The dual of a planar graph

In this section, we will discuss a concept called dual of G. The dual graph is a versatile tool for exploring the interplay between vertices, edges and faces of the planar graph.

DEFINITION 8.5.1 Consider a planar graph G. The *dual graph* of G is denoted by G^* and it is constructed as follows.

There is a corresponding vertex f^* of G^* for every face f of G, and a corresponding edge e^* in G^* for every edge e of G. If an edge e occurs on the boundary between two faces f and g, then the corresponding vertices f^* and g^* in G^* are joined by the edge e^*.

Construction of dual graph: Given a plane drawing of G, place the vertex f^* of G^* inside its corresponding face f. If the edge e lies on the boundary of two faces f and g of G, join the two vertices f^* and g^* by the edge e^* in such a way that it crosses the edge e exactly once and crosses no other edge of G. (This procedure is still possible if the edge e is a bridge.) By the construction, the dual G^* of the planar graph G is also planar as no two edges of G^* intersect. We explain this construction with an example.

EXAMPLE 8.5.2 Consider the planar graph G shown in Figure 8.20. In the graph G, there are seven faces $f_1, f_2, \ldots, f_7$. We select one vertex f_i^* from each face f_i so that there will be seven vertices $f_1^*, f_2^*, \ldots, f_7^*$ in G^*. Now, the edge e in G is a common edge of the two faces f_5 and f_7. So e^* is the edge in G^* joining the vertices f_5^* and f_7^*. We draw the edge e^* in such a way that it crosses the edge e but no other edge of G. The resulting dual graph G^* is shown in the figure, which is clearly planar as no two edges intersect.

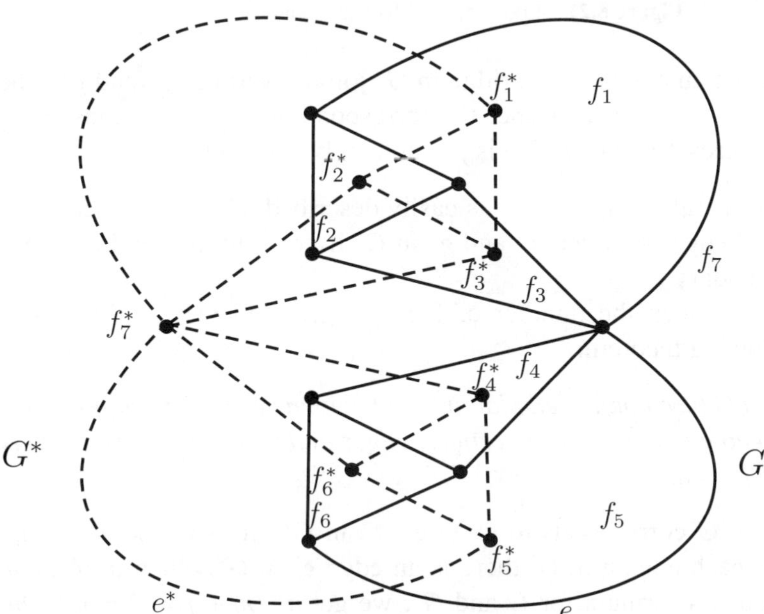

Figure 8.20 Planar graph G and its dual G^*

We prove an interesting result about the relation between loops in a planar graph and bridges in its dual graph.

LEMMA 8.5.3 *Let G be a planar graph. An edge e is a loop in G if and only if the corresponding edge e^* is a bridge in G^*.*

Proof. Suppose an edge l is a loop in G. Then l is the only edge on the common boundary of two faces, say f, and g are the faces that lie inside and outside the loop l, respectively.

Thus, by definition of G^*, l^* is the only edge between vertices f^* and g^* in G^*, and hence l^* is the bridge in G^*, as shown in Figure 8.21.

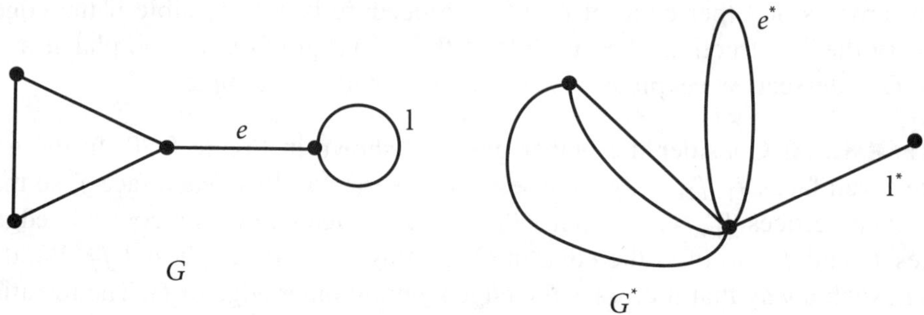

Figure 8.21   Loops and bridges in G and G^*

Conversely, suppose an edge l^* is a bridge in G^* joining vertices f^* and g^*. Then l^* is the only path between f^* to g^* in G^*. Hence the corresponding edge l in G must completely enclose one of these faces f and g, which is possible only if l is a loop in G.   □

The occurrence of parallel edges in G^* is easily described. Given two faces f and g of G, there are k parallel edges between f^* and g^* in G^* if and only if f and g have k edges on their common boundary.

The relationship between the number of faces, edges, and vertices in G and G^* is provided by the following theorem.

THEOREM 8.5.4 *Let G be a connected planar graph with n vertices, m edges and f faces. Let n^*, m^* and f^* denote the number of vertices, edges and faces of G^* respectively. Then $n^* = f, m^* = m$ and $f^* = n$.*

Proof. Each face f in G corresponds to a vertex f^* in G^* and so $n^* = f$. Also, by the definition of G^*, for each edge e in G, there is an edge e^* in G^* which implies $m^* = m$. Then by applying Euler's formula for G and G^*, we get $n - m + f = 2 = n^* - m^* + f^*$. Since $n^* = f$ and $m^* = m$, we have $f^* = n$.   □

DEFINITION 8.5.5 Given a planar graph G, we have its dual G^* as a planar graph. Now, as G^* is a planar graph, we can talk about the dual of G^* which is called as *double dual* of G and it is denoted by G^{**}.

THEOREM 8.5.6 *Let G be a connected planar graph. Then G is isomorphic to its double dual G^{**}.*

Proof. Let G^* be the dual of G with n vertices, m edges and f faces. Let n^*, m^* and f^* denote the number of vertices, edges and faces of G^* respectively. Then $n^* = f, m^* = m$

and $f^* = n$. Now let G^{**} be the dual of G^*. Then $n^{**} = f^* = n, m^{**} = m$ and $f^{**} = n^* = f$. Therefore G and G^{**} have the same number of vertices, edges and faces. Now let v be a vertex in G. Then by the definition of G^*, there exists a face φ containing the vertex v. Since $f^* = n$, φ contains exactly one vertex v. Now for the face φ of G^*, there is a corresponding vertex φ^* in G^{**}. Define a map $\psi : G \to G^{**}$ given by $\psi(v) = \varphi^*$. If $v \neq w$, then there exist two different faces φ and ψ containing v and w so that φ^* and ψ^* are distinct vertices in G^{**}. Thus φ is a graph isomorphism. Hence $G \cong G^{**}$. $\square$

Summary

In this chapter, we have introduced the reader to the fascinating world of planar graphs where the edges never cross each other. We have also discussed Euler's formula, which characterizes planar graphs. The properties of planar graphs are illustrated through various theorems. We have explained Kuratowski's theorem that gives a definitive characterization of a non-planar graph. We have also explored another subclass of planar graphs, the non-Hamiltonian planar graphs through the discussion of Grinberg's theorem which is used primarily to prove that a planar graph is non-Hamiltonian. Additionally we have explained the construction of the dual of a planar graph which offers more insights into planarity.

8.6 Exercises

Section 8.1: Euler's formula

1. Does there exist any 6-regular connected planar graph with an odd number of faces?

2. Is it necessary for a graph to be planar if it has 10 vertices, 10 edges, and an Euler circuit? How many faces would it have?

3. Three neighbors, who do not get along, share access to the same water, oil, and treacle wells. To avoid any encounters, they want to construct non-overlapping paths connecting each of their houses to all three wells. Is it possible to construct these paths without any crossings?

4. Prove that there is no cubic bipartite planar graph with ten vertices.

5. A person is standing in a room where each wall is adorned with a painting. Prove that if the room has at most five walls, there is always a spot where the person can see all the paintings simultaneously. However, prove that if the room has six or more walls, it may not be possible for the person to find such a position.

6. Which of the following four graphs in Figure 8.22 is/are planar? Justify your answers.

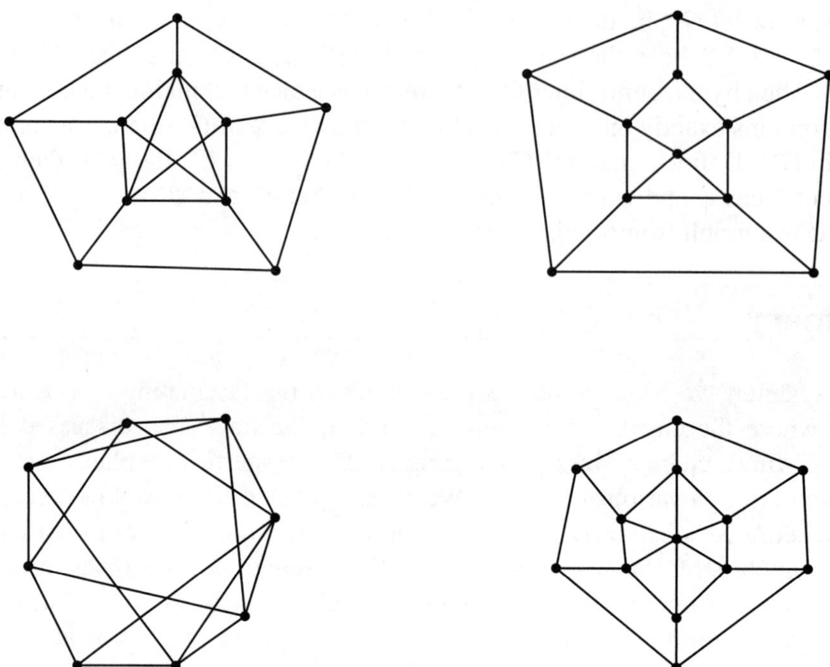

Figure 8.22   Four graphs

7. Design a graph of roads that connects 5 towns such that the graph is planar and minimizes the number of crossings.

8. Consider a football designed using a mix of pentagonal and hexagonal panels (not necessarily regular shape). These panels are stitched together, so that their seams form a cubic graph. Calculate the total number of pentagonal panels present on the football.

Section 8.2: Properties of planar graphs

9. Prove that any subgraph of a planar graph is planar.

10. Show that $K_5 - e$ is planar for each edge e of K_5.

11. Prove that $K_{3,3} - e$ is planar for each edge e of $K_{3,3}$.

12. Does there exist a simple planar graph in which the degree of each vertex is at least 5? Justify your answer.

13. Let G be a simple connected bipartite graph with n vertices and m edges with $m \geq 3$. Show that $m \leq 2n - 4$.

14. Determine the values of k, for which k-cube Q_k graph is planar.

15. Determine the values of r, k and s such that $K_{r,k,s}$ is a planar graph.

16. Let G be a graph with n vertex and m edges. Prove that, if G is a connected graph with girth 5, then $m \leq 5(n-1)/2$. Deduce that the Petersen graph is non-planar. [Hint: Use Euler's formula.]

17. Prove that every connected triangle-free planar graph must have at least one vertex with a degree of three or lower.

18. Prove by induction that every planar graph with at least one edge contains a vertex of degree at most 5.

Section 8.3: Kuratowski's theorem

19. Determine the graph $G * e$ for the graph G and edge e given in Figure 8.23.

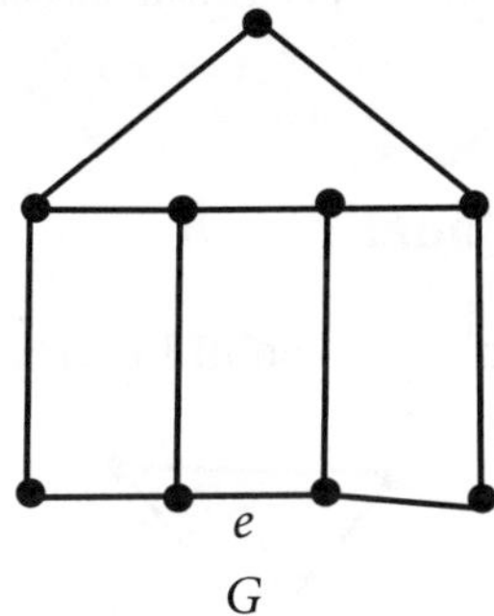

Figure 8.23 The graph G with the edge e

20. Provide an example of a graph G and an edge e such that $\kappa(G) \neq \kappa(G * e)$.

21. Using Kuratowski's theorem, show that the graph shown in Figure 8.24 is non-planar.

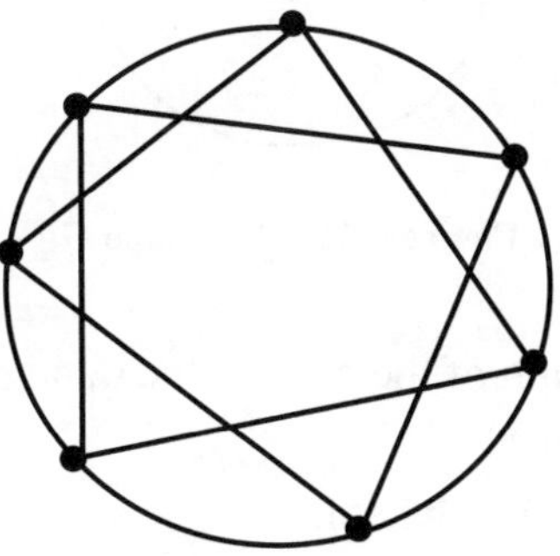

Figure 8.24 A non-planar graph

Section 8.4: Non-Hamiltonian planar graphs

22. Using Grinberg's theorem, show that the planar graph given in Figure 8.25 is not Hamiltonian.

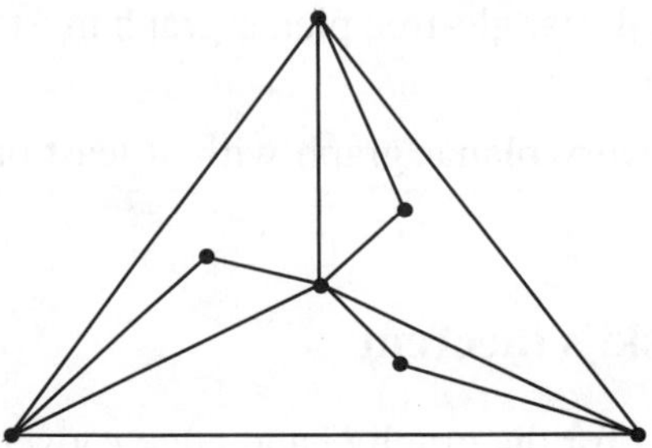

Figure 8.25   A non-Hamiltonian graph

Section 8.5: The dual of a planar graph

23. Draw the dual of the planar graph G given in Figure 8.26.

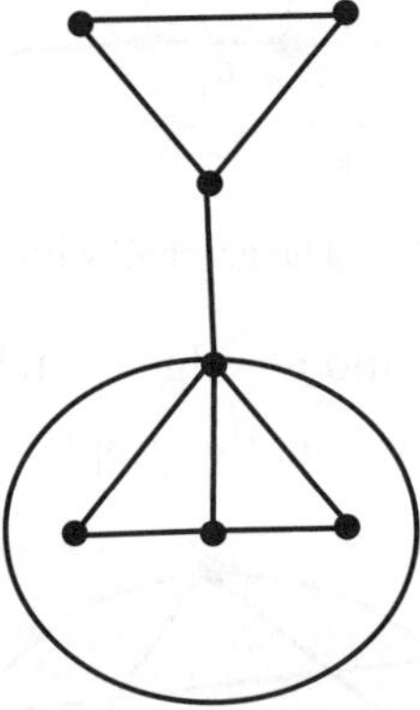

Figure 8.26   The graph G

24. Demonstrate that the dual graph of the cube is the octahedron, and the dual graph of the dodecahedron is the icosahedron.

25. Show that the dual of a wheel is a wheel.

9

Coloring of Graphs

Take a map of your country where the different regions, provinces, or states are clearly shown. How many colors are required to color each region on a map of the country so that the neighboring regions are colored differently? As a fun exercise, let us create a graph by placing a vertex in the middle of each region, and two vertices are adjacent if the states they represent share a border. Now that we have modeled the regions of a map as vertices of a graph, the graph's vertices can be colored in the same way that the regions are colored on the map. The graph's vertices cannot be colored the same when they are adjacent as the neighboring regions cannot have the same color. This interesting problem of coloring has been the source of a great amount of research in graph theory. The applications of graph coloring are numerous, since it is a powerful tool that enables the systematic assignment of resources in situations where certain elements must be distinct or non-overlapping. We will also discuss chromatic number of a graph, which is the minimum number of colors required to have a conflict-free coloring. The parallel concept of edge coloring is essential to frequency assignment in networks, where edges need distinct colors to prevent interference. In this chapter, we will also explore the connection between planarity and coloring, through the five color theorem and four color theorem.

Throughout this chapter, we consider the simple graph without loops and multiple edges.

9.1 Vertex coloring

Let us formally define the concept of vertex coloring of a graph.

DEFINITION 9.1.1 Let G be a graph. A *vertex coloring* (or simply coloring) of G is an assignment of colors to the vertices of G such that no two adjacent vertices are assigned the same color. A *k-coloring* of G is an assignment of exactly k colors to the vertices of G and in this case G is said to be *k-colorable*.

We use the notation $c_1, c_2, c_3, \ldots$ for the colors to differentiate them.

EXAMPLE 9.1.2 Consider the graphs G_1 and G_2 as shown in the Figure 9.1. The graphs G_1 and G_2 are 4-colorable graphs.

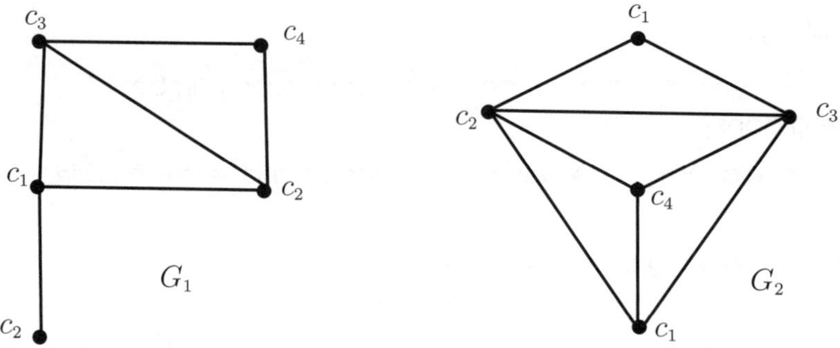

Figure 9.1   4-colorable graphs

It is always possible to provide k-coloring for a given simple graph G for some k, which can be as simple as the number of vertices. Is it possible to have $\ell < k$ such that G is ℓ-coloring for a given k? For example, the graph G_1 has 3-coloring but G_1 cannot be colored by fewer than three colors. This minimum number is known as the chromatic number. Let us formally define it.

DEFINITION 9.1.3 If G is k-colorable but not ℓ-colorable for any $\ell < k$, then k is called the *chromatic number* of G, denoted by $\chi(G)$. If $\chi(G) = k$, then we say that G is *k-chromatic graph*.

For example, the graphs G_1 and G_2 shown in Figure 9.1 are 3-chromatic and 4-chromatic respectively.

If a graph G contains an edge, then $\chi(G) \geq 2$. We shall compute the chromatic number of some standard graphs.

The path P_n: Consider $P_n : v_1, e_1, v_2, e_2, \ldots, e_{n-1}, v_n$. Then assign the color 1 to the vertices $v_1, v_3, v_5, \ldots$ and assign the color 2 to the vertices $v_2, v_4, v_6, \ldots$. Therefore P_n is 2-colorable and so $\chi(P_n) = 2$ for all $n \geq 2$.

The cycle C_n: Consider the n-cycle $C_n : v_1, e_1, v_2, e_2, \ldots, e_{n-1}, v_n, e_n, v_1$ with $n \geq 3$. If n is even, then assign the color 1 to the vertices $v_1, v_3, v_5, \ldots, v_{n-1}$ and assign the color 2 to the vertices $v_2, v_4, v_6, \ldots, v_n$. This yields a proper coloring of G, and C_n is therefore 2-colorable. If n is odd, then assign the color 1 to the vertices $v_1, v_3, v_5, \ldots, v_{n-2}$ and assign the color 2 to the vertices $v_2, v_4, v_6, \ldots, v_{n-1}$. Now the vertex v_{n-1} receives the color 2, and the vertex v_1 receives the color 1. Note that v_{n-1} and v_1 are the neighbors of the vertex v_n. Therefore we have to assign a new color, say color 3 to v_n and so $\chi(C_n) = 3$. Thus

$$\chi(C_n) = \begin{cases} 2 \text{ if } n \text{ is even} \\ 3 \text{ if } n \text{ is odd.} \end{cases}$$

The complete graph K_n: Clearly $\chi(K_n) \leq n$. If we assign any $n-1$ colors to all the vertices of K_n, then two vertices receive the same color. Since any two distinct vertices in K_n are adjacent, the two vertices that receive the same color are adjacent. Therefore $\chi(K_n) \geq n$ and thus $\chi(K_n) = n$.

The complete bipartite graph $K_{m,n}$: Let $X = \{x_1, x_2, \ldots, x_m\}$ and $Y = \{y_1, y_2, \ldots, y_n\}$ be the bipartition of $K_{m,n}$. Assign color 1 to $x_1, x_2, \ldots, x_m$ and color 2 to $y_1, y_2, \ldots, y_n$. Thus $\chi(K_{m,n}) = 2$

The following observations are trivial.

PROPOSITION 9.1.4 *Let G be a simple graph.*

(i). *If H is a subgraph of the graph G, then $\chi(H) \leq \chi(G)$.*

(ii). *If the graph G contains K_ℓ as a subgraph, then $\chi(G) \geq \ell$.*

(iii). *If $G_1, G_2, \ldots, G_m$ are the connected components of G, then*

$$\chi(G) = \max_{1 \leq i \leq m} \chi(G_i).$$

Next, we completely characterize simple graphs with chromatic number 2.

THEOREM 9.1.5 *Let G be a simple graph. Then $\chi(G) = 2$ if and only if G is bipartite.*

Proof. Let G be bipartite with bipartition X and Y. By assigning color 1 to all vertices in X and color 2 to all vertices in Y, we arrive at the conclusion that G is 2-colorable. Since two different colors are required, we get $\chi(G) = 2$.

Conversely, suppose that $\chi(G) = 2$. So we can color G with just two colors. Consider the set X containing all those vertices colored as 1 and the set Y containing all those vertices colored as 2. Since no two adjacent vertices are colored with the same color, no two vertices in X are adjacent, and similarly, no two vertices in Y are adjacent. Thus, any edge in G must have one end vertex in X and another end vertex in Y. Therefore, G is bipartite with bipartition $V(G) = X \cup Y$. $\qquad\qquad\square$

In fact, every tree is bipartite and so it is 2-chromatic.

By Theorem 1.5.6, we know that a graph is bipartite if and only if it has no odd cycle. Therefore we have the following result.

COROLLARY 9.1.6 *Let G be a graph. Then $\chi(G) \geq 3$ if and only if G has an odd cycle.*

We have characterized graphs with a chromatic number of 2. There is no such characterization for graphs whose chromatic number is 3. So we try to get upper bounds on the chromatic number of an arbitrary graph G in terms of the degree of its vertices.

THEOREM 9.1.7 *For any graph G, $\chi(G) \leq \Delta(G) + 1$, where $\Delta(G)$ is the maximum degree of G.*

Proof. Let us prove the result by induction on the number of vertices in a graph. If the number of vertices in a graph is 1, then $G = K_1$ implies $\chi(G) = 1$ and $\Delta(G) = 0$. So the result is true for the case of single vertex graph. Let G be a graph with n vertices. Assume that the result is true for graphs with $n-1$ vertices. We prove that G is $(\Delta(G)+1)$-colorable. Since for any vertex v of G, the subgraph $G - v$ has $n-1$ vertices. By induction, we get $\chi(G - v) \leq \Delta(G - v) + 1$. Thus, a $(\Delta(G - v) + 1)$-coloring of $G - v$ exists. Since $deg(v) \leq \Delta(G)$, there are at most $\Delta(G)$ vertices that are adjacent to the vertex v in G. So, in $G - v$, almost $\Delta(G)$ colors are used in coloring the neighbors of v. If $\Delta(G - v) = \Delta(G)$, then at least one color from $\Delta(G) + 1$ colors used in the coloring for $G - v$ is available to color the vertex v. Thus G has $(\Delta(G)+1)$-coloring. If $\Delta(G - v) < \Delta(G)$, by coloring v with a color which is not used in the coloring of $G - v$, we get $(\Delta(G - v) + 2)$-coloring of G. Since $\Delta(G - v) \leq \Delta(G) - 1$, we have G is $(\Delta(G) + 1)$-colorable. In either case, there exists a $(\Delta(G) + 1)$-coloring of G and hence $\chi(G) \leq \Delta(G) + 1$. $\qquad\square$

Are there graphs for which the equality $\chi(G) = \Delta(G) + 1$ is true? The answer is yes. For example, consider the complete graph K_n. Then $\chi(K_n) = n$ and $\Delta(K_n) = n - 1$ implies $\chi(K_n) = \Delta(K_n) + 1$. Also for the cycle of odd length C_n, we have $\Delta(C_n) = 2$ and $\chi(C_n) = 3 = \Delta(C_n) + 1$. Next, we will prove that these are the only two graphs for which equality holds, as proved by Brook.

THEOREM 9.1.8 (Brook's theorem) *Let G be a simple graph. If G is neither complete nor an odd cycle, then $\chi(G) \leq \Delta(G)$.*

Proof. We first consider the case for G being connected and prove the result using induction on $|V(G)|$. Let $|V(G)| = n$. If $\Delta(G) \leq 2$, then G is either a path P_n or a cycle C_n. Since G is neither complete nor an odd cycle, we get G is either P_n where $n \geq 3$ or C_n where $n = 2k$ with $2 \leq k \in \mathbb{Z}$. Then $\chi(G) = 2 = \Delta(G)$. So assume that $\Delta(G) \geq 3$ and the result holds for graphs having fewer vertices than G. For simplifying the notation, let $\Delta = \Delta(G)$. We have to prove that $\chi(G) \leq \Delta$. Suppose by contradiction, $\chi(G) > \Delta$. This implies that G is not Δ-colorable. For a vertex v of G, consider the subgraph H as the induced subgraph of $G - \{v\}$.

Claim 1: $\chi(H) \leq \Delta$.

Let H' be a connected component of H. If H' is neither complete nor an odd cycle, then by induction hypothesis, $\chi(H') \leq \Delta(H') \leq \Delta$. If H' is either complete or an odd cycle, then $\chi(H') = \Delta(H') + 1$. Also, in this case, as every vertex of H' has the same degree in H', and one of these vertices is adjacent to v in G, we get that $\Delta \geq \Delta(H') + 1$. Thus $\chi(H') \leq \Delta$. Since $\chi(H) \geq \chi(H')$, we get $\chi(H) \leq \Delta$.

Therefore H is Δ-colorable but not G. Then all the Δ colors in Δ-coloring of H must be used to color the neighbors of v in G, otherwise, v can be colored by the unused colors among one to Δ implying G is Δ-colorable, a contradiction. In particular, $deg(v) = \Delta$. For given any Δ-coloring of H, let v_i, $1 \leq i \leq \Delta$, be the vertex adjacent to v in G which is colored with i. For $1 \leq i \neq j \leq \Delta$, consider the subgraph $H_{i,j}$ of H induced by the vertices colored by either i or j.

Claim 2: For $1 \leq i \neq j \leq \Delta$, the vertices v_i and v_j belong to the same component $C_{i,j}$ of $H_{i,j}$.

Suppose not, then the colors i and j can be interchanged in the component of $H_{i,j}$ containing the vertex v_j. This yields a Δ-coloring of H in which both v_i and v_j have the same coloring i, a contradiction to v_i receives color i and v_j receives color j.

Claim 3: The component $C_{i,j}$ is a $v_i - v_j$ path.

Since $C_{i,j}$ is a component, there exists a $v_i - v_j$ path P in $C_{i,j}$. Since v_i is adjacent to v in G, $deg_H(v_i) \leq \Delta - 1$. Note that the neighbors of v_i have pairwise different colors. Hence the neighbor of v_i on the path P is its only neighbor in $C_{i,j}$. Similarly the neighbor of v_j on P is the only neighbor of v_j in $C_{i,j}$. If $C_{i,j} \neq P$, then there exists an inner vertex on P with at least three neighbors having the same color. Let u be the first vertex on P with at least three neighbors. Without loss of generality, assume that the color of u is j (proof goes similar if it is colored with i). We now recolor the vertex u with a color different from i and j, say with the color k, refer Figure 9.2. Let w be the preceding vertex of u in the path P. Then in this new coloring of H, the path from v_i to w becomes a component of $H_{i,j}$ and so the vertices v_i and v_j are in different components of $H_{i,j}$, a contradiction to Claim 2.

Claim 4: $C_{i,j} \cap C_{i,k} = \{v_i\}$ for $1 \leq i \neq j \neq k \leq \Delta$.

Clearly $v_i \in C_{i,j} \cap C_{i,k}$. Suppose $u \in C_{i,j} \cap C_{i,k}$ such that $u \neq v_i$. Then u is adjacent to two vertices colored with j in $C_{i,j}$ and adjacent to two vertices colored with k in $C_{i,k}$, as shown in Figure 9.3. Now we recolor the vertex u with a new color r which is different from k and

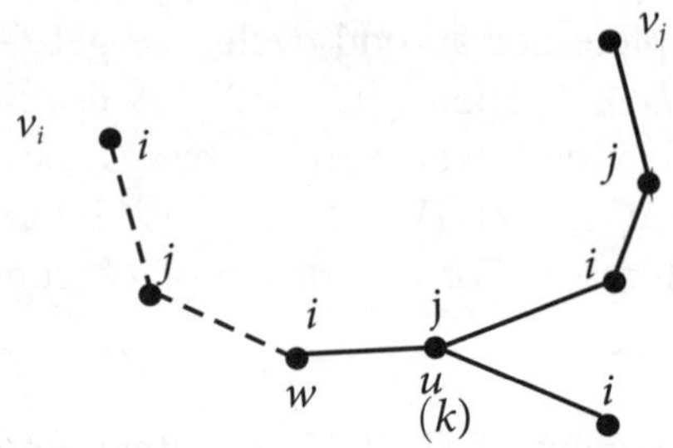

Figure 9.2 Diagrammatic representation for Claim 3 of Theorem 9.1.8

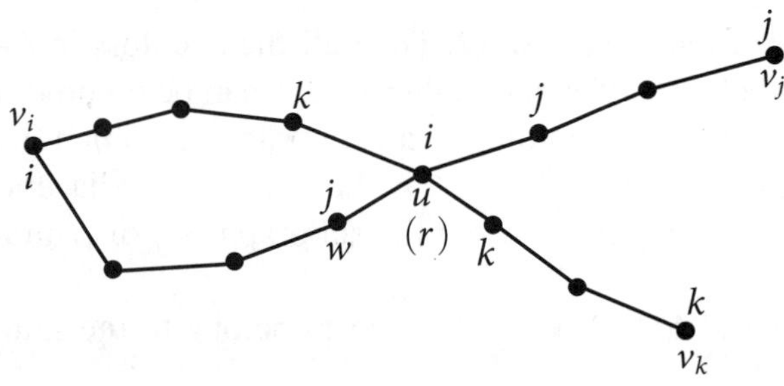

Figure 9.3 Diagrammatic representation for Claim 4 of Theorem 9.1.8

j. Then similar to Claim 3, in this new coloring of H, the vertices v_i and v_j are in different components, a contradiction. We have completed all our claims.

Since $deg(v) = \Delta$ and G is not complete, there exists a pair of non-adjacent vertices in the neighbors of v, say v_1 and v_2. Then a $v_1 - v_2$ path $C_{1,2}$ contains a vertex v which is adjacent to v_1. Since v_1 has color 1, v assigns color 2. Also, since $\Delta \geq 3$, there exists a neighbor v_3 of v in G. Let u be the vertex adjacent to v_1 in the $C_{1,2}$ path. That is u receives the color 2. Now we interchange the colors of 1 and 3 in $C_{1,3}$ of $H_{1,3}$, as shown in Figure 9.4. In the new coloring of H, we denote a $v_i - v_j$ path by $C'_{i,j}$. As u has retained its previous color 2, we get $u \in C'_{1,2}$. Since v_1 receives the color 3 in the new coloring, we have $u \in C'_{2,3}$. Therefore $u \in C'_{1,2} \cap C'_{2,3}$, a contradiction to claim 4. Hence $\chi(G) \leq \Delta(G)$.

Suppose G has connected components $G_1, G_2, \ldots, G_k$. Then by applying the above case to each G_i, we get $\chi(G_i) \leq \Delta(G_i) \leq \Delta(G)$ for all $1 \leq i \leq k$.

Therefore, $\chi(G) = \max_{1 \leq i \leq n} \chi(G_i) \leq \Delta(G)$.

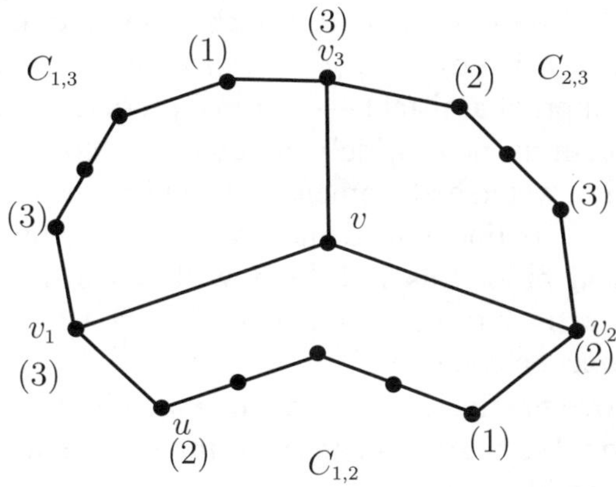

Figure 9.4   Swapping colors of v_1 and v_3 in $C_{1,3}$

9.2 Coloring of planar graphs

Let us start with a simple question: How many colors are needed to color any map, regardless of how many faces it contains, so that no two adjacent faces have the same color? It is fascinating to learn that just four colors are enough to color the entire map, under certain conditions. This is known as the *four color problem*, and it is one of the most fundamental results in graph theory, especially with regard to planar graphs.

The four color problem was discovered by *Francis Guthrie* while attempting to color a map of England. He discovered that only four colors are sufficient to color the map so that no two adjacent faces have the same color. He then began coloring various maps and discovered that this appears to be true for any other map, establishing the four color conjecture: His brother *Frederick Guthrie* showed Francis's attempts to prove this result to De Morgan, who was unable to respond and wrote a letter to Hamilton, to which Hamilton replied that he was unlikely to attempt this problem soon.

In 1879, *Alfred Bray Kempe* used Kempe chains to prove this conjecture. However, in the year 1889, *Percy John Heawood* demonstrated that Kempe's proof was incorrect. Following this, *Kenneth Appel* and *Wolfgang Haken* produced a proof in 1976 after more than 80 years. Mathematicians *George David Birkhoff*, *Kempe* and *Heinrich Heesch* made significant contributions to proving this conjecture. This is one of the first major mathematical theorems to be proved using computers. As a result, many mathematicians rejected these proofs and those who proved it, faced harsh criticism. Appel and Haken spent 1200 hours on the computer working out the details of the final proof. Over the years, the proof has been simplified and made more efficient, but it still requires a computer for verification of some cases. In 1997, *Robertson, Sanders, Seymour* and *Thomas* presented a more refined version of the proof, using modern techniques and more efficient algorithms. In 2005, the theorem was verified by *Georges Gonthier* using a general-purpose theorem-proving software (Coq is a proof assistant — a formal proof-checking software).

The four color theorem is now widely accepted as correct by the mathematical community, but it remains a watershed moment in the philosophical debate over computer-assisted proofs. The proof continues to demonstrate how computers have changed the landscape of mathematical proofs, as well as how they can be integrated with formal verification systems. There is currently no non-computational proof of the theorem, but this does not diminish its acceptance.

In this section, we prove the five-color problem, which is simpler and does not require a computer for verification. We begin with an even simpler one, the six-color theorem, and then prove the five-color problem.

THEOREM 9.2.1 (The six color theorem) *Every planar graph G is 6-colorable.*

Proof. We prove the result by induction on n, the number of vertices of G. If $n \leq 6$, then clearly G is 6-colorable. So, assume that $n > 7$ and the induction hypothesis that any plane graph with less than n vertices is 6-colorable. Since G is planar, by Corollary 8.2.2, there exists a vertex v with $deg(v) \leq 5$. Now consider the subgraph $H = G - v$ of G. Then H is a planar graph with $n - 1$ vertices and so, by induction hypothesis, H is 6-colorable. Since $deg(v) \leq 5$, the adjacent vertices can be colored with at most five colors. So there is at least one color which is not assigned to the vertex v. Now we color the vertex v in G with one of the missing colors at v to get a 6-coloring of G. $\qquad\qquad\square$

Heawood proved that every map can be 5-colored.

THEOREM 9.2.2 (The five color theorem, Heawood) *Every planar graph G is 5-colorable.*

Proof. We prove the result by induction on n, the number of vertices of G. If $n \leq 5$, then G is clearly 5-colorable. So assume that $n \geq 6$ and the induction hypothesis that any planar graph with less than n vertices is 5-colorable. Since G is a planar graph, by Corollary 8.2.2, there exists a vertex v with $deg(v) \leq 5$. Then by induction hypothesis, the subgraph $H = G - v$ is 5-colorable. We have the following cases:

Case 1: If $deg(v) \leq 4$ or $deg(v) = 5$ with the vertices adjacent to v are colored with 4 colors or less, then we can color the vertex v with any one of the missing color to get a 5-coloring of G.

Case 2: Suppose $deg(v) = 5$ with the vertices adjacent to v are colored with exactly five different colors. Let v_i be the adjacent vertex of v colored with i for $i \in \{1, 2, \ldots, 5\}$. Now consider the induced subgraph $H_{1,3}$ of H consisting of all the vertices colored with 1 and 3 together with all the edges connecting these vertices. Again we have two sub cases:

Case 2.1: Suppose there is no walk between v_1 and v_3 in $H_{1,3}$. Let S be the set of all vertices u in $H_{1,3}$ such that $v_1 - u$ walk exists in $H_{1,3}$. Then $v_1 \in S$ but $v_3 \notin S$. Now swap the colors

1 and 3 for the vertices in S so that the color 1 is not used to color the vertices v_1 to v_5. Therefore we can color the vertex v with color 1 to get a 5-coloring of G.

Case 2.2: Suppose there is a walk from v_1 to v_3. Then we consider another subgraph $H_{2,4}$ (or $H_{2,5}$) of H. A path from v_1 to v_3 gives a closed cycle as shown in Figure 9.5. Since G is

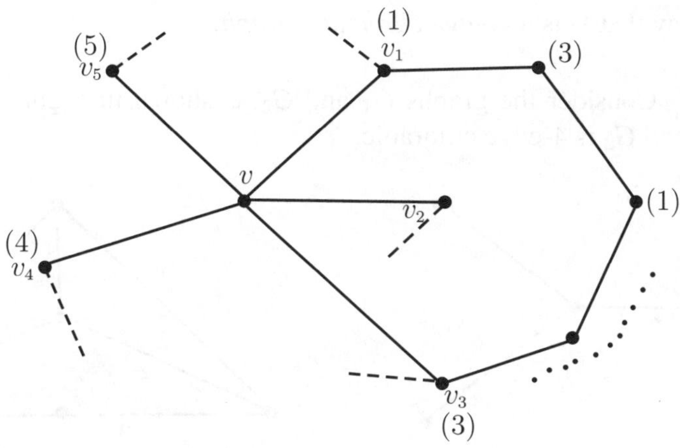

Figure 9.5   Illustration of Case 2.2

planar, the vertex v_2 cannot be connected to v_4 in $H_{2,4}$, that is, there is no walk from v_2 to v_4. Therefore, as in Case 2.1, we can swap the colors of 2 and 4 in the portion of $H_{2,4}$ which is connected to v_2. Then, we see that the color 2 is not used to color any of the vertices v_1 to v_5. Therefore we can color the vertex v with the color 2 to get a 5-coloring of G. $\square$

We end this section by stating the four color theorem without proof

THEOREM 9.2.3 (The four color theorem) *Every planar graph G is 4-colorable.*

9.3   Edge coloring

A graph's chromatic number indicates the color of its vertices, but we could also inquire about the color of its edges. As with vertex coloring, we may insist that adjacent edges be colored differently. We consider two edges to be adjacent if they are incident at the same vertex. The minimum number of colors required to color a graph's edges is known as the edge chromatic number.

DEFINITION 9.3.1 Let G be a graph. An *edge coloring* of G is an assignment of colors to the edges of G such that no two incident edges are assigned the same color. A *k-edge coloring* of G is an assignment of exactly k-colors to the edges of G and in this case G is said to be *k-edge colorable.*

NOTE 9.3.2 If an edge is colored with a color c, then we call that edge as *c-edge*. Also, if a color c is assigned to any edge incident to a vertex v, then we say that c *is assigned at v*.

DEFINITION 9.3.3 The minimum number n for which there is an n-edge coloring of G is called the *edge chromatic number* (or edge chromatic index) of G and is denoted by $\chi'(G)$. If $\chi'(G) = k$, we say that G is a *k-edge chromatic graph*.

EXAMPLE 9.3.4 Consider the graphs G_1 and G_2 as shown in Figure 9.6. Then G_1 is 3-edge colorable and G_2 is 4-edge colorable.

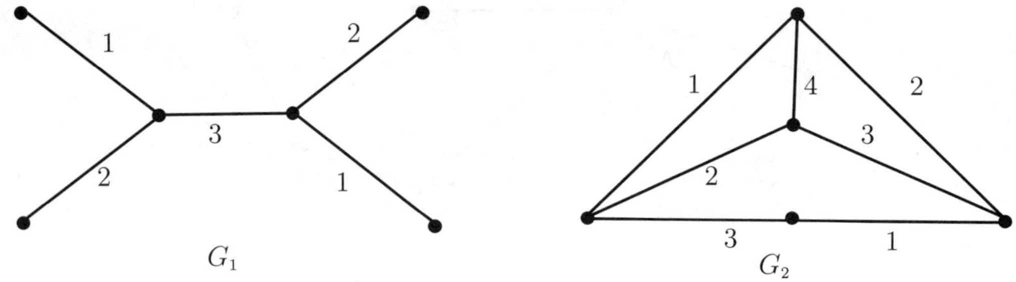

Figure 9.6   3-edge colorable and 4-edge colorable graphs

EXAMPLE 9.3.5 Six friends are playing chess. Everyone will play each other once. They have only one chess set. Games will last one hour. How many hours will the tournament last?

To answer this question, represent each player with a vertex and place an edge between two players if they will be playing each other. In this case, we obtain the complete graph K_6. We must color the edges with each color representing a different hour. Since no player will be able to play two games (edges) concurrently, different edges incident to the same vertex will be colored differently. Thus we must know the edge chromatic number of K_6. Since $\chi'(K_6) = 5$, the friends will play for 5 hours.

Let us find the edge chromatic number of path graph and cycle graph.

The path P_n: Consider the $P_n : v_1, e_1, v_2, e_2, \ldots, e_{n-1}, v_n$. Then assign the color 1 to the edges $e_1, e_3, e_5, \ldots$ and assign the color 2 to the edges $e_2, e_4, e_6, \ldots$. Therefore P_n is 2-edge colorable and so $\chi'(P_n) = 2$ for all $n \geq 2$.

The cycle C_n: Consider the n-cycle $C_n : v_1, e_1, v_2, e_2, \ldots, e_{n-1}, v_n, e_n, v_1$ with $n \geq 3$. If n is even, then assign the color 1 to the edges $e_1, e_3, e_5, \ldots, e_{n-1}$ and assign the color 2 to the edges $e_2, e_4, e_6, \ldots, e_n$. This yields a proper coloring of G and so C_n is 2-colorable. If n is odd, then assign the color 1 to the edges $e_1, e_3, e_5, \ldots, e_{n-2}$ and assign the color 2 to the edges $e_2, e_4, e_6, \ldots, e_{n-1}$. Now the edge e_{n-1} receives the color 2 and the edge e_1 receives the color 1. Note that e_n is incident with both e_1 and e_{n-1}. Therefore we have to assign a new color, say color 3 to e_n and so $\chi'(C_n) = 3$. Thus

$$\chi'(C_n) = \begin{cases} 2 \text{ if } n \text{ is even} \\ 3 \text{ if } n \text{ is odd.} \end{cases}$$

Next, we will discuss the edge chromatic number for bipartite graphs.

THEOREM 9.3.6 *If G is a nonempty bipartite graph, then $\chi'(G) = \Delta(G)$.*

Proof. We prove the result by induction on the number of edges of G. Clearly the result is true if G has one edge as $\chi'(G) = 1 = \Delta(G)$. So assume that G has more than one edge and the result is true for all non-empty bipartite graphs having edges less than G. Let $\Delta = \Delta(G)$ and $e = uv$ be an edge of G. Then $G - e$ is non-empty bipartite graph and thus by induction hypothesis, there exists a Δ-edge coloring of $G - e$. Since the vertices u and v in $G - e$ are incident with at most $\Delta(G) - 1$ edges, u is not incident with an α-edge and v is not incident with a β-edge for some $\alpha, \beta \in \{1, 2, \ldots, \Delta\}$. Now if $\alpha = \beta$, then the edge e can be colored with the color α and so G is Δ-colorable. So assume that $\alpha \neq \beta$. If u is not incident with β-edge, then e can be colored with β as v is not incident with β-edge. So assume that u is incident with some β-edge, say e'. Let W be the largest β-edge and α-edge alternating walk starting from the edge e'. If any vertex is repeated in W, then the particular vertex has at least three incident edges from W. That is, two incident edges must have the same color, a contradiction. Thus no vertex is repeated in W implies W is a path. Suppose v belongs to W. Since v is not incident with β-edge, W must end with v and the incident edge of v in W is α-edge. Since W starts with β-edge and ends with α-edge, it is an even path. Then $W + e$ is an odd cycle in G, a contradiction to G is a bipartite graph. Thus v is not on W.
Now recoloring W by swapping the colors α and β gives a new Δ-edge coloring of $G - e$,

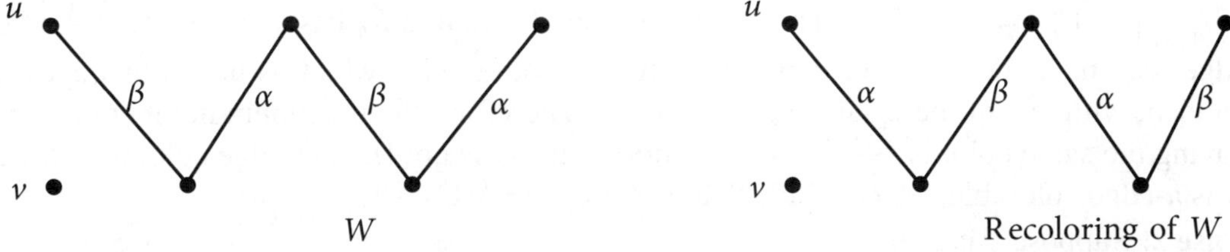

Figure 9.7   Representation of W and its recoloring

as shown in Figure 9.7. In this new Δ-edge coloring of $G - e$, neither u nor v is incident with β-edge, so the edge e can be colored with β to get Δ-coloring of G. $\square$

COROLLARY 9.3.7 $\chi'(K_{m,n}) = \max\{m, n\}$.

We now discuss the edge chromatic number of complete graphs.

THEOREM 9.3.8 *For the complete graph on n vertices,*

$$\chi'(K_n) = \begin{cases} n-1 & \text{if } n \text{ is even} \\ n & \text{if } n \text{ is odd.} \end{cases}$$

Proof. Note that $|E(K_n)| = \frac{n(n-1)}{2}$ and when $G = K_n$, we have $\Delta(G) = n-1$.

Case 1: Suppose n is odd.

We can draw K_n such a way that the outer edges of K_n form a regular n-gon. We color these outer edges with different colors, say from color 1 to color n. Since each of the remaining internal edges of G is parallel to exactly one of the outer edge, we assign it the same color as assigned to its parallel outer edge. For illustration, we have explained the coloring for K_5 in Figure 9.8. So two edges have same color only if they are parallel edges. Since parallel edges are not incident, we have n-edge coloring of G.

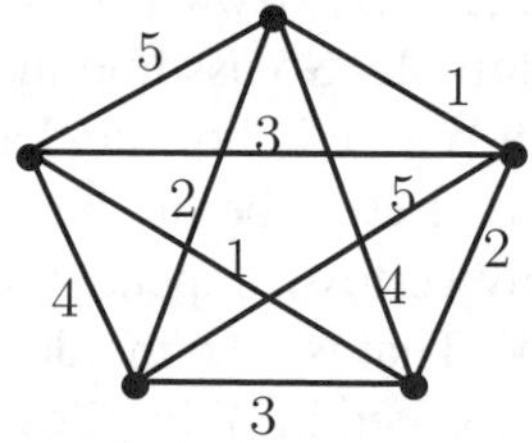

Figure 9.8   Same coloring to parallel edges

Suppose there exists $(n-1)$-edge coloring for G. Since there are $\frac{n(n-1)}{2}$ edges in K_n and only $n-1$ colors are available, by pigeonhole principle, a color will be used for at least $\left\lceil \frac{n(n-1)}{2(n-1)} \right\rceil = \left\lceil \frac{n}{2} \right\rceil = \frac{n+1}{2}$ edges. The maximum length cycle in K_n has n edges and $\frac{n-1}{2}$ of its edges can have the same coloring. Note that any other edge which is not in this cycle is incident with one of these same colored edges. Therefore, the maximum number of edges having the same color is $\frac{n-1}{2}$, a contradiction. Thus G is not $(n-1)$-edge colorable. Since G is n-edge colorable, we conclude that $\chi'(G) = n = \Delta(G) + 1$.

Case 2: Suppose n is even.

We know that for any vertex w of G, $G' = G - w$ is complete with $n-1$ vertices. Since $n-1$ is odd, $\chi'(G') = n-1$. Since $\Delta(G') = n-2$, each vertex v in G' is incident with $n-2$ edges and so there exists a color, say c_v, which is not assigned to any edge incident with v. Now we color G from G' by coloring the edge wv by the color c_v for all $v \in V(G')$ and thus G is $(n-1)$-edge colorable. Hence $\chi'(G) = n-1 = \Delta(G)$. $\square$

We now prove one of the important results of graph theory, which establishes the lower and upper bounds for the edge chromatic index of any nontrivial graph.

THEOREM 9.3.9 (Vizing theorem(1964)) *Let G be a nontrivial graph. Then $\chi'(G)$ is either $\Delta(G)$ or $\Delta(G)+1$.*

Proof. Let v be a vertex of G such that $deg(v) = \Delta(G)$. Then v is incident with $\Delta(G)$ edges and hence at least $\Delta(G)$ different colors are required to color the edges incident with v in G, which implies that $\Delta(G) \leq \chi'(G)$.

We claim that $\chi'(G) \leq \Delta(G)+1$. Let $\Delta = \Delta(G)$. We prove the result by contradiction. Suppose G is not $(\Delta+1)$-edge colorable. We choose a subgraph H of G having maximum number of edges such that H is $(\Delta+1)$-edge colorable. We show the existence of a subgraph H' of G which is $(\Delta+1)$-edge colorable having more edges than H, a contradiction to the choice of H.

Since H is $(\Delta+1)$-edge colorable but not G, H is a proper subgraph of G and so there exists an edge uv_1 in G but not in H. Since $deg(u) \leq \Delta$, there exists a color, say c from $(\Delta+1)$ colors, such that the color c is not used to color any edge incident with u. For simplicity, we say c is not assigned at u. As $deg(v_1) \leq \Delta$, by the same argument, there exists a color c_1 not assigned at v_1, refer Figure 9.9. In Figure 9.9, we have explained this by enclosing the color c_i at the vertex v_i. Now there exists a vertex v_2 such that the

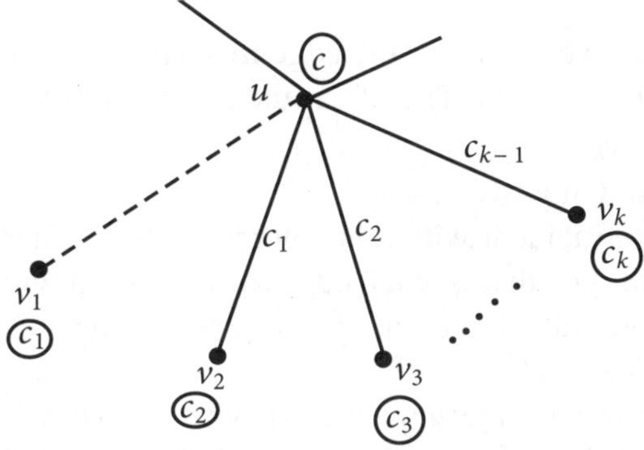

Figure 9.9   Colors not represented are encircled at the vertices

edge uv_2 of H is colored with c_1, if not, then, the edge uv_1 can be colored with c_1 and hence $H \cup \{uv_1\}$ is $(\Delta+1)$-edge colorable, a contradiction to the choice of H. Again as $deg(v_2) \leq \Delta$, there exists a color c_2 not assigned at v_2. Then by same argument as above, there exists an edge uv_3 of H colored with c_2. This process can be continued to get a sequence of edges $uv_1, uv_2, \ldots, uv_k$ such that

(i). the color c_i is not assigned at the vertex v_i for all $1 \leq i \leq k$ and

(ii). the edge uv_{j+1} is colored with c_j for all $1 \leq j \leq k-1$.

Since the $deg(u)$ is finite, this process stops say at v_k.

Next, we claim that the color c is assigned at every vertex $v_1, v_2, \ldots, v_k$. Suppose c is not assigned at v_j for some $1 \le j \le k$. Then we shift(cascade) the colors c_i from uv_{i+1} to uv_i for all $1 \le i \le j-1$, as shown in Figure 9.10. Then in this new coloring, c is not assigned at u

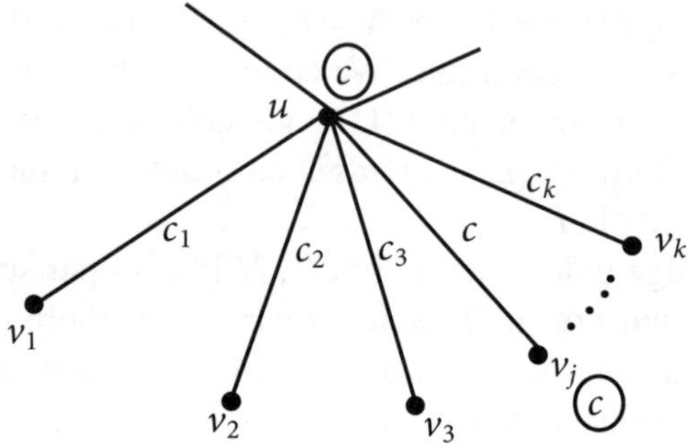

Figure 9.10   Cascading of colors c_i

as well as at v_j and hence we can color the edge uv_j with c to get a $(\Delta+1)$-edge coloring to $H \cup \{uv_1\}$, a contradiction again. Therefore, the claim holds true, that is, c is assigned at all the vertices $v_1, v_2, \ldots, v_k$.

Now we have the following two cases:

Case 1: Suppose no edge incident with u is colored with c_k. Then we can apply the same cascading of colors technique, that is, each edge uv_i is colored with c_i for $i = 1, 2, \ldots, k-1$ and the edge uv_k is colored with c_k to get a $(\Delta+1)$-edge coloring of $H \cup \{uv_1\}$, a contradiction to the choice of H.

Case 2: Suppose there is an edge uv_j colored with c_k. That is $c_j = c_k$ for some $1 \le j \le k-1$. Then we shift the colors c_i from uv_{i+1} to uv_i for all $1 \le i \le j$, leaving uv_{j+1} and keeping other edge colors as it is, as shown in Figure 9.11. Consider the subgraph $H' = (H \cup \{uv_1\}) - uv_{j+1}$. Then H' and H have the same number of edges. Now consider $H'(c, c_j)$ the subgraph of H' induced by the edges colored with c and c_j. Then as discussed earlier, any component of $H'(c, c_j)$ is an even cycle or a path having edges colored with c and c_j alternatively. As c is assigned at each vertex v_1 to v_k, in particular, it is assigned at v_{j+1} and v_k. On the other hand, c_j is not assigned at v_{j+1} and v_k. Also c_j is assigned at u and c is not assigned at u. Therefore, u, v_{j+1} and v_k are of degree 1 and hence can't be in the same component of $H'(c, c_j)$, as the path has exactly two vertices of degree 1.

Case 2.1: Suppose u and v_{j+1} are in different components of $H'(c, c_j)$. Then we swap the colors c and c_{j+1} in the component which contains the vertex v_{j+1}. As a result c is not assigned at v_{j+1}. As c is already not assigned at u, coloring the edge uv_{j+1} with c to get a $(\Delta+1)$-edge coloring to $H' \cup \{uv_{j+1}\}$, a contradiction.

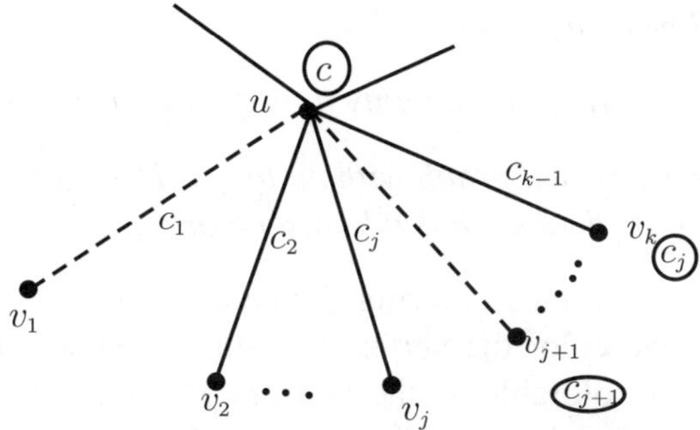

Figure 9.11 Recoloring for Case 2

Case 2.2: Suppose u and v_{j+1} are in the component of $H'(c, c_j)$. Then v_k lies in different component. We swap c and c_j in this component and shift the colors c_i from from uv_{i+1} to uv_i for all $1 \le i \le k-1$. Since c is not assigned at u and v_k, we can color the edge uv_k with c and so $(\Delta + 1)$-edge coloring exists for $H' \cup \{uv_{j+1}\}$, a contradiction. Hence $\chi'(G) \le \Delta(G) + 1$. $\qquad\square$

Vizing's theorem says that the edge chromatic number of any nontrivial graph G is either $\Delta(G)$ or $\Delta(G) + 1$. So graphs can be classified into two classes as per the chromatic index, as follows:

DEFINITION 9.3.10 A graph G is called a *class 1* graph if $\chi'(G) = \Delta(G)$ and is called a *class 2* graph if $\chi'(G) = \Delta(G) + 1$.

EXAMPLE 9.3.11 By Theorem 9.3.6, all bipartite graphs are class 1 graphs and by Theorem 9.3.8, all complete graphs are class 2 graphs. We leave to the reader to check that Petersen graph is a class 2 graph.

9.4 Face coloring

Now that we are familiar with the coloring of vertices and edges let us begin to color the faces of planar graphs.

DEFINITION 9.4.1 A planar connected graph with no bridges is called a *map*.

DEFINITION 9.4.2 A map G is said to be *k-face colorable* if faces of G can be colored with at most k colors such that no two adjacent faces have the same color.

So we have seen three types of coloring: to vertices, to edges and to faces. As there is a one to one correspondence between faces of G and vertices of the dual graph G^*, we can establish relation between face coloring and vertex coloring as we prove in the next result.

THEOREM 9.4.3 *Let G be a simple graph.*

(i). *A map G is k-face colorable if and only if its dual G^* is k-vertex colorable.*

(ii). *Let G be a planar connected graph without loops. Then G has a vertex coloring of k colors if and only if its dual G^* has a k-face coloring.*

Proof. (i). Suppose G has a k-face coloring. Since each vertex f^* in G^* corresponds to unique face f in G, we can color the vertex f^* with the same color used for the face f of G. Then G^* is k-vertex colorable as the vertices f^* and g^* in G^* are adjacent in G^* if the corresponding faces f and g are adjacent in G. Conversely, suppose G^* is k-vertex colorable. By similar argument as above, we color the face f in G with the same color used for the vertex f^* in G^*, which implies G is k-face colorable.

(ii). The dual G^* has no bridges as G has no loops. Implies G^* is a map. Then, applying (i) to G^*, we have G^* is k-face colorable if and only if the double dual G^{**} is k-vertex colorable. Since G is isomorphic to G^{**}, we have the result. $\square$

Next we will characterize 2-face colorable graphs in terms of Euler graphs. First, let us prove a lemma.

LEMMA 9.4.4 *A planar connected graph G is bipartite if and only if the dual graph G^* is Eulerian.*

Proof. Since G is planar, every face of G is surrounded by a cycle. Suppose G is bipartite. Then every cycle in G is even. Therefore every face of G has an even number of edges in its boundary. Note that for each face f in G there exists a unique vertex f^* in G^* with the degree of f^* is the degree of f. Therefore the degree of each vertex in G^* is even. Hence G^* is Eulerian. Conversely, suppose G^* is Eulerian. Then the degree of each vertex f^* in G^* is even and thus the degree of the corresponding face f in G is even. Hence G has only even cycles which implies that G is bipartite. $\square$

THEOREM 9.4.5 *A map G is 2-face colorable if and only if it is an Euler graph.*

Proof. Suppose G is 2-face colorable. Then G^* is 2-vertex colorable implies $\chi(G^*) = 2$. By Theorem 9.1.5, G^* is bipartite. Thus the double dual G^{**} is an Euler graph, by Lemma 9.4.4. Since G and G^{**} are isomorphic, we conclude that G is Euler. Conversely, suppose G is an Euler graph. Then its double dual G^{**} is Euler and so, again by Lemma 9.4.4, G^* is bipartite. Thus $\chi(G^*) = 2$ implies G^* is 2-vertex colorable. Hence G is 2-face colorable. $\square$

Summary

In this chapter, we introduced the concept of vertex coloring and demonstrated the coloring process for some well-known graphs such as paths, cycles and complete graphs. We examined the relation between chromatic number of a graph and the maximum degree of a graph. We have explored coloring in a planar graph through the six color theorem, the five color theorem and finally the "famous" four color theorem. The parallel concept of edge coloring is introduced and demonstrated through examples. The relation between edge chromatic number and the maximum degree of a graph is established through Vizing's theorem. We then moved on to face coloring of a graph and presented theorems that characterize k-face colorability in a graph.

9.5 Exercises

Section 9.1: Vertex coloring

1. If W_n is the wheel graph on n vertices, then show that
$$\chi(G) = \begin{cases} 3 & \text{if } n \text{ is odd} \\ 4 & \text{if } n \text{ is even.} \end{cases}$$

2. Show that there exist at least n vertices of degree at least $n-1$, if $\chi(G) = n$.

3. Consider a graph G which contains exactly one odd cycle. Then show that $\chi(G) = 3$.

4. If G is a graph such that any pair of odd cycles has a common vertex. Then prove that $\chi(G) \leq 5$.

5. Prove that $\chi(G) = \triangle(G) + 1$ if and only if G is a cycle of odd length or a complete graph.

6. Let G be a connected graph. If G is not regular, then $\chi(G) \leq \triangle(G)$.

7. Prove that $\chi(G) + \chi(\bar{G}) \leq n+1$, where n is the number of vertices of G.

8. If the graph G contains complete graph with n vertices as a subgraph, then prove that $\chi(G) \geq n$.

9. Find the chromatic number of a complete bipartite graph $K_{m,n}$.

10. If any two odd cycles of G have a vertex in common, then prove that $\chi(G) \leq 5$.

11. Show that a simple graph with n vertices and more than $\lfloor \frac{n}{4} \rfloor$ edges cannot be a bipartite graph.

Section 9.2: Coloring of planar graphs

12. Note that $\chi(K_5) = 5$. Does it contradict the four color theorem? Explain.

13. Prove that the chromatic number of planar graph is not greater than four.

14. Consider the graph G_5 as given in 9.12:

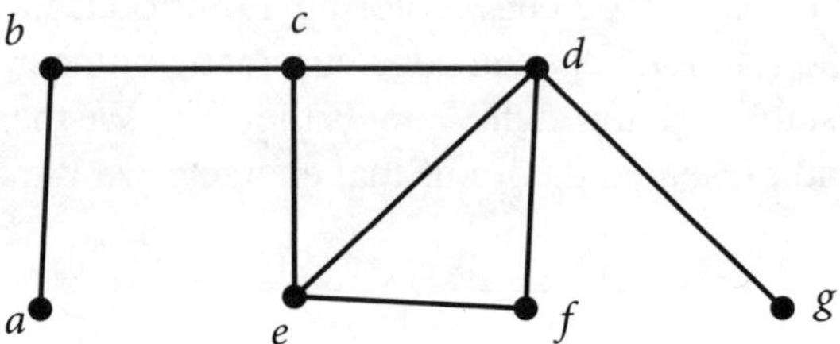

Figure 9.12   The graph G_5

Is G_5 3- colorable? If yes, give the color assignment to each vertex. Otherwise, briefly justify why G_5 is not 3-colorable.

15. In a garden, there are 6 flower beds, where each flower bed is adjacent to no more than 3 others. Show that it is possible to assign at most 4 different colors to the flower beds such that no two adjacent beds have the same color.

Section 9.3: Edge coloring

16. Find the vertex chromatic number $\chi(G)$ and edge chromatic number $\chi'(G)$ of the graphs G_1 and G_2 given in Figure 9.13 and Figure 9.14 respectively.

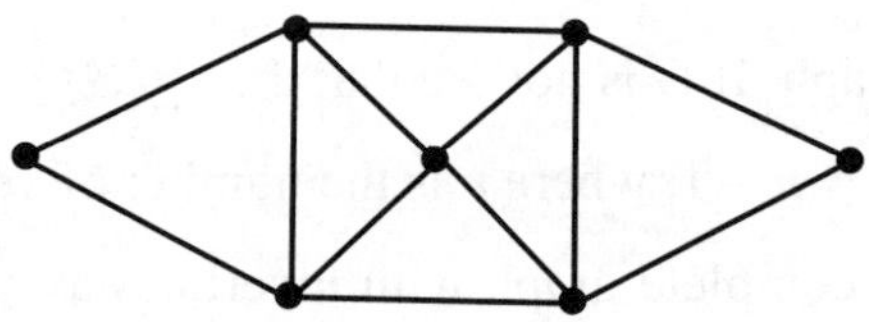

Figure 9.13   The graph G_1

17. Let G denote the Petersen graph. Prove that $\chi'(G) = 4$.

18. Let G be a 3-regular Hamiltonian graph then show that $\chi'(G) = 3$.

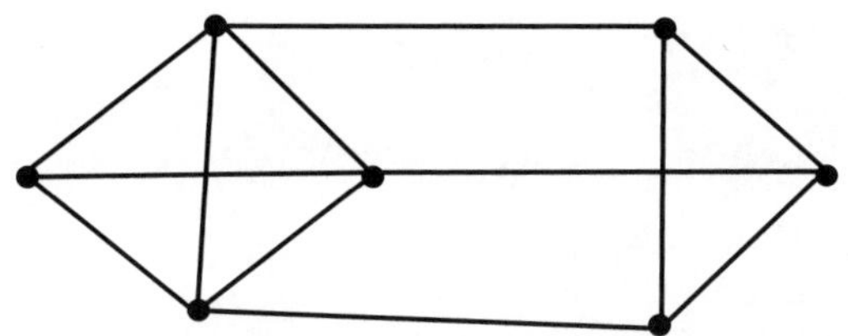

Figure 9.14   The graph G_2

19. Does $\chi'(G) = \Delta(G)$ imply that G is bipartite? Explain.

20. What is the edge chromatic index of the Petersen graph?

Section 9.4: Face coloring

21. Is it possible to color the faces of K_4 with three colors?

10

Independent Sets and Ramsey's Theory

In graph theory, independent sets represent collections of vertices that are pairwise non-adjacent, meaning no two vertices within an independent set share an edge. The study of independent sets is often linked to cliques (sets of mutually adjacent vertices) and covering numbers (the smallest set of vertices and edges that cover the entire graph) as they provide contrasting perspectives on how elements within a graph relate to each other.

In this chapter we will also investigate edge independence and edge covering numbers which extend these concepts to edges instead of vertices. A vital element of this chapter is Ramsey's theorem, a key principle in combinatorics that provides a deep insight that within large enough graphs, specific structures such as cliques or independent sets of a specific size must emerge. The determination of Ramsey numbers, the smallest number of vertices needed to guarantee a specific clique or independent set, introduces a fascinating challenge. This is because Ramsey numbers grow quickly and are notoriously difficult to compute. Calculating Ramsey numbers involves advanced combinatorial reasoning and is essential for understanding the inherent order and structure that arise in large and complex networks.

10.1 Independence, clique and covering number

DEFINITION 10.1.1 Two vertices that are not adjacent in a graph G are said to be independent. A set S of vertices is called as an *independent set* of vertices if every two vertices of S are independent.

A single vertex in any simple graph constitutes an independent set.

EXAMPLE 10.1.2 The independent sets of the graph in Figure 10.1 are $\{v_1\}$, $\{v_2\}$, $\{v_3\}$, $\{v_4\}$, $\{v_1, v_4\}$ and $\{v_2, v_3\}$.

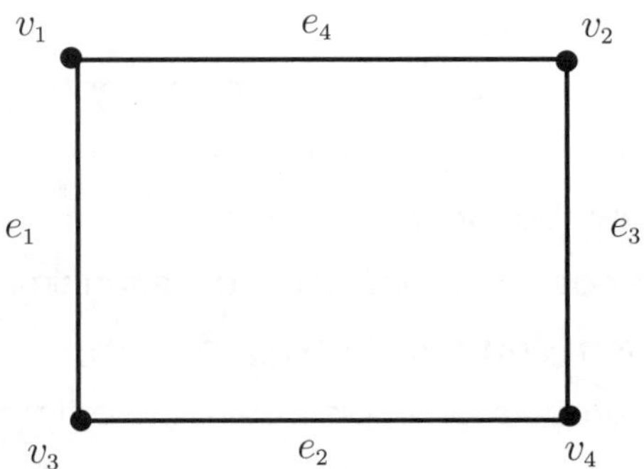

Figure 10.1   Graph for finding independent sets

DEFINITION 10.1.3 An *independent set of edges* in a graph G is a set of edges such that each two of which are not incident. This is also called as *matching*.

Note that Chapter 7 contains a detailed discussion of the concept of matching.

EXAMPLE 10.1.4 The independent edges of graph G in Figure 10.1 are $\{e_1\}$, $\{e_2\}$, $\{e_3\}$, $\{e_4\}$, $\{e_1, e_3\}$ and $\{e_2, e_4\}$.

DEFINITION 10.1.5 A vertex independent set $S \subseteq V(G)$ is called a *maximum independent set* of G if G has no independent set S_0 with $|S_0| > |S|$.

DEFINITION 10.1.6 A *maximal independent set* of G is an independent set that is not a proper subset of any other independent set in G.

DEFINITION 10.1.7 The number of vertices in the maximum independent set is called the *independence number* of G and is denoted by $\alpha(G)$. That is $\alpha(G) = \max\{|A| : A$ is an independent set in $G\}$.

Note that a maximum independent set is maximal but the converse is not always true.

EXAMPLE 10.1.8 For the Petersen graph given in Figure 10.2, $S_1 = \{v_2, v_3, v_{10}\}$ is a maximal independent set and $S_2 = \{v_2, v_3, v_6, v_9\}$ is a maximum independent set. So the independence number is 4. Here S_1 is not a maximum independent set of the Petersen graph.

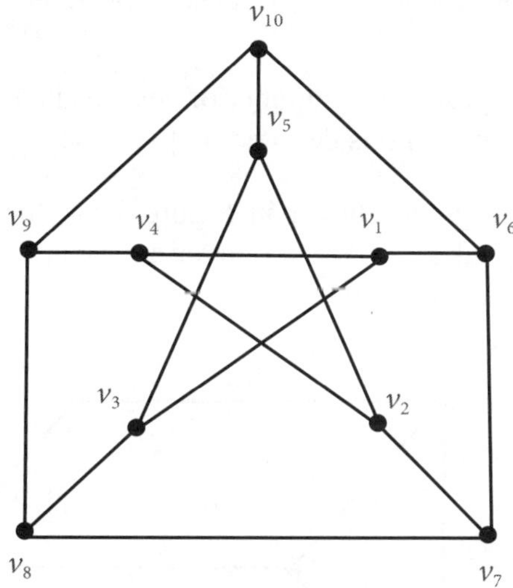

Figure 10.2 Petersen graph

DEFINITION 10.1.9 A subset S of vertices of a graph G is called a *clique* if every pair of vertices in S is joined by at least one edge, and no proper superset of S has this property. Thus, a clique of a graph G is a maximal induced complete subgraph in G. The *clique number* of a graph G, denoted by $\omega(G)$, is the number of vertices in the largest clique in G.

EXAMPLE 10.1.10 There are four cliques in the graph G shown in Figure 10.3 which are $\{v_1, v_2\}, \{v_2, v_3, v_4\}, \{v_2, v_4, v_5\}$ and $\{v_5, v_6\}$. Here, the cardinality of the largest clique is 3. So, the clique number of G is 3.

The relationship between cliques and independent sets is that, in a graph G, the cliques exactly match the independent sets in the complement graph $\overline{G}$. Likewise, cliques in $\overline{G}$ exactly correspond to independent sets in G. In particular, $\alpha(G) = \omega(\overline{G})$ and $\omega(G) = \alpha(\overline{G})$.

The concept of vertex covering of a graph has been explained in Chapter 7, Section 7.3. It is important to recall that a vertex cover is a collection of vertices that has at least one

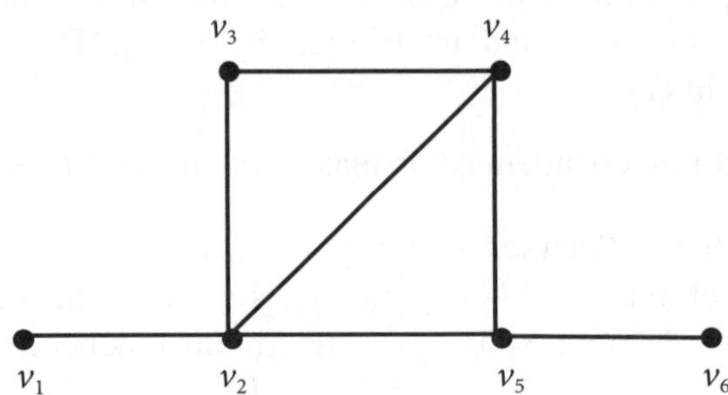

Figure 10.3   Graph for finding clique sets

endpoint of each edge in the graph. The number of vertices in a minimum covering of G is called the *covering number* of G and is denoted by $\beta(G)$.

EXAMPLE 10.1.11 In the graph shown in Figure 10.4, the set $W = \{v_2, v_4, v_6\}$ is a covering set and no proper subset of W is a covering. Hence, W is a minimal covering, which makes the covering number 3.

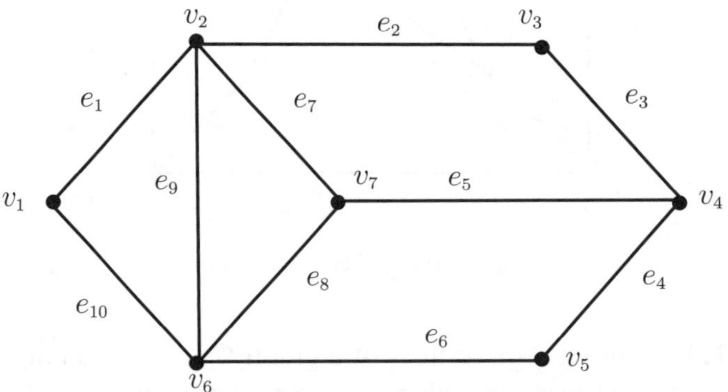

Figure 10.4   Graph for finding covering number

Table 10.1 displays the independence numbers and vertex-covering numbers for different standard graphs.

LEMMA 10.1.12 A subset S of $V(G)$ is independent if and only if $V(G) \setminus S$ is a covering of G.

Proof. Consider a graph G, having set of vertices $V(G)$ and set of edges $E(G)$. A subset S of $V(G)$ is independent when no pair of vertices within S are adjacent in G. That is, each

Table 10.1 Independence and vertex-covering numbers of some standard graphs

	P_n	C_n	K_n	$\overline{K_n}$	$K_{r,s}$
α	$\lceil \frac{n}{2} \rceil$	$\lfloor \frac{n}{2} \rfloor$	1	n	$\max\{r,s\}$
β	$\lfloor \frac{n}{2} \rfloor$	$\lceil \frac{n}{2} \rceil$	$n-1$	0	$\min\{r,s\}$

edge in G connects two vertices that are not in S. In other words, the set $V(G) \setminus S$ forms a cover for the graph G. This means that every edge in G is connected to at least one vertex in the set $V(G) \setminus S$. Hence, the set $V(G) \setminus S$ is a vertex cover of the graph G. $\qquad\square$

THEOREM 10.1.13 For any graph G of order n, $\alpha(G) + \beta(G) = n$.

Proof. Let S be a maximum independent set of G. Then by Lemma 10.1.12, $V(G) \setminus S$ is a covering of G. Thus $|V(G) \setminus S| = n - \alpha \geq \beta$, which implies that $\alpha + \beta \leq n$. Now, it is enough to prove that $\alpha + \beta \geq n$. Let X be a minimum covering of G. Then $V(G) \setminus X$ is an independent set in G and so $|V(G) \setminus X| = n - \beta \leq \alpha$. Thus $\alpha + \beta \geq n$. Thus, we conclude that $\alpha + \beta = n$. $\qquad\square$

REMARK 10.1.14 For any graph G, the vertex chromatic number $\chi(G) \leq \frac{|V(G)|}{\alpha(G)}$. Clearly the vertex set can be partitioned into independent sets and one color can be assigned to all the vertices in a partition set. Since the number of vertices in each of these partitions is at most $\alpha(G)$, we get $\chi(G) \leq \frac{|V(G)|}{\alpha(G)}$.

10.2 Edge independence and edge covering number

Chapter 7 addressed the topic of matchings. It is important to keep in mind that an independent set of edges is a collection of edges where no two of them share a vertex. Typically, we are referring to the matching of a graph, when discussing edge independence. The term *edge independence number* refers to the total number of edges in a maximum edge independent set, and it is represented by the symbol $\alpha'(G)$. Alternatively, $\alpha'(G) = \max\{|M| : M \text{ is a matching in } G\}$.

EXAMPLE 10.2.1 Consider the graph in Figure 10.5, the edge subset $\{e_2, e_3, e_7\}$ is a maximum matching of the graph and thus the edge independence number is 3.

DEFINITION 10.2.2 The number of edges in a minimum edge covering of G is called the edge covering number of G and is denoted by $\beta'(G)$.

EXAMPLE 10.2.3 In the graph given in Figure 10.4, the set $S = \{e_1, e_3, e_6, e_8\}$ is a minimal edge covering and so the edge covering number is 4.

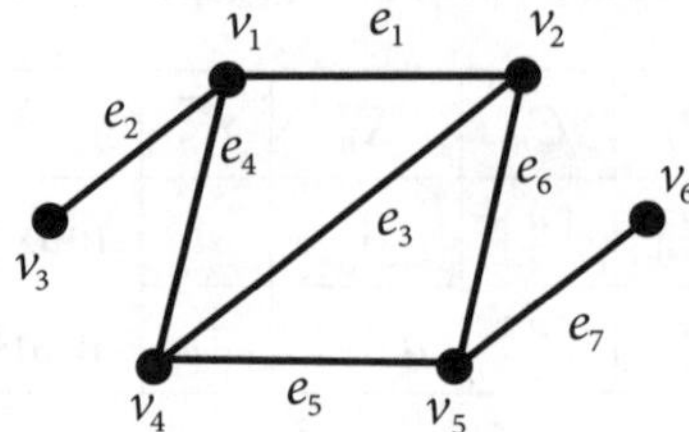

Figure 10.5   Graph for finding edge independence number

Table 10.2 displays the cardinality of edge-independence and edge-covering numbers for different standard graphs.

Table 10.2   Edge independence and edge-covering numbers of some standard graphs

	P_n	C_n	K_n	$K_{r,s}$
α'	$\lfloor \frac{n}{2} \rfloor$	$\lfloor \frac{n}{2} \rfloor$	$\lfloor \frac{n}{2} \rfloor$	$\min\{r,s\}$
β'	$\lceil \frac{n}{2} \rceil$	$\lceil \frac{n}{2} \rceil$	$\lceil \frac{n}{2} \rceil$	$\max\{r,s\}$

LEMMA 10.2.4 If G is a graph of order n having no isolated vertices, then $\beta(G) \geq \alpha'(G)$.

Proof. Consider a vertex cover S in G. Let X be an independent set of edges such that $|X| = \alpha'(G)$. Therefore corresponding to every edge e of X, S has a vertex say v_e, that is incident with e. Additionally, in set X, for any pair of distinct edges e and f, the vertices v_e and v_f are also distinct. Thus $|S| \geq |X|$, which implies that $\beta(G) \geq \alpha'(G)$. $\qquad\square$

THEOREM 10.2.5 For any bipartite graph $G, \alpha'(G) = \beta(G)$.

Proof. From Lemma 10.2.4, we get that $\beta(G) \geq \alpha'(G)$. So it is enough to prove that $\beta(G) \leq \alpha'(G)$. Let U and W be the bipartite sets of G. Let M be a maximum matching in G. That is $\alpha'(G) = |M|$. Let A be the set of all M-unmatched vertices in U. Note that $|M| = |U| - |A|$. Consider S to be the set comprising of all vertices of G that are connected to at least one vertex in A by an M-alternating path. Define $U' = S \cap U$ and $W' = S \cap W$. We get that $U' - A$ is matched to W' and $N(U') = W'$ from the proof of Hall's theorem, refer Theorem 7.2.5. So we can conclude that $|U'| - |W'| = |A|$. Therefore either $C = (U \setminus U') \cup W'$ is a vertex cover in G or there exists an edge vw in G such that $v \in U'$ and $w \notin W'$, which implies that $|C| = |U| - |U'| + |W'| = |U| - |A| = |M|$. Thus $\beta(G) \leq |C| = |M| = \alpha'(G)$. Hence the proof. $\qquad\square$

THEOREM 10.2.6 For any graph G with minimum degree $\delta(G) > 0$, $\alpha'(G) + \beta'(G) = n$.

Proof. Let G be a graph with $\delta(G) \geq 1$. Consider an independent set of edges E_1 in the graph G with $|E_1| = \alpha'(G)$. Then E_1 covers $2\alpha'(G)$ vertices of G. For each vertex not covered by E_1, select an incident edge and form a new set, say E_2, by taking the union of these edges with E_1. Now E_2 is necessarily an edge cover in G, ensuring that $|E_2| \geq \beta'(G)$. Now we get $2|E_1| + |E_2| = n$ which implies $\alpha'(G) + \beta'(G) \leq n$.

Next, let's consider an edge cover E' in G with $|E'| = \beta'(G)$. Because of the minimality of E', every element of $G[E']$ is a tree. Choose one edge from each component of $G[E']$ to create a new set of edges, E''. Consequently, $|E''| \leq \alpha'(G)$. Now by Corollary 3.1.5, if $G[E']$ is a forest with k components, then the size of $G[E']$ is $n - k$. Therefore $\alpha'(G) + \beta'(G) \geq |E''| + |E'| = k + (n - k) = n$. Thus $\alpha' + \beta' = n$. $\qquad\qquad\square$

Note that by Theorems 10.1.13, 10.2.6 and 10.2.4, we get $n - \alpha(G) = \beta(G) \geq \alpha'(G) = n - \beta'(G)$. So we can conclude that $\beta'(G) \geq \alpha(G)$. Let us mention it in the following statement.

COROLLARY 10.2.7 *If G is a graph of order n having no isolated vertices, then $\beta'(G) \geq \alpha(G)$.*

COROLLARY 10.2.8 If G is a bipartite graph with no isolated vertices, then $\alpha(G) = \beta'(G)$.

Proof. By Theorem 10.1.13, $\alpha(G) + \beta(G) = n$ and by Theorem 10.2.6, we have $\alpha'(G) + \beta'(G) = n$. Therefore $\alpha(G) + \beta(G) = \alpha'(G) + \beta'(G)$. By Theorem 10.2.5, we know that $\alpha'(G) = \beta(G)$. Combining these equations, we get $\alpha(G) = \beta'(G)$. $\qquad\qquad\square$

Now consider a specific question: Can we characterize the simple graphs for which the parameter value is 1 for each of the four values that follow: α, α', β and β'? The answer is discussed in the following example.

EXAMPLE 10.2.9 Let G be a simple graph.

- $\alpha(G) = 1$ if and only if any two vertices of G are adjacent; which happens when the graph under consideration is complete. Thus the independence number is 1 if and only if G is a complete graph.

- $\alpha'(G) = 1$ if and only if G is a triangle or a nontrivial star with/without isolated vertices. For, suppose that G is a nontrivial star with isolated vertices. Then deleting all the endpoints in the graph leaves no edges remaining. Therefore $\alpha'(G) = 1$. Also $\alpha'(K_3) = 1$. Conversely, suppose that $\alpha'(G) = 1$ and ignore the isolated vertices. Let v be a vertex of maximum degree. If every edge is incident to v, then G is a star. Otherwise, there exists an edge e not incident to v that shares an endpoint with every edge incident to v. Since e has only two endpoints, $deg(v) = 2$, and there is only one such edge e. Thus G is a triangle.

- $\beta(G) = 1$ if and only if one vertex is incident to all edges. Thus the vertex covering number is 1 if and only if G is a nontrivial star with/without isolated vertices.

- $\beta'(G) = 1$ if and only if $G = K_2$ since every edge covers two vertices, $\beta'(G) = 1$ requires that $|V(G)| = 2$ and indeed $\beta'(K_2) = 1$.

Let us discuss some other interesting behaviors with regard to independence numbers and covering numbers.

THEOREM 10.2.10 *If T is a tree, then $\alpha'(T) = |V(T)| - \alpha(T)$.*

Proof. Let T be a tree with n vertices. Note that the vertices outside a maximum independent set form a vertex covering, refer to Lemma 10.1.12. Therefore $\beta(T) = n - \alpha(T) = n - \alpha(T)$. Since tree is a bipartite graph, by Theorem 10.2.5, we have $\alpha'(T) = \beta(T) = n - \alpha(T)$. $\qquad\square$

THEOREM 10.2.11 *A graph G is bipartite if and only if $\alpha(H) = \beta'(H)$ for every subgraph H of G with no isolated vertices.*

Proof. If G is bipartite, then every subgraph H of G is also bipartite. Consequently, according to Corollary 10.2.8, we find that for every subgraph H of G with no isolated vertices, the independence number $\alpha(H)$ equals the vertex covering number $\beta'(H)$, that is $\alpha(H) = \beta'(H)$.

Conversely, if $\alpha(H)$ equals $\beta'(H)$ for every subgraph H of G with no isolated vertices, and G is not bipartite, then G must contain an odd cycle C. This implies that to cover the vertices of C, it requires $\alpha(C) + 1$ edges, contradicting the assumption. $\qquad\square$

THEOREM 10.2.12 *Every maximal matching in a graph G has at least $\frac{\alpha'(G)}{2}$ edges.*

Proof. When M is a maximal matching, the set of vertices saturated by M forms a vertex cover. Otherwise, if both end vertices of an edge have not been included in the set, then the edge could be added to M. Thus $\beta(G) \leq 2|M|$. Since every vertex cover has size at least $\alpha'(G)$, we obtain $\beta(G) \geq \alpha'(G)$. Thus from above equations we get $2|M| \geq \beta(G) \geq \alpha'(G)$. Thus $|M| \geq (\alpha'(G))/2$. Hence every maximal matching in a graph G has at least $\frac{\alpha'(G)}{2}$ edges. $\qquad\square$

THEOREM 10.2.13 *Let G be a nontrivial simple graph. Then $\alpha(G) \leq |V(G)| - \frac{|E(G)|}{\Delta(G)}$. In particular, $\alpha(G) \leq \frac{n(G)}{2}$ when G is regular.*

Proof. Let S be an independent set of size $\alpha(G)$. By Lemma 10.1.12, $V(G) \setminus S$ is a vertex covering. Summing the vertex degrees in $V(G) \setminus S$ provides an upper bound on $|E(G)|$. Thus $|E(G)| \leq [|V(G)| - \alpha(G)]\Delta(G)$. Hence $\alpha(G) \leq |V(G)| - \frac{|E(G)|}{\Delta(G)}$. Assume that G is regular. Hence by the Fundamental theorem of graph theory, refer Theorem 2.1.4, we have $|E(G)| = \frac{n(G)\Delta(G)}{2}$. Thus $\alpha(G) \leq \frac{n(G)}{2}$. $\qquad\square$

THEOREM 10.2.14 *Let G be a bipartite graph. Then $\alpha(G) = \frac{|V(G)|}{2}$ if and only if G has a perfect matching.*

Proof. Let G be a bipartite graph. Suppose that G has a perfect matching. Then $\alpha'(G) = \frac{|V(G)|}{2}$. By Theorem 10.1.13, we have $\alpha(G) = |V(G)| - \beta(G)$. Since G is bipartite, by Theorem 10.2.5, we have $\beta(G) = \alpha'(G)$. Therefore, $\alpha(G) = |V(G)| - \alpha'(G) = \frac{|V(G)|}{2}$. Conversely, suppose that $\alpha(G) = \frac{|V(G)|}{2}$. By Theorem 10.1.13, $\beta(G) = \frac{|V(G)|}{2}$. Since G is bipartite, by Theorem 10.2.5, we get $\alpha'(G) = \frac{|V(G)|}{2}$. Thus G has a perfect matching. $\qquad\square$

10.3 Ramsey's theorem

Frank Ramsey's (1930) theorem laid the foundation for Ramsey theory, a branch of mathematics studying patterns in complex systems. A prime example of this theorem is the Ramsey puzzle, which highlights the inevitability of specific relationships within groups of individuals. For instance, within a gathering of six people, there will always emerge either a trio who all know each other or a trio who are all strangers to each other.

THEOREM 10.3.1 *In any simple graph G with six vertices, either there are three mutually adjacent vertices or there are three mutually non-adjacent vertices. Equivalently, either G or its complement $\overline{G}$ contains K_3 as a subgraph.*

Proof. Consider a graph G with six vertices. Let $u \in V(G)$. Then, either the degree of u in G, denoted by $deg_G(u) \geq 3$ or $deg_{\overline{G}}(u) \geq 3$, because $\deg_G(u) + deg_{\overline{G}}(u) = 5$. Now, without loss of generality, let's assume that $\deg_G(u) \geq 3$, due to the symmetry of the theorem statement concerning G and $\overline{G}$. Consider x, y and z as three neighbors of u. If any two of x, y, and z are adjacent, then those two along with u form a complete K_3 as a subgraph in G. Otherwise, if x, y and z are not adjacent to each other, then they constitute a complete K_3 as a subgraph in $\overline{G}$. $\qquad\square$

Statement of Ramsey's theorem is given below.

THEOREM 10.3.2 *For any $k+1 \geq 3$ positive integers $t, n_1, n_2, \ldots, n_k$, there exists a positive integer n such that if each of the t-element subsets of the set $\{1, 2, \ldots, n\}$ is colored with one of the k colors $1, 2, \ldots, k$, then for some integer i with $1 \leq i \leq k$, there is a subset S of $\{1, 2, \ldots, n\}$ containing n_i elements such that every t-element subset of S is colored i.*

To understand Ramsey's theorem in the context of graph theory, let us consider the complete graph K_n with vertex set $\{1, 2, \ldots, n\}$. In Theorem 10.3.2, when $t = 2$, each edge of K_n corresponds to a 2-element subset of $\{1, 2, \ldots, n\}$. We assign one of the colors $1, 2, \ldots, k$ to each edge, forming a k-edge coloring of K_n (though not necessarily a proper one). This particular case of Ramsey's theorem holds special significance in the field of graph theory.

THEOREM 10.3.3 *For any $k \geq 2$ positive integers $n_1, n_2, \ldots, n_k$, there exists a positive integer n such that for every k-edge coloring of K_n (that is not necessarily a proper edge coloring), there is a complete subgraph K_{n_i} of K_n for some i $(1 \leq i \leq k)$ such that every edge of K_{n_i} is colored i.*

10.4 Determination of Ramsey numbers

In Ramsey's theory, we focus on the scenario where $k = 2$. In a red-blue edge coloring of a graph G, each edge is colored either red or blue, and neighboring edges may share the same color. The graph could be entirely red, entirely blue, or a mix of both colors.

DEFINITION 10.4.1 For two graphs F and H, the *Ramsey number $R(F, H)$* is the smallest number n such that in any red-blue coloring of the complete graph K_n, there will either be a subgraph isomorphic to F with all red edges or a subgraph isomorphic to H with all blue edges. Then, $R(F, H) = R(H, F)$ for any two graphs F and H.

DEFINITION 10.4.2 The *Ramsey number $R(F, F)$* is the smallest number n such that in any red-blue coloring of K_n, there will always be a monochromatic subgraph isomorphic to F. This number $R(F, F)$ is often referred to as the Ramsey number of the graph F.

Let's discuss an example that might assist us in comprehending the Ramsey numbers.

EXAMPLE 10.4.3 Let us find the value of $R(K_3, K_3)$.

Let $V(K_5) = \{v_1, v_2, \ldots, v_5\}$. We can color the edges of K_5 using red and blue such that the edges forming the outer 5-cycle are red, and all other edges are blue, refer to Figure 10.6. In Figure 10.6, solid lines indicate red edges, while dashed lines indicate blue edges. This specific coloring ensures that there is no red K_3 or blue K_3, thus proving $R(K_3, K_3) \geq 6$.

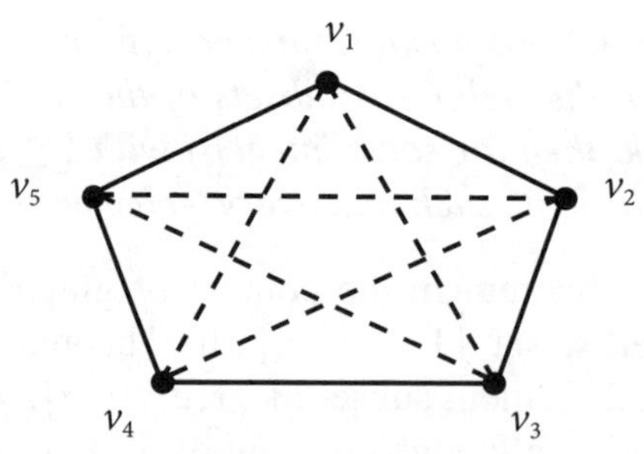

Figure 10.6   The graph K_5

To establish that $R(K_3, K_3) \leq 6$, let's examine any red-blue edge coloring of K_6. Consider a vertex v_1 in K_6. This vertex connects to five other vertices, implying that at least three incident edges must be with the same color. Suppose these three edges v_1v_2, v_1v_3, and v_1v_4 are red. If any of the edges v_2v_3, v_2v_4, or v_3v_4 is red, it forms a red K_3. If none of these edges are red (and thus blue), it forms a blue K_3, refer Figure 10.7. In Figure 10.7, solid lines indicate red color edges, dashed lines indicate blue color edges and dotted lines indicate the remaining edges. Consequently, we conclude that $R(K_3, K_3) \leq 6$ and therefore $R(K_3, K_3) = 6$.

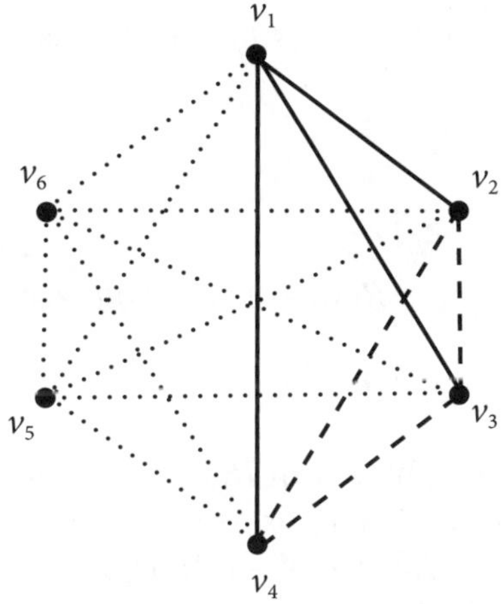

Figure 10.7   The graph K_6

The numbers $R(K_s, K_t)$ are commonly expressed as $R(s,t)$ and are called as the classical Ramsey numbers. In fact, if F has order s and H has an order t, then

$$R(F, H) \leq R(s, t).$$

DEFINITION 10.4.4 Let s and t be positive integers. The *Ramsey number* $R(s,t)$ is the minimum positive number n such that for every simple n-vertex graph G, either G contains a K_s-subgraph or its complement graph $\overline{G}$ contains a K_t-subgraph.

THEOREM 10.4.5 $R(s,t) = R(t,s)$ *for all* $s, t \geq 1$.

Proof. The Ramsey number $R(s,t)$ is the minimum number n such that for every edge-coloring of K_n with colors red and blue, there exists either a red K_s-subgraph or a blue K_t-subgraph. This version of the definition generalizes to arbitrarily many edge colors and to arbitrary subgraphs in each color. Here we observe that without loss of generality, we can interchange s and t. Hence $R(s,t) = R(t,s)$ for all $s, t \geq 1$. $\qquad\square$

LEMMA 10.4.6 $R(s,1) = 1$ *for all $s \geq 1$.*

Proof. The edge-complement of a nontrivial graph always contains K_1. This means that in any edge-coloring of a complete graph with at least two vertices, there will always be a monochromatic complete subgraph with only one vertex. $\qquad\square$

LEMMA 10.4.7 $R(s,2) = s$ *for all $s \geq 1$.*

Proof. If a simple graph G with s vertices is not complete, then its edge complement $\overline{G}$ contains K_2. Therefore $R(s,2) \leq s$. Since K_{s-1} obviously does not contain a K_s-subgraph and $\overline{K}_{s-1}$ contains no edges, the reverse inequality also holds. $\qquad\square$

Let us explore the upper bounds for $R(s,t)$, which was established by *Paul Erdös* and *George Szekeres* in 1935.

THEOREM 10.4.8 *For all $s,t \geq 3$,*

(i). $R(s,t) \leq R(s-1,t) + R(s,t-1)$.

(ii). *Furthermore, if $R(s-1,t)$ and $R(s,t-1)$ are both even, then*

$$R(s,t) < R(s-1,t) + R(s,t-1).$$

Proof. (i). Consider a simple graph G with $R(s-1,t) + R(s,t-1)$ vertices, and let $u \in V(G)$. Then, either $\deg_G(u) \geq R(s-1,t)$ or $\deg_{\overline{G}}(u) \geq R(s,t-1) = R(t-1,s)$, because $\deg_G(u) + \deg_{\overline{G}}(u) = R(s-1,t) + R(s,t-1) - 1$. If $\deg_G(u) \geq R(s-1,t)$, then the subgraph H of G induced by the open neighborhood of u has at least $R(s-1,t)$ vertices. Therefore, either the subgraph H contains a K_{s-1} subgraph, which together with vertex u forms a K_s subgraph of G, or $\overline{H}$ (and hence $\overline{G}$ contains a K_t subgraph). Then, we can apply the same argument to $\overline{G}$ with the roles of s and t reversed if $\deg_{\overline{G}}(u) \geq R(t-1,s)$.

(ii). Now suppose $R(s-1,t)$ and $R(s,t-1)$ are both even, and that graph G has $R(s-1,t) + R(s,t-1) - 1$ vertices. Then $\deg_G(u) + \deg_{\overline{G}}(u) = R(s-1,t) + R(s,t-1) - 2$. There exists a vertex u in G with an even degree because the total number of vertices is odd. If $\deg_G(u) \geq R(s-1,t)$, then the proof follows as in part (i). If $\deg_G(u) < R(s-1,t)$, then $\deg_G(u) \leq R(s-1,t) - 2$ and so $\deg_{\overline{G}}(u) \geq R(t-1,s)$. The result follows as in part (i). $\qquad\square$

The next result shows that Ramsey numbers exist for all pairs of positive integers.

COROLLARY 10.4.9 *For all positive integers s and t, $R(s,t) \leq \binom{s+t-2}{s-1}$.*

Proof. We will prove the statement by induction on the sum $s+t$. The statement is true for $s+t = 2$, as shown by Lemma 10.4.6 . Assume that for some $k \geq 3$, the inequality holds for all positive integers s and t whose sum is less than k. Now consider $s+t = k$. Then, by Theorem 10.4.8,

$$R(s,t) \leq R(s-1,t) + R(s,t-1).$$

By the induction hypothesis, we get

$$R(s,t) \le \binom{(s-1)+t-2}{(s-1)-1} + \binom{s+(t-1)-2}{s-1}$$
$$= \binom{s+t-3}{s-2} + \binom{s+t-3}{s-1}.$$

By Pascal's recursion formula,

$$\binom{n-1}{k} + \binom{n-1}{k-1} = \binom{n}{k} \quad \text{for } n,k \in \mathbb{N}.$$

Thus

$$R(s,t) \le \binom{s+t-2}{s-1}.$$

$\square$

Ramsey number calculations

Determining values of the Ramsey numbers $R(s,t)$ for $3 \le s \le t$ has proved to be quite difficult. Next, we determine the values of $R(3,t)$ for an arbitrary integer t, followed by an upper bound for $R(4,4)$. From Corollary 10.4.9, we observe that the bound for $R(s,t)$ is sharp if one of s and t is 1 or 2. The bound is also sharp for $s = t = 3$. By Corollary 10.4.9, we get

$$R(3,t) \le \frac{t^2+t}{2}.$$

An improved bound for $R(3,t)$ is provided in the next theorem.

THEOREM 10.4.10 *For every integer $t \ge 3$,*

$$R(3,t) \le \frac{t^2+3}{2}.$$

Proof. We prove this by using induction on t. For $t = 3$, $R(3,t) = 6 = \frac{t^2+3}{2}$ so that $R(3,t) \le \frac{t^2+3}{2}$ holds for $t = 3$. Now, assume that

$$R(3,t-1) \le \frac{(t-1)^2+3}{2},$$

for some integer $t \ge 4$, and consider $R(3,t)$. From Theorem 10.4.8,

$$R(3,t) \le t + R(3,t-1). \tag{10.1}$$

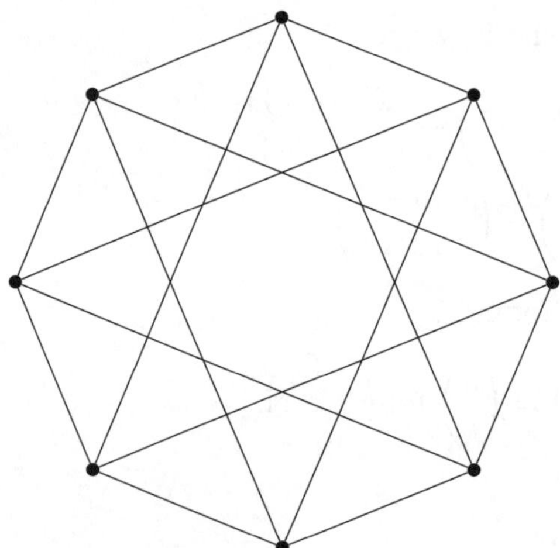

Figure 10.8   Graph showing $R(3,4) \geq 9$

Furthermore, strict inequality holds when t and $R(3,t-1)$ are even. Combining Equation 10.1 and the inductive hypothesis, we have

$$R(3,t) \leq \frac{(t-1)^2+3}{2} = \frac{t^2+4}{2}. \tag{10.2}$$

In order to complete the proof, it remains to show that the inequality given in Equation 10.2 is strict. When t is odd, then $R(3,t) < \frac{t^2+4}{2}$ since t^2+4 is odd. Thus we may assume that t is even. If $R(3,t-1) < \frac{(t-1)^2+3}{2}$, then clearly the inequality in Equation 10.2 is strict. If

$$R(3,t-1) = \frac{(t-1)^2+3}{2} = \frac{t^2}{2} - t + 2,$$

then $R(3,t-1)$ is even since t is even. Therefore, the inequality in Equation 10.2 is strict, which provides us the result. $\qquad\square$

Based on Theorem 10.4.10, we have $R(3,4) \leq 9$ and $R(3,5) \leq 14$. In fact, these inequalities are equalities in both cases. The equality $R(3,4) = 9$ is demonstrated by the existence of a graph with 8 vertices that contains neither a triangle nor an independent set of four vertices. Similarly, $R(3,5) = 14$ is proven by the existence of a graph with 13 vertices that contains neither a triangle nor an independent set of five vertices. These graphs are illustrated in Figures 10.8 and 10.9. However, while Theorem 10.4.10 gives $R(3,6) \leq 19$, but the actual value is $R(3,6) = 18$.

By Theorem 10.4.8 and Theorem 10.4.5 together with the fact that $R(3,4) = 9$, we have $R(4,4) \leq R(3,4) + R(4,3) = 9 + 9 = 18$. Let us write it as a statement.

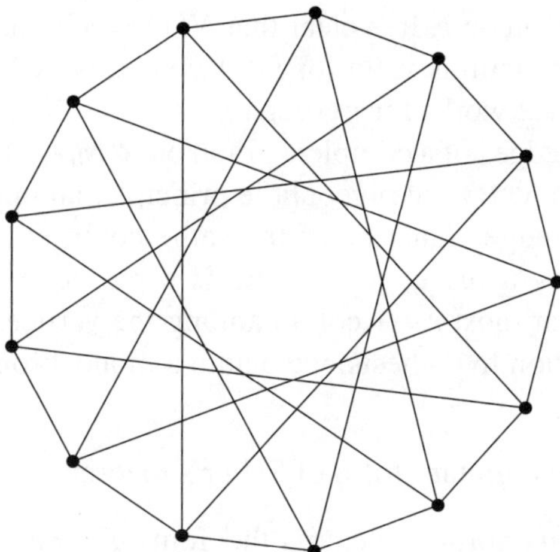

Figure 10.9   Graph showing $R(3,5) \geq 14$

COROLLARY 10.4.11 $R(4,4) \leq 18$.

Known classical Ramsey numbers

The following table provides the known values of $R(s,t)$ for integers s and t with $s,t \geq 3$.

Table 10.3   Known classical Ramsey numbers

s	3	3	3	3	3	3	3	4	4
t	3	4	5	6	7	8	9	4	5
$R(s,t)$	6	9	14	18	23	28	36	18	25

10.5 Schur's theorem

Schur's theorem is an important application of Ramsey's theorem. In the beginning of 19th century, Schur tried to prove Fermat's last theorem by reducing the equation $X^n + Y^n = Z^n$ mod p where p is a prime. Despite his efforts, he did not succeed. It was later shown that for every positive integer n, the equation has nontrivial solutions modulo p for all sufficiently large primes p.

The following result is a special case of Ramsey's theorem.

THEOREM 10.5.1 *For every positive integer r, there is some integer $N = N(r)$ such that if the edges of K_N, the complete graph on N vertices, are colored with r colors, then there is always a monochromatic triangle.*

Proof. We will use induction on r. It is clear that $N(1) = 3$ is sufficient for $r = 1$. Now, let $r \geq 2$ and assume that the claim is valid for $r - 1$ colors, with $N = N_0$. We will show that choosing $N = R(N_0 - 1) + 2$ works for r colors.

Consider coloring the edges of a complete graph on $R(N_0 - 1) + 2$ vertices with r colors. Choose an arbitrary vertex v. By the pigeonhole principle, among the $R(N_0 - 1) + 1$ edges incident to v, at least N_0 edges must be of the same color, say red. Let v_0 be the set of vertices connected to v by a red edge. If there is a red edge within v_0, we have a red triangle. If not, there are at most $r - 1$ colors among the vertices in v_0, which has at least N_0 vertices. By the induction hypothesis, we obtain a monochromatic triangle.

$\square$

The next result gives the finitary form of Schur's theorem.

THEOREM 10.5.2 (Finitary form of Schur's theorem) *For every positive integer r, there exists a positive integer $N = N(r)$ such that if the elements of $[N]$ are colored with r colors, then there is a monochromatic solution to $x + y = z$ with $x, y, z \in [N]$.*

Proof. Consider a function $f : [N] \to [r]$ representing a coloring. We color the edges of a complete graph with vertices $\{1, \ldots, N+1\}$ by assigning the edge $\{i, j\}$ with $i < j$ the color $f(j - i)$. According to 10.5.1, if N is sufficiently large, then there exists a monochromatic triangle, let's say on vertices $i < j < k$. Therefore, $f(j - i) = f(k - j) = f(k - i)$. We take $x = j - i$, $y = k - j$, and $z = k - i$. This implies $f(x) = f(y) = f(z)$ and $x + y = z$. Hence the proof.

$\square$

Its application to Fermat's last theorem is given in the following result.

THEOREM 10.5.3 *Let n be a positive integer. For all sufficiently large primes p, there exists $X, Y, Z \in \{1, \ldots, p-1\}$ such that $X^n + Y^n \equiv Z^n \pmod{p}$.*

Proof. We denote by $\mathbb{Z}_p^*$ the group of nonzero residues modulo p under multiplication. Let H be the subgroup of n-th powers in $\mathbb{Z}_p^*$. The index of H in $\mathbb{Z}_p^*$ is at most n, so the cosets of H partition $\{1, 2, \ldots, p-1\}$ into at most n sets. According to the finite statement of Schur's theorem (Theorem 10.5.2), for sufficiently large primes p, there exists a solution to $x + y = z$ in $\mathbb{Z}$ within one of the cosets of H, denoted by aH for some $a \in \mathbb{Z}_p^*$. Since H consists of n-th powers, we can express $x = aX^n$, $y = aY^n$, and $z = aZ^n$ for some $X, Y, Z \in \mathbb{Z}_p^*$. Consequently, we have $aX^n + aY^n \equiv aZ^n \pmod{p}$, leading to $X^n + Y^n \equiv Z^n \pmod{p}$ as desired.

$\square$

Summary

We have introduced the reader a set containing only pairwise non-adjacent vertices, known as the independent set. Maximal and maximum independent sets are also introduced. The interesting relation between cliques, independent sets and vertex covering numbers have been explored thoroughly. The edge independent sets and edge coverings have been illustrated through definitions and examples. The relations between edge independent sets, edge covering numbers are found to be similar to those existing in vertices. The characteristics of vertex and edge independence numbers in various types of graphs have been described through theorems and examples. The reader is introduced to Ramsey's theory and the process of determination of Ramsey numbers. A special case of Ramsey's theorem, Schur's theorem has been discussed and its application to Fermat's last theorem has also been explained.

10.6 Exercises

Section 10.1: Independence, clique and covering number

1. Prove or disprove: In any tree, there is a maximal independent set of vertices containing all the vertices of degree one.

2. Let G be a graph of order $n \geq 2$. Prove that

$$\alpha(G) + \alpha'(G) \leq n \leq \alpha(G) + 2\alpha'(G)$$

and also show that the bounds are sharp.

3. Prove that a graph G is bipartite if and only if $\beta(H) \geq \frac{1}{2}n$ for every subgraph H of G, where n is the number of vertices of H.

4. Determine the clique number of cycle C_{10}.

Section 10.2: Edge independence and edge covering number

5. Determine the values of $\alpha(G)$, $\alpha'(G)$, $\beta(G)$, and $\beta'(G)$ for each of the following graphs G:

    (i). the path P_n, $n \geq 2$,

    (ii). the cycle C_n, $n \geq 3$,

(iii). the complete graph K_n, $n \geq 2$,

(iv). the complete bipartite graph $K(p,q)$, $p \geq q \geq 1$,

(v). the wheel W_n, $n \geq 4$,

(vi). the Petersen graph, and

(vii). the n-cube graph Q_n, $n \geq 1$.

6. Consider two graphs G and H. Let $G+H$ denote the join of G and H.

   (i). Express $\alpha(G+H)$ in terms of $\alpha(G)$ and $\alpha(H)$.

   (ii). What can be said about $\beta(G+H)$?

7. Consider a graph G of order n with $\delta(G) \geq 1$. Prove that

   (i). $\beta'(G) \leq \beta(G)\Delta(G)$ and

   (ii). $\beta(G)+\beta'(G) \geq n$ and that the bound is sharp for each n.

8. Using theorem 10.2.5, prove Hall's theorem.

Section 10.3: Ramsey's theorem

9. Prove that $R(P_3, K_3) = 5$.

10. Prove that $R(K_{1,3}, K_3) = 7$.

11. Find $R(2K_2, 3K_2)$.

12. Find $R(C_4, C_4)$.

13. Show that $R(K_3, K_n) \leq \binom{n+1}{2} = \frac{n^2+n}{2}$ for every integer $n \geq 2$.

14. Prove that if G is a graph of order $R(s,t) - 1$, then

$$K_{s-1} \subseteq G \text{ or } K_t \subseteq \overline{G},$$
$$K_s \subseteq G \text{ or } K_{t-1} \subseteq \overline{G}.$$

15. Show that $r(s', t') \leq r(s,t)$ if $2 \leq s' \leq s$ and $2 \leq t' \leq t$. Also, show that $s = s'$ and $t = t'$ if and only if equality holds.

Section 10.5: Schur's theorem

16. Consider a 2-coloring of the integers $\{1, 2, \ldots, 15\}$. Show that a monochromatic solution to $x+y = z$ must exist.

11

Directed Graphs

LEARNING OBJECTIVES

- Define directed graphs and distinguish them from undirected graphs.
- Calculate indegree and outdegree in a sample directed graph.
- Define tournaments and orientability.
- Explain the modes of traversal in directed graphs.
- Apply the Bellman–Ford algorithm to find the shortest path between a pair of vertices in a directed graph with negative weights.
- Apply Floyd–Warshall algorithm to find shortest paths between all pairs of vertices in a directed graph.
- Define networks within the context of directed graphs.
- State and prove the max-flow min-cut cut theorem.
- Use the Ford-Fulkerson algorithm to calculate maximum flows in sample networks.

Directed graphs, also known as digraphs are a fundamental structure in graph theory, where the relations between two vertices are allowed to be asymmetrical. An edge as we have known so far, has always signified a two-way relationship between two vertices. But in directed graphs, a directed edge has a designated direction in which you may go from say a to b but not always from b to a. This directionality introduces the concept of ordered pairs where each edge has a distinct starting point (source) and an ending point (target). Understanding their properties and behavior is crucial for solving problems related to reachability, hierarchy and optimization. This chapter explores the core concepts of directed graphs including their representation, traversal techniques and properties while also providing a succinct introduction to the world of networks.

Throughout this book, there have been references to how certain concepts are beneficial for navigating networks, allocating resources within networks, solving assignment problems in networks, and more. But it is only now that we introduce the reader to networks, which are models of relationships and interactions across various types of real world systems such as transportation systems, communication channels and biological pathways. While a network can be represented as a graph, it differs from an abstract graph by often including additional information, such as weights, directions or types of connections, that help capture the complexity of real-life systems. In this chapter, we will also introduce a pivotal theorem in the world of networks: the max-flow min-cut theorem that became the basis for the cornerstone algorithms such as the Ford–Fulkerson and Edmonds–Karp algorithms. We will also be discussing the Ford–Fulkerson algorithm that provides practical solutions to maximize flow in any flow network, such as telecommunications and data network, transportation and logistics network, job assignments, project and resource allocation and image segmentation. In addition to all of the above, the Ford–Fulkerson algorithm laid the groundwork for future research, making it one of the essential algorithms in optimization and graph theory.

11.1　Definitions

Due to the asymmetrical nature of the *directed edges*, we will define some terminologies and notations that are unique to directed graphs.

DEFINITION 11.1.1 A *directed graph* is a tuple $D = (V, E)$ where V is the vertex set and E is the directed-edge set such that for each directed edge $\vec{e}$ in E, an ordered pair of vertices (u, v) is assigned to $\vec{e}$. The edge $\vec{e}$ is said to join u to v, u is called the origin or initial vertex or tail of the edge $\vec{e}$ and v is called the terminus or terminal vertex or the head of $\vec{e}$. The directed graph is also called as the *digraph* and the directed edges are called *arcs*.

EXAMPLE 11.1.2 Consider the Figure 11.1. Here D is a directed graph $D = (V, A)$ where the vertex set $V = \{u_1, u_2, \ldots, u_5\}$ and the arc set $E = \{\vec{e_1}, \vec{e_2}, \ldots, \vec{e_8}\}$. Each directed edge is associated with an ordered pair. For example, $\vec{e_1}$ is associated with (u_5, u_1), $\vec{e_4}$ is associated with (u_3, u_5) and $\vec{e_8}$ is associated with (u_2, u_2).

Although we use the word "arc" or "directed edge" to differentiate it from the usual terminology of edge as used with regards to a simple graph G, we may use the word "edge" in directed graphs as well, when coupled with the notation, edge $\vec{e}$. Instead of strictly adhering to an ordered pair representation, we may also use the usual notation of edges to denote a relationship between vertices in order. For example, in the digraph shown in Figure 11.1, edge $\vec{e_4}$ is $\overrightarrow{u_3 u_5}$, instead of being strictly represented as (u_3, u_5). The reader

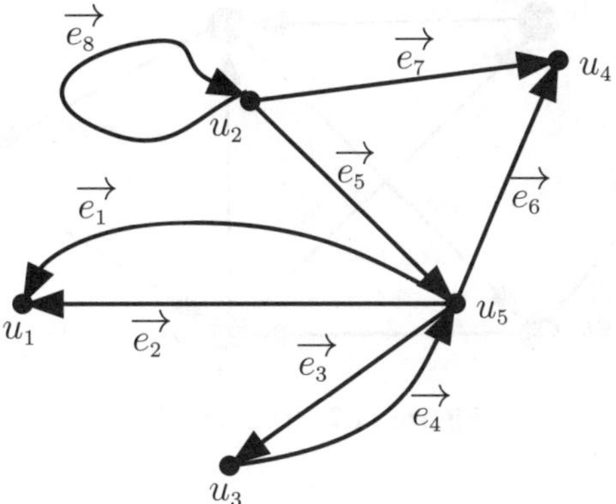

Figure 11.1   A digraph D

can understand from the notation that we are discussing a directed edge and therefore $\overrightarrow{u_3u_5}$ signifies that u_3 is the origin and u_5 is the terminus.

DEFINITION 11.1.3 Given a digraph D, we can obtain a graph G called the *underlying graph* of D by replacing each ordered pair (u,v) associated with some arc $\overrightarrow{e}$ with an edge $e = uv$. In other words, we get the underlying graph G by simply replacing arcs with edges without any direction.

DEFINITION 11.1.4 Let D be a digraph. Then a *directed walk* in D is given by $W = u_0\overrightarrow{e_1}u_1 \ldots \overrightarrow{e_k}u_k$, where the arc $\overrightarrow{e_i}$ has origin u_{i-1} and terminus u_i. The number of arcs in W is called the length of W.

DEFINITION 11.1.5 Let D be a digraph and $W = u_0\overrightarrow{e_1}u_1 \ldots \overrightarrow{e_k}u_k$ be a directed walk. Then W is called a *directed trail* if all the arcs $\overrightarrow{e_1}, \overrightarrow{e_2}, \ldots, \overrightarrow{e_k}$ are distinct.

DEFINITION 11.1.6 Let D be a digraph and $W = u_0\overrightarrow{e_1}u_1 \ldots \overrightarrow{e_k}u_k$ be a directed walk. Then W is called as a *directed path* if all the vertices $u_0, u_1, \ldots, u_k$ are distinct. Clearly every directed path is a directed trail because if the vertices are distinct then the arcs are distinct too.

DEFINITION 11.1.7 Let D be a digraph and $W = u_0\overrightarrow{e_1}u_1 \ldots \overrightarrow{e_k}u_k$ be a directed walk. Then W is called as a *directed cycle* if W is a directed path and $u_0 = u_k$. In other words, a closed directed path is called a directed cycle.

DEFINITION 11.1.8 A vertex u of the digraph D is said to be *reached* from a vertex v if there is a directed path in D from v to u.

EXAMPLE 11.1.9 Consider the digraph D as shown in Figure 11.2. Then directed walk $W_1 = u_1u_2u_5u_4u_1u_2u_3$ of length 6 is not a directed trail since the arc from u_1 to u_2 repeats

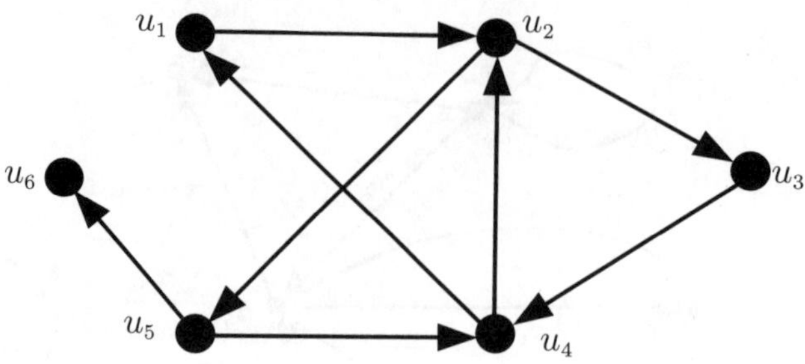

Figure 11.2 Digraph D

twice and $W_2 = u_1u_2u_3u_4u_2u_5$ is a directed trail of length 5 which is not a directed path, since the vertex u_2 is repeated twice whereas $W_3 = u_1u_2u_5u_4u_3$ is a directed path of length 4, which is not a cycle. We have a directed cycle $W_4 = u_2u_5u_4u_1u_2$ of length 4. Also the vertex u_6 is reachable from all the other vertices u_1 to u_5, but no vertex is reachable from u_6.

DEFINITION 11.1.10 Let $D_1 = (V_1, E_1)$ and $D_2 = (V_2, E_2)$ be two digraphs. Then D_1 and D_2 are said to be *isomorphic* if there is a one-to-one correspondence between V_1 and V_2 and also a one-to-one correspondence between E_1 and E_2 such that if $\vec{e_1} = (u_1, u_2)$ is an arc in D_1 and the vertices u_1 and u_2 correspond to the vertices v_1 and v_2 in D_2, then the arc $\vec{e_1} = (u_1, u_2)$ corresponds to the arc $\vec{e_2} = (v_1, v_2)$ of D_2.

EXAMPLE 11.1.11 Consider the digraphs D_1 and D_2 as shown in Figure 11.3. Then the

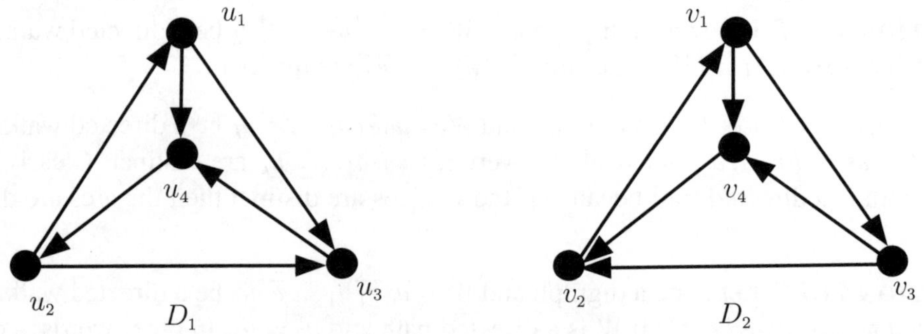

Figure 11.3 Isomorphic digraphs

digraphs D_1 and D_2 are isomorphic by the correspondence $u_1 \mapsto v_3$, $u_2 \mapsto v_1$, $u_3 \mapsto v_4$ and $u_4 \mapsto v_2$.

In digraphs, connectedness can be defined in two forms, namely weakly connected and strongly connected.

DEFINITION 11.1.12 A digraph D is called as *weakly connected* if its underlying graph G is connected.

DEFINITION 11.1.13 A digraph D is said to be *strongly connected* if for any pair of vertices u and v in D, there exists a directed path from u to v. It is sometimes called *di-connected*.

EXAMPLE 11.1.14 The digraph as shown in Figure 11.2 is weakly connected but not strongly connected as u_5 is not reachable from u_6. However, it is easy to check that the digraph $D - \{u_6\}$ is strongly connected.

DEFINITION 11.1.15 A digraph D is called *simple* if, for any pair of vertices u and v of D, there is at most one arc from u to v and there is no arc from u to itself.

EXAMPLE 11.1.16 The digraph D shown in Figure 11.2 is simple but that of Figure 11.1 is not simple as it contains more than one arc from u_5 to u_1.

Given a digraph D, we have an underlying graph G as discussed above. Similarly given a graph G, we can associate a digraph D, called the orientation of G, as we define below.

DEFINITION 11.1.17 Let G be a graph. For each edge $e = uv$ in G, consider an arc either from u to v or from v to u, but not both. Then we get a digraph D, called an *orientation* of G.

EXAMPLE 11.1.18 Consider the graph G as shown in Figure 11.4. This graph has several orientations, two of which are shown in the figure.

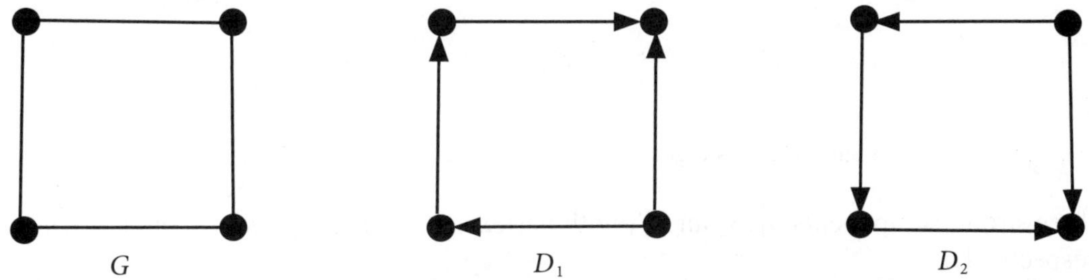

Figure 11.4 Graph G and two of its orientations D_1 and D_2

DEFINITION 11.1.19 A *subdigraph* (or subgraph for directed graphs) S of a digraph D refers to a subgraph of D that consists of some of the vertices and edges of D.

DEFINITION 11.1.20 Let D be a digraph and S be a subdigraph of D. Then S is said to be a *strong component* of D if S is strongly connected and is not a proper subdigraph of any other strongly connected subdigraph of D. In other words, S is a maximal strongly connected subdigraph.

REMARK 11.1.21 Note that if S is a strong component of a digraph D, then S consists of all those vertices of D such that any vertex is reachable from any other vertex and contains all those arcs in D which join these vertices.

EXAMPLE 11.1.22 Consider the digraph as shown in Figure 11.5. Then there are

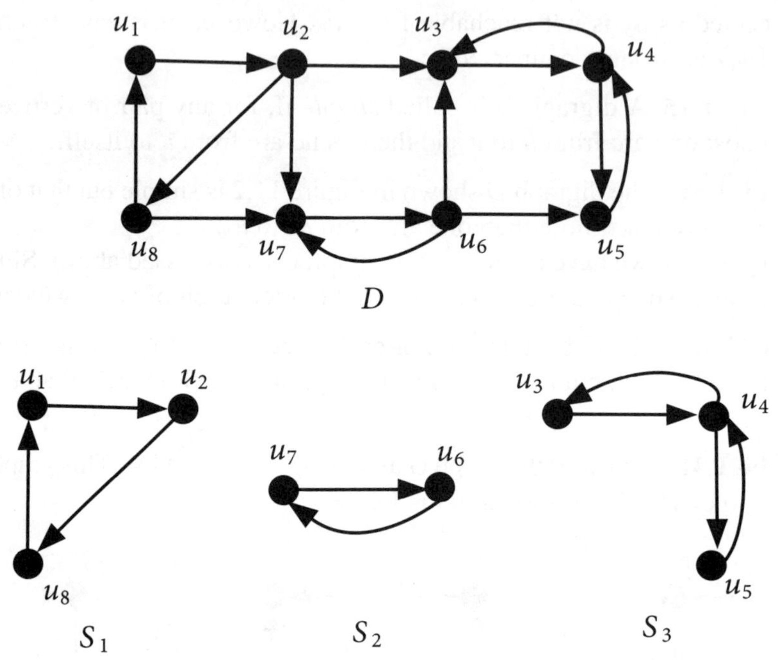

Figure 11.5   A digraph D and its strong components S_1, S_2 and S_3

three strong components S_1, S_2 and S_3 with vertex set $\{u_1, u_2, u_8\}$, $\{u_6, u_7\}$ and $\{u_3, u_4, u_5\}$ respectively.

11.2 Indegree and outdegree

In a graph G, the degree of a vertex v is the number of edges incident to v. The concept of the degree of a vertex in a digraph is slightly different due to the directions of edges. For each vertex v in D, we have indegree and outdegree of v as defined below.

DEFINITION 11.2.1 Let v be a vertex in the digraph D. The *indegree* $id(v)$ of v is defined as the number of arcs of D that have v as its head, which is the number of arcs that "go to" v.

DEFINITION 11.2.2 Let v be a vertex in the digraph D. The *outdegree $od(v)$* of v is defined as the number of arcs of D that have v as its tail, that is, the number of arcs that "go out" of v.

EXAMPLE 11.2.3 Consider the digraph D as shown in Figure 11.5. Then $id(u_1) = 1$, $id(u_2) = 1, id(u_3) = 3, id(u_4) = 2$ and $od(u_1) = 1, od(u_2) = 3, od(u_3) = 1, od(u_4) = 2$.

In the next result, we give the relation between indegree, outdegree and number of arcs in a digraph.

THEOREM 11.2.4 (The first theorem of digraph theory) *Let D be a digraph with n vertices* $\{u_1, u_2, \ldots, u_n\}$ *and q arcs. Then* $\sum_{i=1}^{n} id(u_i) = \sum_{i=1}^{n} od(u_i) = q$.

Proof. Since every arc goes to exactly one vertex and every arc is going to some vertex v, we get that

$$\sum_{i=1}^{n} id(v_i) = \text{number of arcs} = q.$$

Similarly, as each arc goes out of exactly one vertex, we get

$$\sum_{i=1}^{n} od(v_i) = q.$$

$\square$

DEFINITION 11.2.5 Let D be a weakly connected digraph. Then a directed trail W in D is said to be an *Euler trail*, if W contains all the arcs of D. A directed closed Euler trail is called the *Euler tour* of D. A digraph D is called an *Euler digraph* if it contains a directed Euler tour.

EXAMPLE 11.2.6 Consider the digraphs D_1 and D_2 as shown in Figure 11.6. Then the digraph D_1 is Euler as it contains an Euler tour $u_1u_2u_3u_4u_2u_5u_4u_1$ whereas D_2 is not Euler but it has Euler trail $v_1v_2v_3v_4v_5v_2$.

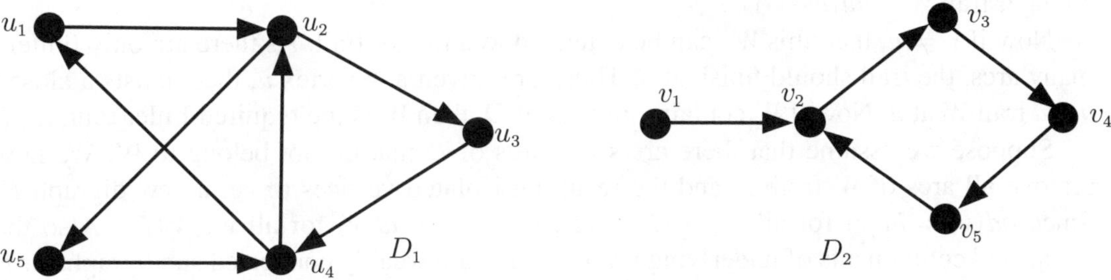

Figure 11.6 D_1 is an Euler digraph, but D_2 has a Euler trail

If we observe the above example, in D_1, $id(u) = od(u)$ for each vertex u of D_1 which is not true in D_2. So do we have a relation between Euler digraph and indegree and outdegree of its vertices? The answer is yes as proved in the following result.

THEOREM 11.2.7 *A weakly connected digraph D with at least one arc is Euler if and only if* $od(v) = id(v)$ *for each vertex v of D.*

Proof. Assume that D is Euler and let T be a directed Euler tour of D at the vertex v. If u is any vertex that appeared k times in the Euler tour T, then for each of it's occurrences, it is entered by an arc and left by a different arc. Therefore $id(u) = k = od(u)$ for all $u \in V$.

Conversely, suppose that $od(v) = id(v)$ for all $v \in V$. We prove the results by induction on q, the number of arcs in D. Let $q = 1$. If $a = (u, v), u \neq v$, then $od(u) = 1$ and $id(v) = 0$ or $od(u) = 0$ and $id(v) = 1$, which is not true in either case. Therefore the arc should be a loop. Thus the directed loop is the required Euler tour. Suppose $q = 2$. Then the only possibilities are either both arcs are loops at the same vertex or both arcs between the same two vertices with opposite directions, as shown in Figure 11.7. In both cases, we have an Euler tour.

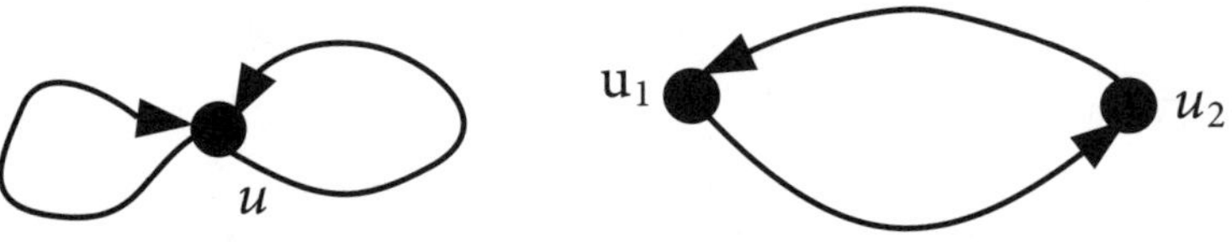

Figure 11.7　　The only Euler digraphs with two arcs

Now let $q \geq 3$, and assume the induction hypothesis for all weakly connected digraphs D' with less than q arcs and $od(v) = id(v)$ for all $v \in V(D')$. Since D is weakly connected, there is no isolated vertex in D implies that the outdegree of every vertex is non-zero. Now start with any vertex u in D. Since $od(u) > 0$, there exists a trial W' in D starting at u. If W' finishes at u, then we have a closed $u - u$ trail in D. If W' finishes at $v \neq u$, say $W' : uu_1u_2 \ldots v$, then there exists an arc going out of v which is not part of W'. So we have a longer trail $W'' : uu_1u_2 \cdots vv'$.

Now if $v' \neq u$, then this W'' can be extended to a longer trail. As there are only finitely many arcs, the trail should finish at u. Therefore, given any vertex u, there exists a closed $u - u$ trail W at u. Now if W contains all arcs of D, then W is the required Euler tour.

Suppose we assume that there are some arcs of D that do not belong to W. We now remove all arcs of W from D and the resulting isolated vertices to get a new digraph F. Since $od(v) = id(v)$ for all $v \in V(D)$, we get $od(v) = id(v)$ for all $v \in V(F)$. Also the connected components of underlying graph G of F are weakly connected subdigraphs of F having less than q arcs. Thus by induction, each subdigraph contains an Euler tour. Since D is weakly connected, each of these subgraphs has a vertex in common with W. Now by attaching the Euler tours of subdigraphs at these common vertices of W, we get an Euler tour of D. Hence D is Euler. $\square$

The next result characterizes the digraphs having Euler trail.

THEOREM 11.2.8 *Let D be a weakly connected digraph with at least two vertices. Then D has a directed Euler trail which begins at u and ends at v if and only if $od(u) = id(u) + 1$ and $id(v) = od(v) + 1$ and $od(w) = id(w)$ for all $w \neq u, v$. Furthermore, in this case the trail begins at u and ends at v.*

Proof. Suppose that D contains an Euler trail $W : uu_1u_2 \cdots u_n v$ that begins at u and ends at v. If the vertex u appears K_1 times in the Euler trail W and v appears K_2 times in the same Euler trail, then $od(u) = K_1 + 1$, $id(u) = K_1$, $od(v) = K_2$ and $id(v) = K_2 + 1$. Also, for any vertex $w \neq u, v$, in W, we have $od(w) = id(w)$. Therefore $od(u) = id(u) + 1$, $id(v) = od(v) + 1$ and $id(w) = od(w)$ for all $w \neq u, v$.

Conversely, let D be a weakly connected digraph containing vertices u & v satisfying the given condition. Now consider the digraph F as D together with a new arc from v to u. Then in F, $od(w) = id(w)$ for all $w \in F \setminus \{u, v\}$. Also in F, $id(u) = (od(u) - 1) + 1 = od(u)$. Similarly, $od(v) = (id(v) - 1) + 1 = id(v)$. Thus every vertex w in F satisfies $od(w) = id(w)$. Hence F has an Euler tour T, by Theorem 11.2.7. Then $T \setminus \{(v, u)\}$ is the required Euler trail from u to v. $\square$

11.3 Tournaments

In this section, we discuss a specific type of digraph called a tournament. The reason for this name is because this digraph is useful to record the results of games in a round-robin tournament where draws are not allowed, for example, tennis.

DEFINITION 11.3.1 An orientation of a complete graph is called a *tournament*, that is, a tournament is a digraph with no directed loops in which any two distinct vertices are joined by exactly one arc.

EXAMPLE 11.3.2 Non-isomorphic tournaments on n vertices for $n = 1, 2, 3$ are shown in Figure 11.8.

EXAMPLE 11.3.3 A tournament on 5 vertices is shown in Figure 11.9. Here any arc from u_i to u_j indicates that u_i has beaten u_j. Thus in this tournament, u_1 has beaten every other competitor, whereas u_5 lost all matches. The vertex u_3 has won three matches, u_2 has won two and u_4 has won one.

In the above example, the vertex u_1 has maximum outdegree 4 and all other vertices can be reached from u_1 through a directed path of length at most 2. This is in fact true for any tournament having a vertex with maximum outdegree.

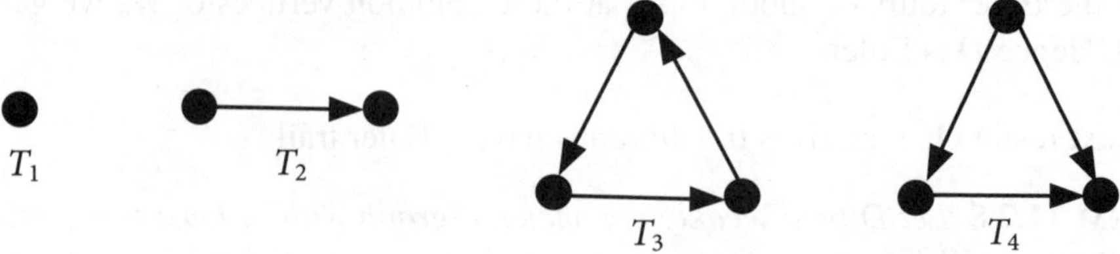

Figure 11.8   Non-isomorphic tournaments up to 3 vertices

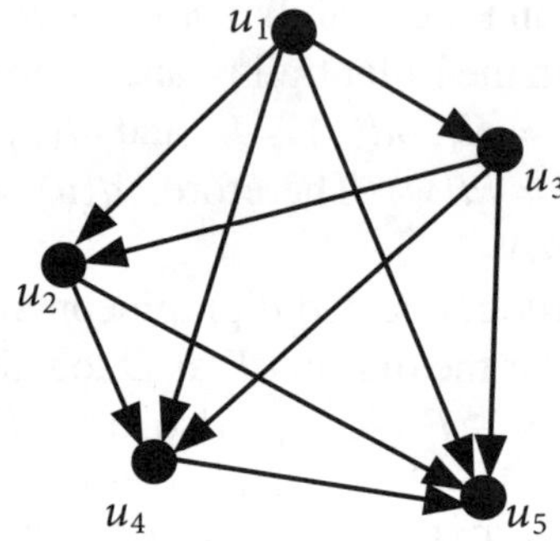

Figure 11.9   Tournament on 5 vertices

THEOREM 11.3.4 *Let T be a tournament and v be a vertex with maximum outdegree. Then for every vertex u of T, there is a directed path from v to u of length at most 2.*

Proof. Suppose T has n vertices and $od(v) = m$. Let the m arcs from v be joined to the vertices $v_1, v_2, \ldots, v_m$. Since T is a tournament, the remaining $n - m - 1$ vertices, say, $u_1, u_2, \ldots, u_{n-m-1}$ are joined to the vertex v, as shown in Figure 11.10. Then each arc from v to v_i for $1 \leq i \leq m$ gives a directed path from v to v_i of length 1. Now we show that for each $u_j, 1 \leq j \leq n - m - 1$, there is a directed path of length 2 from v to u_j.

For given vertex u_j, if there is an arc from v_i to u_j for some i, then vv_iu_j is the required directed path of length 2. Suppose there exists a vertex u_k such that there is no arc from v_i to u_k for all $i = 1, 2, \ldots m$. Since T is a tournament, there exists an arc from u_k to each v_i, for all $i = 1, 2, \ldots m$. We also have an arc from u_k to v. Therefore, the outdegree of u_k is $od(u_k) = m + 1$, a contradiction to the maximum outdegree of T. Hence for each u_j, there exists a directed path from v to u_j of length 2. $\qquad\square$

DEFINITION 11.3.5 A *directed Hamiltonian path* of a digraph D is a directed path in D that includes every vertex of D exactly once.

The next result proves that each tournament has a directed Hamiltonian path.

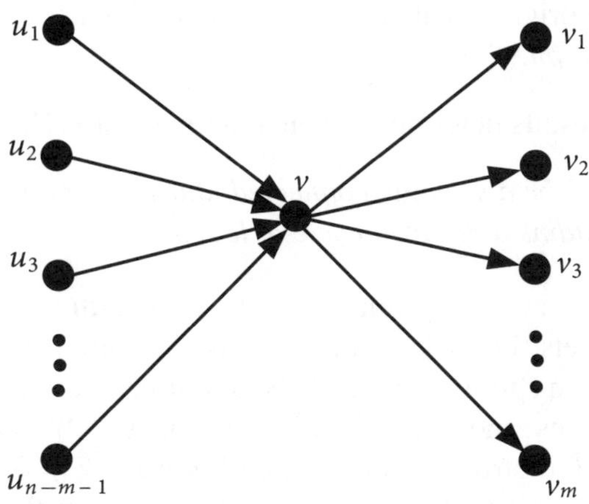

Figure 11.10   Vertex v has maximum outdegree

THEOREM 11.3.6 *Every tournament T has a directed Hamiltonian path.*

Proof. Suppose T has n vertices. We prove the result by induction on n. For $n = 1, 2$ or 3, T has a directed Hamiltonian path P as shown in Figure 11.8. So assume that $n \geq 4$ and the result is true for all tournaments with less than n vertices. Let v be a vertex of T and consider the directed graph $T - v$. Then $T - v$ is a tournament with $n - 1$ vertices, since given any $u, w \in V(T - v)$, there is exactly one arc joining u and w, the same arc as in T. Thus by induction hypothesis, there exists a directed Hamiltonian path $P = v_1 v_2 \ldots v_{n-1}$ in $T - v$. Now if there is an arc from v to v_1 or from v_{n-1} to v, then $P' = v v_1 v_2 \ldots v_{n-1}$ or $P'' = v_1 v_2 \ldots v_{n-1} v$ is the required Hamiltonian path in T. So assume that there is no arc from v to v_1 and from v_{n-1} to v. Since $T - v$ is a tournament, there exists an arc from v_1 to v. Let v_i be the last vertex on path P having an arc from v_i to v. Then there is no arc from v_{i+1} to v, which implies that there exists an arc from v to v_{i+1}. Then the path $Q = v_1 v_2 \ldots v_i v v_{i+1} v_{i+2} \ldots v_{n-1}$ is the required Hamiltonian path in T, as shown in Figure 11.11. $\square$

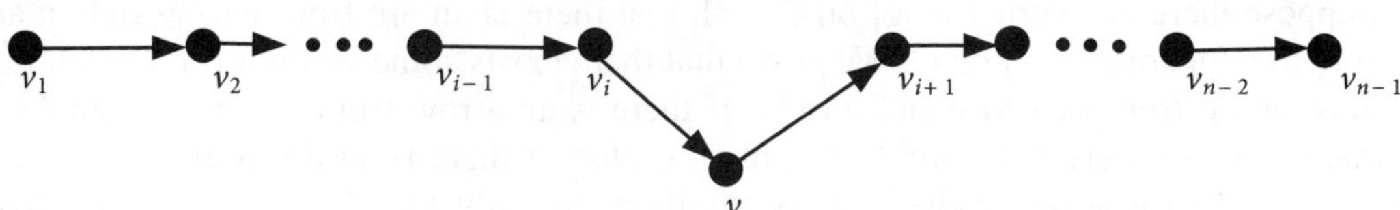

Figure 11.11   Hamiltonian path Q in T

DEFINITION 11.3.7 Let D be a digraph. A directed cycle which includes each vertex of D exactly once except the origin is called a *directed Hamiltonian cycle* and in this case D is called *Hamiltonian digraph*.

The following two results determine when a tournament is Hamiltonian.

THEOREM 11.3.8 *Let T be a strongly connected tournament on n vertices. Then for each k with $3 \leq k \leq n$, T contains a directed cycle of length k.*

Proof. First we prove that T contains a cycle of length 3. Given any vertex v of T, consider the two sets $W = \{w :$ there exists some arc from v to $w\}$ and $Z = \{z :$ there exists some arc from z to $v\}$. Since T is a tournament, there is exactly one arc between any two vertices and thus $W \cap Z = \emptyset$. (if $x \in W \cap Z$, then $v \to x \to v$, a contradiction). Also as T is strongly connected, $W \neq \emptyset$ and $Z \neq \emptyset$, for if $W = \emptyset$, then there is no way to go from v to any other vertex, which is a contradiction to T being strongly connected. Again as T is strongly connected, there exists an arc from some $w' \in W$ to some $z' \in Z$, since for $w \in W$ and $z \in Z$, there exists a directed path $ww_1 \ldots w_k z_1 \ldots z$, choose w_k as w' and z_1 as z'. Thus we have a directed cycle $vw'z'v$ of length 3. So by induction, assume

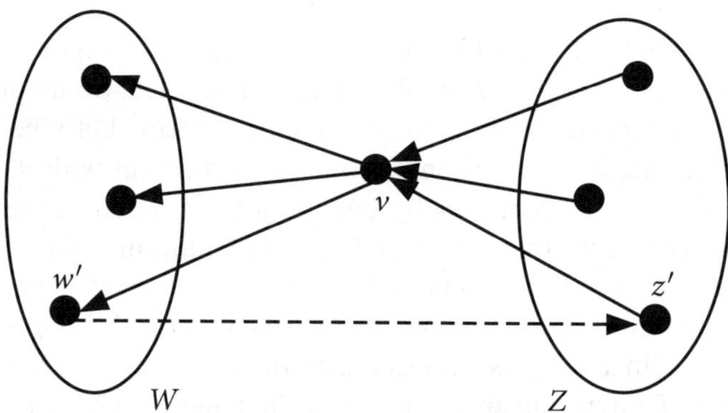

Figure 11.12 T containing a 3 cycle

that T has a directed cycle $C : v_1 v_2 \ldots v_k v_1$ of length $k < n$. We will prove that T has a cycle of length $k+1$.

Suppose there is a vertex v not on C such that there is an arc from v to v_i and an arc from v_j to v for some v_i, v_j on C. We prove that there exists some vertex v_l on C such that there is an arc from v_{l-1} to v and v to v_l. If there is an arrow from v_{l-1} to v, then $l = i$ works, otherwise there is an arc from v to v_{i-1}. Now if there is an arrow from v_{i-2} to v, take $l = i - 1$ or else repeat the process. As there are only finitely many vertices, there exists such a vertex v_l. Then the cycle $C' = v_1 v_2 \ldots v_{l-1} v v_{l+1} \ldots v_k v_1$ is a cycle of length $k+1$, as shown in Figure 11.13. So assume that no vertex outside C which satisfies this

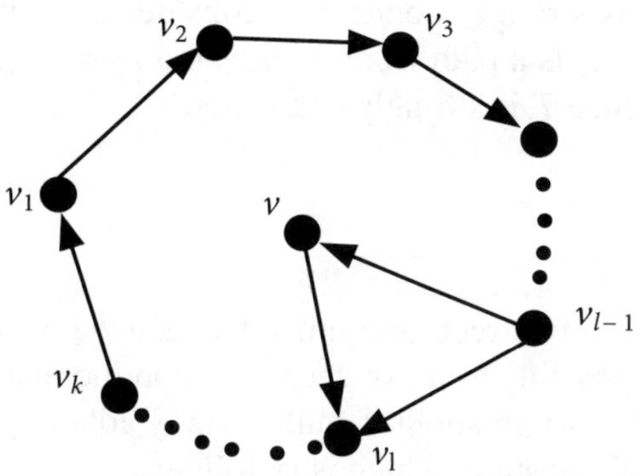

Figure 11.13   Existence of $k+1$ cycle in T

property. Then given any vertex v not on C, we have either for each i, $1 \leq i \leq k$, there is an arc from v_i to v or an arc from v to v_i. Therefore the set of vertices not belonging to C can be divided into two sets. Let $W = \{w \notin C : \text{for each } i, \text{ there exists an arc from } v_i \text{ to } w\}$ and $Z = \{z \notin C : \text{ for each } i, \text{ there exists an arc from } z \text{ to } v_i\}$. Then as in the case for $n = 3$, we have $W \cap Z = \emptyset$, $W \neq \emptyset$ and $Z \neq \emptyset$. Since T is strongly connected, there exists an arc from some $w' \in W$ to some $z' \in Z$. Then we have a cycle $C' = v_1 w' z' v_3 v_4 \ldots v_k v_1$ of length $k+1$, as shown in Figure 11.14. Hence by induction, there exists a cycle of length $n \geq 3$. $\square$

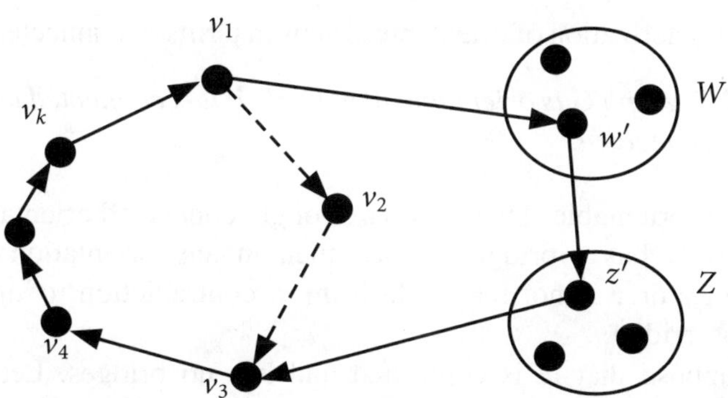

Figure 11.14   Existence of $k+1$ cycle in T

COROLLARY 11.3.9 *A tournament T is Hamiltonian if and only if it is strongly connected.*

Proof. Suppose T has n vertices. If T is strongly connected, then, by Theorem 11.3.8, T must have a directed cycle of length n. Such a cycle is a directed Hamiltonian cycle as it includes each vertex of T. Hence T is Hamiltonian.

Conversely, suppose T is Hamiltonian with a directed Hamiltonian cycle $C = v_1 v_2 \ldots v_n v_1$. To show T is strongly connected, consider any two vertices v_i and v_j with $i > j$. Then $P_1 = v_j v_{j+1} \ldots v_i$ is a path from v_j to v_i and $P_2 = v_i v_{i+1} \cdots v_{n-1} v_n v_1 v_2 \cdots v_{j-1} v_j$ is a path from v_i to v_j. Hence T is strongly connected. $\qquad\square$

11.4 Traffic flow

Cities frequently use one-way street assignments to alleviate traffic congestion. Given a street map of a city, is it possible to make each street on the map one-way so that one can travel from one part of the city to another while strictly adhering to traffic rules? This can be rewritten as a graphical question, which is as follows:

We can consider a graph G with vertex being a street intersection. Two vertices x and y of G will be joined by an edge if one can reach from x to y without any vertex in between x and y. Then the resulting graph G is a street map of the city.

Now making the streets a one-way is nothing but assigning a direction to each of the edges of G, namely, orienting each edge of G. By doing this, we get a digraph D, an orientation of G. So now the original question is the same as asking, "Can we find a digraph D in which every vertex u is reachable from every other vertex v?"

DEFINITION 11.4.1 A *strongly connected orientation* of a digraph is an assignment of directions to the edges of an undirected graph such that the resulting directed graph is strongly connected. A graph G is called *orientable* if it has a strongly connected orientation.

We give the characterization of orientable graphs in terms of connectedness and bridges.

THEOREM 11.4.2 *A graph G is orientable if and only if it is connected and has no bridges, that is G is 2-edge connected.*

Proof. Suppose G is orientable. Then G has a strongly connected orientation, in particular, G is connected. If G has a bridge $e = uv$, then in any orientation of G either u is not reachable from v or v is not reachable from u, contradiction to strongly connected. Therefore G has no bridge.

Conversely, suppose that G is connected and has no bridges. Let $H = \langle \{v\} \rangle$ be a subgraph of G induced by a single vertex v. Then H is orientable since if v contains a loop, then the corresponding arcs make H strongly connected. So we can choose a largest possible subset $U \subseteq V$ such that $H = \langle U \rangle$ is orientable. If $H = G$, then G is orientable. If $H \neq G$, then $U \not\subseteq V$ implies there exists $v \in V$ but $v \notin U$. Since H is orientable, the orientation of the edges of H gives a strongly connected digraph D. Since G is connected with no bridges, there are two edge disjoint paths from u to v in G, say $P = u_0 u_1 \ldots u_{m-1} v$ and $P' = v_0 v_1 \ldots v_{n-1} v$ with $u_0 = v_0 = u$. Since $u \in H$ but $v \notin H$, there exists a vertex $u_i \in P$ which is the last vertex in P belonging to H. Similarly, let $v_j \in P'$ be the last vertex in P'

belonging to H. Now consider the following two paths

$$Q = u_i u_{i+1} \cdots u_{m-1} v \text{ and } Q' = v_j v_{j+1} \cdots v_{n-1} v.$$

Orient the edges in Q as $u_i \to u_{i+1} \to u_{i+2} \to \ldots \to v$ and orient the edges in Q' as $v \to v_{n-1} \to v_{n_2} \to \ldots \to v_j$, as shown in Figure 11.15.

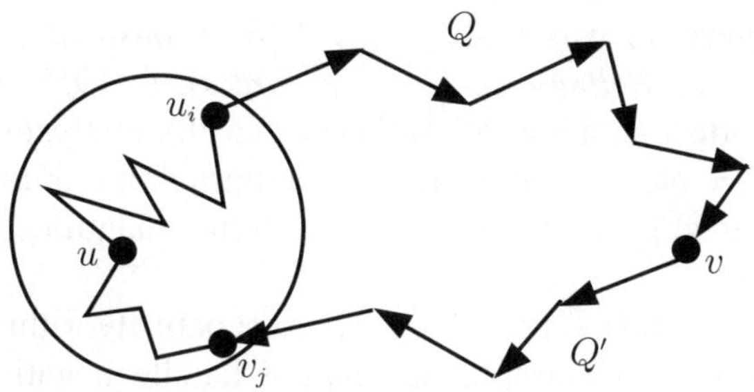

Figure 11.15   Orientation of paths Q and Q'

Now consider the digraph D' induced by the vertex set

$$U' = U \cup \{u_i, u_{i+1}, \ldots, u_{m-1}, v\} \cup \{v_j, v_{j+1}, \ldots, v_{n-1}\}.$$

To prove D' is strongly connected, let x and y be two vertices in D'. Then we have the following cases:

(i.) If $x, y \in U$, since D is strongly connected, there exists a directed path from x to y.

(ii.) If $x \in U$ and $y = u_k$ for some k, then there exists a directed path $x \to u_j \to u_k$.

(iii.) If $x \in U$ and $y = v_r$ for some r, then there exists a directed path $x \to u_j \to v \to v_r$.

(iv.) If $x = u_k$ and $y = v_r$, then there exists a directed path $u_k \to v \to v_r$.

Thus D' is strongly connected and $U' \supsetneq U$, a contradiction to the maximality of U. Hence $H = G$ implies G is orientable.

$\square$

The above theorem guarantees the existence of strongly connected orientation for 2-edge connected graphs.

11.5 Traversal in a directed graph

After being introduced to a directed graph, one usually wonders about the ways in which one can traverse across the vertices and the arcs of the directed graph or digraph. We are naturally led to the question of how shortest paths can be discovered between vertices of a digraph. To that end, we introduce the reader to two algorithms, one of which is a single-source shortest path algorithm, much like Dijkstra's algorithm and another algorithm which helps to find the shortest path between any two vertices of a digraph.

11.5.1 Bellman–Ford algorithm

The Bellman–Ford algorithm was first proposed by *Alfonso Shimbel* in 1955, but was later published by *Richard Bellman* and *Lester Ford Jr.* in 1958 and 1956 respectively. It is a single-source shortest path algorithm that works efficiently, even with negative edge weights or arc weights. Note that *arc weights* in a digraph are numerical values assigned to the directed edges (arcs) between nodes, representing quantities such as distance, cost, capacity or other metrics.

We are familiar with another single-source shortest path algorithm, Dijkstra's algorithm that also works on a weighted graph, but cannot handle negative edge weights. The Bellman–Ford algorithm works in a similar manner to Dijkstra's algorithm except for some key points. A relaxation step is run $|V(G)| - 1$ times in which the distance of each vertex from the source vertex is progressively improved until a solution emerges. The Bellman–Ford algorithm also helps to identify negative-weight cycles in a graph. A directed cycle whose arc weights sum to a negative weight is called a *negative-weight cycle*.

The Bellman–Ford algorithm follows the steps:

Step 1: For a directed weighted graph G with weights $w(u,v)$ for each arc, we initialize the distance function for all vertices in $V(G)$ except the source vertex s as ∞. The source vertex is initialized with a distance function of 0, that is $dist(v) = \infty$ if $v \neq s$, $dist(s) = 0$.

Step 2: For each arc $\vec{e} = \vec{uv} \in E(G)$ if $dist(u) + w(u,v) < dist(v)$, update $dist(v)$ to $dist(u) + w(u,v)$. Repeat this step $|V(G)| - 1$ times.

Step 3: For each arc $\vec{e} = \vec{uv} \in E(G)$ if $dist(u) + w(u,v) < dist(v)$, state that a negative cycle is present and that a shortest path cannot be found.

Step 3 is the same as Step 2 except that we do it more than $|V(G)| - 1$ times. If a relaxation is possible even after $|V(G)| - 1$ times, then it is an indicator of a negative-weight cycle.

The pseudocode for the Bellman–Ford algorithm is written to ensure that the relaxation step occurs $|V(G)| - 1$ times and during each relaxation, the distance function is updated if it meets the stipulated condition.

The main operation of this algorithm is the relaxation step where the algorithm iterates $|V(G)|$-1 times. For each edge e, the algorithm checks if $dist[u] + w(u,v) < dist[v]$ and updates $dist[v]$ and $pred[v]$ if the condition is met. Since this nested loop iterates $|V(G)|$-1 times each time checking $|E(G)|$ edges, the time complexity for the relaxation step is

---

Algorithm 1: Bellman–Ford Algorithm

Data: A weighted digraph $G = (V(G), E(G))$, source vertex s
with $V(G) = \{v_1, v_2 \ldots v_n\}$ and weights of edges $w(u, v)$ for all pairs of vertices.
Result: $dist[\,]$, an array in which each entry represents the distance/weight of the
        shortest path from s, an array $pred[\,]$ which stores the predecessor of v in
        the shortest path from s to v

Initialization:
for (*Each vertex v in* $V(G)$) {

    $dist[v] = \infty$;   /* Set initial distances to infinity */
    $pred[v] = 0$;   /* No predecessors initially */
    $dist[s] = 0$;   /* Distance to the source is 0 */

}

Relaxation step:
for ($i = 1$; $i < |V(G)| - 1$; $i = i + 1$) {

    for (*each edge* $\vec{e} = \vec{uv} \in E(G)$) {

        if $dist[u] + w(u, v) < dist[v]$ **then**
            $dist[v] = dist[u] + w(u, v)$
            $pred[v] = u$
        end

    }

}

Checking for negative cycles:
for (*each edge* $\vec{e} = \vec{uv} \in E(G)$) {

    if $dist[u] + w(u, v) < dist[v]$ **then**
        return Graph contains a negative-weight cycle
    end

}
return $dist[\,], pred[\,]$

$O(|V(G)| - 1) \times |E(G)| \approx O(|V(G)| \times |E(G)|)$. In the negative cycle checking step, the algorithm makes a pass over each edge to detect negative-weight cycles. Hence the overall complexity of Bellman–Ford algorithm is $O(|V(G)|.|E(G)|)$.

The correctness of the Bellman–Ford Algorithm can be stated as follows: "Let the Algorithm 1 be run on a weighted directed graph G with vertex set $V(G)$ and edge set $E(G)$ with source s and weight function $w : E \to R$. If G contains no negative-weight cycles that are reachable from s then the algorithm returns $TRUE$ and the $dist(v)$ function returns the distance of the vertex v from the source s. If G contains a negative-weight cycle reachable from s, then the algorithm returns $FALSE$". For the proof of this theorem, the reader may refer to [9, Theorem 24.4].

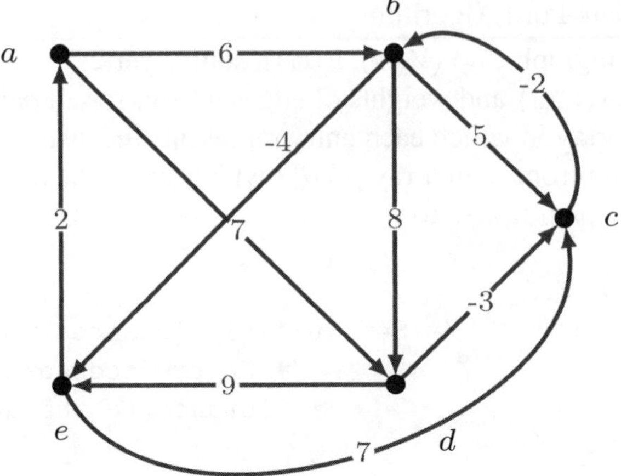

Figure 11.16   A directed weighted graph G

EXAMPLE 11.5.1 Determine the shortest path from a to all other vertices of the weighted graph G in Figure 11.16 [15].

In this graph, the distance function is initialized as $dist(a) = 0$ while the distances of all other vertices from a are ∞.

Initialization: We initialize the distances of all the vertices in this step.

Table 11.1   Initialization of $dist(v)$

Vertex	**Distance** $dist(v)$
a	0
b	∞
c	∞
d	∞
e	∞

Iteration 1: The algorithm checks if $dist(u) + w(u,v) < dist(v)$ for each edge $\vec{uv}$ and updates $dist(v)$ to the value of $dist(u) + w(u,v)$ if it is true. The value of $dist(v)$ does not change if the statement is false.

Table 11.2   Iteration 1 for directed graph G

Edge	Distance $dist(v)$
$\vec{ab}$	$dist(b) = min(\infty, 0+6) = 6$
$\vec{ad}$	$dist(d) = min(\infty, 0+7) = 7$
$\vec{bc}$	$dist(c) = min(\infty, 6+5) = 11$
$\vec{bd}$	$dist(d) = min(7, 6+8) = 7$
$\vec{be}$	$dist(e) = min(\infty, 6-4) = 2$
$\vec{cb}$	$dist(b) = min(6, 11-2) = 6$
$\vec{dc}$	$dist(c) = min(11, 7-3) = 4$
$\vec{de}$	$dist(e) = min(2, 7+9) = 2$
$\vec{ea}$	$dist(a) = min(0, 2+2) = 0$
$\vec{ec}$	$dist(c) = min(4, 2+7) = 4$

Table 11.3 $dist(v)$ after Iteration 1

Vertex	Distance $dist(v)$
a	0
b	6
c	4
d	7
e	2

Iteration 2: The iteration must run $|V(G)| - 1$ times, which is 4 times, since there are 5 vertices in this graph.

Table 11.4   Iteration 2 for directed graph G

Edge	Distance $dist(v)$
$\overrightarrow{ab}$	$dist(b) = min(6, 0+6) = 6$
$\overrightarrow{ad}$	$dist(d) = min(7, 0+7) = 7$
$\overrightarrow{bc}$	$dist(c) = min(4, 6+5) = 4$
$\overrightarrow{bd}$	$dist(d) = min(7, 6+8) = 7$
$\overrightarrow{be}$	$dist(e) = min(2, 6-4) = 2$
$\overrightarrow{cb}$	$dist(b) = min(6, 4-2) = 2$
$\overrightarrow{dc}$	$dist(c) = min(4, 7-3) = 4$
$\overrightarrow{de}$	$dist(e) = min(2, 7+9) = 2$
$\overrightarrow{ea}$	$dist(a) = min(0, 2+2) = 0$
$\overrightarrow{ec}$	$dist(c) = min(4, 2+7) = 4$

Table 11.5 $dist(v)$ after Iteration 2

Vertex	Distance $dist(v)$
a	0
b	2
c	4
d	7
e	2

Iteration 3: We can see that the distance function has been updated for some of these vertices, while the others remain the same. We will run through another iteration to see if there is any change.

Table 11.6   Iteration 3 for directed graph G

Edge	Distance $dist(v)$
$\overrightarrow{ab}$	$dist(b) = min(2, 0+6) = 2$
$\overrightarrow{ad}$	$dist(d) = min(7, 0+7) = 7$
$\overrightarrow{bc}$	$dist(c) = min(4, 2+5) = 4$
$\overrightarrow{bd}$	$dist(d) = min(7, 2+8) = 7$
$\overrightarrow{be}$	$dist(e) = min(2, 2-4) = -2$
$\overrightarrow{cb}$	$dist(b) = min(2, 4-2) = 2$
$\overrightarrow{dc}$	$dist(c) = min(4, 7-3) = 4$
$\overrightarrow{de}$	$dist(e) = min(-2, 7+9) = -2$
$\overrightarrow{ea}$	$dist(a) = min(0, -2+2) = 0$
$\overrightarrow{ec}$	$dist(c) = min(4, -2+7) = 4$

Table 11.7 $dist(v)$ after Iteration 3

Vertex	Distance $dist(v)$
a	0
b	2
c	4
d	7
e	-2

Iteration 4: Most values for the distance function of the vertices should have settled by now. There is not likely to be much change in the last iteration. Still we will do the fourth iteration to finalize the distance of each vertex from a.

Table 11.8   Iteration 4 for directed graph G

Edge	Distance $dist(v)$
$\vec{ab}$	$dist(b) = min(2, 0+6) = 2$
$\vec{ad}$	$dist(d) = min(7, 0+7) = 7$
$\vec{bc}$	$dist(c) = min(4, 2+5) = 4$
$\vec{bd}$	$dist(d) = min(7, 2+8) = 7$
$\vec{be}$	$dist(e) = min(-2, 2-4) = -2$
$\vec{cb}$	$dist(b) = min(2, 4-2) = 2$
$\vec{dc}$	$dist(c) = min(4, 7-3) = 4$
$\vec{de}$	$dist(e) = min(-2, 7+9) = -2$
$\vec{ea}$	$dist(a) = min(0, -2+2) = 0$
$\vec{ec}$	$dist(c) = min(4, -2+7) = 4$

Table 11.9 $dist(v)$ after Iteration 4

Vertex	Distance $dist(v)$
a	0
b	2
c	4
d	7
e	-2

Iteration 5: Usually 4 iterations are sufficient to calculate the distance of the each vertex from the source vertex a. But to check for negative-weight cycles, we must iterate once more to see if the values change.

Table 11.10   Iteration 5 for directed graph G

Edge	Distance $dist(v)$
$\vec{ab}$	$dist(b) = min(2, 0+6) = 2$
$\vec{ad}$	$dist(d) = min(7, 0+7) = 7$
$\vec{bc}$	$dist(c) = min(4, 2+5) = 4$
$\vec{bd}$	$dist(d) = min(7, 2+8) = 7$
$\vec{be}$	$dist(e) = min(-2, 2-4) = -2$
$\vec{cb}$	$dist(b) = min(2, 4-2) = 2$
$\vec{dc}$	$dist(c) = min(4, 7-3) = 4$
$\vec{de}$	$dist(e) = min(-2, 7+9) = -2$
$\vec{ea}$	$dist(a) = min(0, -2+2) = 0$
$\vec{ec}$	$dist(c) = min(4, -2+7) - 4$

Table 11.11 $dist(v)$ after Iteration 5

Vertex	Distance $dist(v)$
a	0
b	2
c	4
d	7
e	-2

Since the values do not change between Iteration 4 and 5, we can safely conclude that there are no negative cycles and the distances given by Table 11.11 are indeed the required shortest distance from a to other vertices.

EXAMPLE 11.5.2 Determine the shortest path from a to all other vertices of the weighted graph G in Figure 11.17. Note that while the graph is similar to Figure 11.16, the weights have changed.

The reader can try to solve this problem. Since we have 5 vertices, we must do 4 iterations to find the distances of each vertex from the source vertex a. We will arrive at the following table after 4 iterations.(Verify!)

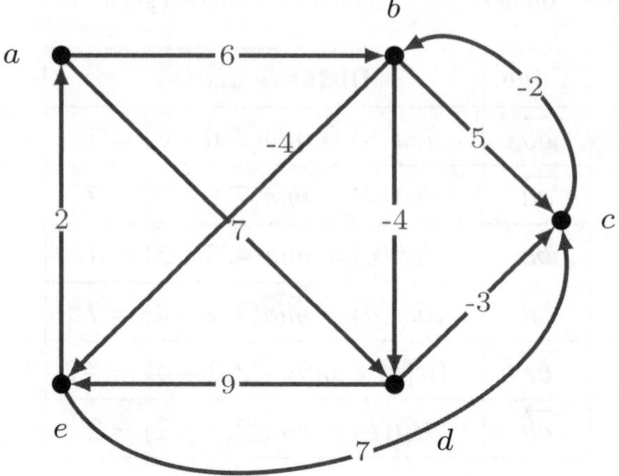

Figure 11.17   A directed weighted graph G

Table 11.12 $dist(v)$ after Iteration 4

Vertex	Distance $dist(v)$
a	0
b	-12
c	-10
d	-7
e	-7

Let us iterate one more time to check for a negative cycle.

Table 11.13 $dist(v)$ after Iteration 5

Vertex	Distance $dist(v)$
a	0
b	-12
c	-19
d	-16
e	-16

We can see that there is a constant change in the distance values with regard to the arc $\overrightarrow{bd}$. In other words, if we continue iterating, we will be getting newer values that never stabilize. This is due to the fact that $\overrightarrow{bd}$, $\overrightarrow{dc}$, $\overrightarrow{cb}$ form a negative-weight cycle and the Bellman–Ford algorithm fails to deliver the measure of distance for the vertices b, d, c from a in this graph.

NOTE 11.5.3 It is important to note that the distance function or $dist(u)$ of the source vertex is always 0. While iterating, we may get a negative value for $dist(u)$, which in theory is less than 0, but we should not update the distance function of the source vertex to such values. The reasoning behind this rule is that the distance of the source vertex to itself is always 0.

11.5.2 Floyd–Warshall algorithm

We will introduce the reader to an all-pairs-shortest-path algorithm, which as the name itself suggests, provides the length of the shortest path or the total weight of the shortest path between all the pairs of vertices in a graph. This algorithm in its present form was published by an American computer scientist *Robert Floyd* in 1962. In the same year, another American computer scientist *Stephen Warshall* published this algorithm to determine the transitive closure of a graph. Hence, this algorithm is named after both of them, although there were similar algorithms, such as *Kleene's algorithm* which followed the same principles but was published much earlier in 1956. The Floyd–Warshall algorithm works on undirected and unweighted graphs but we will mainly use it for directed graphs. It can also work efficiently on weighted directed graphs allowing for negative weights. But we will assume that negative-weight cycles are not present in the weighted digraph.

The Floyd-Warshall algorithm uses dynamic programming formulation to arrive at the desired result. Dynamic programming is a powerful algorithmic technique that is used to solve complex problems by breaking them down into simpler overlapping subproblems. If we can construct an optimal solution to the main problem by constructing optimal solutions to the subproblems, then the problem becomes suitable for dynamic programming. Floyd–Warshall algorithm is an excellent candidate for dynamic programming because we will use recursion at each stage to obtain an all-pairs-shortest-path matrix. *Recursion* is a method where a process calls itself, either directly.

The steps of the algorithm are as follows:

Step 1: For each vertex pair, we initialize a distance matrix $D[i][j]$ with the weights of each edge $\overrightarrow{ij}$ as d_{ij} where $i \neq j$, i.e., if v_i and v_j are vertices of the weighted digraph with an edge $\overrightarrow{ij}$ between them, then the weight of the edge $\overrightarrow{ij}$ is updated as d_{ij} in $D[i][j]$ (We deliberately use edge $\overrightarrow{ij}$ instead of $\overrightarrow{v_i v_j}$ for simplicity). The distance matrix entry d_{ij} is 0, if $i = j$ (there are no self-loops). If an edge $\overrightarrow{ij}$ does not exist between the vertex pair (v_i, v_j), $d_{ij} = \infty$.

Step 2: An intermediate vertex k where $k \neq i$ and $k \neq j$ is introduced between each pair of vertices (v_i, v_j) and the value of d_{ij} is updated to the weight $d_{ik} + d_{kj}$ if it is less than the original value of d_{ij}.

In other words, $D[i][j] = min(D[i][j], D[i][k] + D[k][j])$

Step 3: Repeat Step 2 for all values of k.

The distance matrix keeps changing at each iteration when we introduce a new intermediate vertex k. To keep track of these different matrices, the initialized distance matrix is $D^0[i][j]$ and the subsequent distance matrices are $D^1[i][j], D^2[i][j] \ldots$ until we exhaust the list of intermediate vertices.

EXAMPLE 11.5.4 Determine the shortest path of all pairs of vertices in the weighted digraph G given in Figure 11.18 [16].

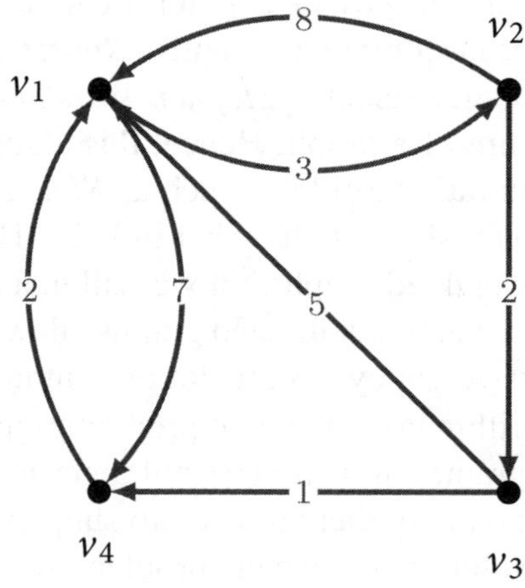

Figure 11.18   A directed weighted graph G

We have not given any negative weights in the graph, although Floyd–Warshall would work just as efficiently in the presence of negative weights.

Initialization: We will initialize the distance matrix $D^0[i][j]$ of all the vertices in graph G.

$$\mathbf{D^0[i][j]} = \begin{array}{c} \\ v_1 \\ v_2 \\ v_3 \\ v_4 \end{array} \begin{array}{cccc} v_1 & v_2 & v_3 & v_4 \\ \left(\begin{array}{cccc} 0 & 3 & \infty & 7 \\ 8 & 0 & 2 & \infty \\ 5 & \infty & 0 & 1 \\ 2 & \infty & \infty & 0 \end{array} \right) \end{array}$$

Since there are no edges between the pairs of vertices $(v_1, v_3), (v_2, v_4), (v_3, v_2), (v_4, v_2)$ and (v_4, v_3) in the graph, the values of d_{ij} corresponding to these edges in the distance matrix is ∞.

Iteration 1: We must now introduce intermediate vertices to re-calculate the values of d_{ij}. Let us introduce the intermediate vertex v_1. Since v_1 cannot be introduced into paths or edges containing v_1 as the start or end vertex such as (v_1, v_2) or (v_4, v_1), we can easily retain all the values of the row and column corresponding to v_1. $D^1[i][j]$ is initially constructed as follows:

$$\mathbf{D^1[i][j]} = \begin{array}{c} \\ v_1 \\ v_2 \\ v_3 \\ v_4 \end{array} \begin{array}{cccc} v_1 & v_2 & v_3 & v_4 \\ \left(\begin{array}{cccc} 0 & 3 & \infty & 7 \\ 8 & 0 & - & - \\ 5 & - & 0 & - \\ 2 & - & - & 0 \end{array}\right) \end{array}$$

The zeroes in the main diagonal of the matrix are also retained since there are no self-loops. Let us start with (v_2, v_3). After the introduction of the intermediate vertex v_1, we see that $d_{21} = 8$ and $d_{13} = \infty$ in $D^0[i][j]$. By step 2 of the algorithm, $d_{23} = d_{21} + d_{13} = 8 + \infty = \infty$. Clearly this value is greater than the original value of $d_{23} = 2$ in $D^0[i][j]$ and since we aim for the minimum of the two, the value of d_{23} is retained as 2 in $D^1[i][j]$.

Let us consider (v_2, v_4). After the introduction of the intermediate vertex v_1, we can see that $d_{21} = 8$ and $d_{14} = 7$ in $D^0[i][j]$. Hence $d_{24} = d_{21} + d_{14} = 8 + 7 = 15$ is found to be lesser than the original value of ∞ in $D^0[i][j]$. Hence d_{24} is updated to 15 in $D^1[i][j]$. In a similar manner the intermediate vertex v_1 is introduced for (v_3, v_2), $(v_3, v_4), (v_4, v_2)$ and (v_4, v_3) and the distance values are calculated. The distance matrix $D^1[i][j]$ is found to be:

$$\mathbf{D^1[i][j]} = \begin{array}{c} \\ v_1 \\ v_2 \\ v_3 \\ v_4 \end{array} \begin{array}{cccc} v_1 & v_2 & v_3 & v_4 \\ \left(\begin{array}{cccc} 0 & 3 & \infty & 7 \\ 8 & 0 & 2 & 15 \\ 5 & 8 & 0 & 1 \\ 2 & 5 & \infty & 0 \end{array}\right) \end{array}$$

Iteration 2: We will introduce the intermediate vertex v_2. As outlined in Iteration 1, the values of d_{ij} in the second row and second column remain unchanged. The lead diagonal values are 0. In each case, we will compare the values with that of $D^1[i][j]$. Following usual methods, the distance matrix $D^2[i][j]$ is found to be as follows:

$$
\mathbf{D^2[i][j]} = \begin{array}{c} \\ v_1 \\ v_2 \\ v_3 \\ v_4 \end{array}
\begin{array}{cccc}
v_1 & v_2 & v_3 & v_4 \\
\left(\begin{array}{cccc}
0 & 3 & 5 & 7 \\
8 & 0 & 2 & 15 \\
5 & 8 & 0 & 1 \\
2 & 5 & 7 & 0
\end{array}\right)
\end{array}
$$

Iteration 3: In this iteration, the intermediate vertex v_3 is introduced and the distance matrix is updated with new d_{ij} if it is less than the existing d_{ij} values found in $D^2[i][j]$. We do not disturb the values of the lead diagonal nor the third row and column. The distance matrix $D^3[i][j]$ is as follows:

$$
\mathbf{D^3[i][j]} = \begin{array}{c} \\ v_1 \\ v_2 \\ v_3 \\ v_4 \end{array}
\begin{array}{cccc}
v_1 & v_2 & v_3 & v_4 \\
\left(\begin{array}{cccc}
0 & 3 & 5 & 6 \\
7 & 0 & 2 & 3 \\
5 & 8 & 0 & 1 \\
2 & 5 & 7 & 0
\end{array}\right)
\end{array}
$$

Iteration 4: The intermediate vertex is v_4 in this iteration and we follow the same steps to update the distance matrix.

$$
\mathbf{D^4[i][j]} = \begin{array}{c} \\ v_1 \\ v_2 \\ v_3 \\ v_4 \end{array}
\begin{array}{cccc}
v_1 & v_2 & v_3 & v_4 \\
\left(\begin{array}{cccc}
0 & 3 & 5 & 6 \\
5 & 0 & 2 & 3 \\
3 & 6 & 0 & 1 \\
2 & 5 & 7 & 0
\end{array}\right)
\end{array}
$$

Since there are no more vertices to be introduced as intermediate vertex, the algorithm terminates and $D^4[i][j]$ gives the length of the shortest path between all pairs of vertices in the digraph G of Figure 11.18.

The pseudocode for the Floyd–Warshall algorithm is easy to write if one has identified that there are three *nested for-loops* involved in the iterations. In fact, this was formulated by *Peter Inger* in 1962 and led to more clarity on the execution of the algorithm. Since there are triply nested for-loops in the Floyd–Warshall algorithm, for the variables k, i, j it is easy to conclude that the complexity of the Floyd–Warshall algorithm is $O(n^3)$, where n is the number of vertices in G or alternatively the number of rows of $D[i][j]$.

Algorithm 2: Floyd–Warshall Algorithm

Data: A 2-dimensional array $D[i][j]$
with each entry d_{ij} representing the weight of the edge between vertices v_i and v_j, provided $i \neq j$. If $i = j$, d_{ij} is initialized as 0 and if there is no edge between v_i and v_j, $d_{ij} = \infty$

Result: $D[i][j]$, an array in which each entry represents the distance/weight of the
shortest path

```
n ← |V(G)|
for ( k = 1; k ≤ n; k = k + 1 ) {
    for ( i = 1; i ≤ n; i = i + 1 ) {
        for ( j = 1; j ≤ n; j = j + 1 ) {
            | D[i][j] = min(D[i][j], D[i][k] + D[k][j])
        }
    }
}
```

11.6 Networks

11.6.1 Introduction to networks

We are aware, by now, that a lot of real life situations can be modeled using networks. One good example would be the transportation networks. Let us suppose we are running an e-commerce company that connects various businesses with consumers. Everyday we deliver a variety of goods directly from businesses to customer's doorstep. To streamline this, we have warehouses in major cities where goods are received, sorted and stored before being dispatched across the city based on online orders. A dedicated collection team picks up goods from the businesses and brings them to the warehouses for further distribution.

For instance, walnuts grown in Northern India, in Himachal Pradesh, might be transported over long distances to a warehouse in a southern city like Chennai. However each delivery route has a maximum load capacity. When this limit is exceeded, the collection team becomes overloaded, leading to potential waste and loss. There are different ways to transport goods from the businesses. For example, a truck carrying walnuts might travel as far as Maharashtra, where the load is transferred to two smaller trucks for the journey to Chennai. This creates a network of routes and transport methods to move goods from one point to another.

This logistics challenge can be visualized as a directed graph with one point s as the farm pickup location and another point t as the warehouse. Each route has a maximum load capacity, representing the maximum amount it can handle. As owners of this e-commerce business, we would certainly be interested in determining the maximum amount of goods that can be transported from s to t through this network without overloading any route. We can imagine our network to look something like the Figure 11.19.

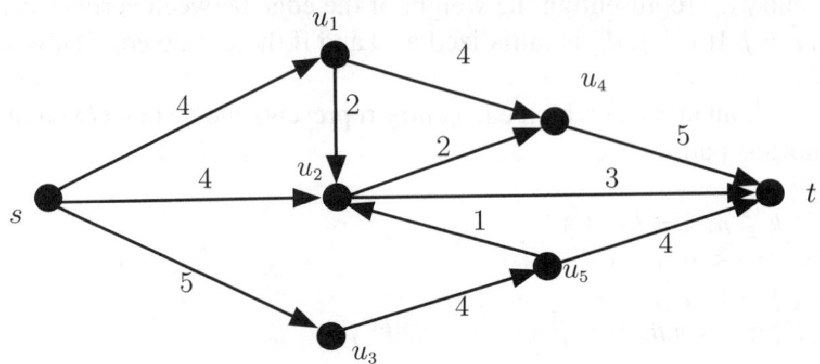

Figure 11.19   E-commerce network

Before formally defining a network, consider how the e-commerce model offers a clear picture of how goods move through various channels. This model can also apply to other scenarios such as a drip irrigation system of a large farm. Here each pipeline has a maximum capacity and must be regulated to ensure even water distribution across the farm.

Another example is arranging flights for a large group of guests to attend a destination wedding at a scenic island location far from their home. Planning the logistics involves managing each flight's capacity and routes to transport guests efficiently. Determining the maximum number of people who can reach the island and the routes they should take is a problem that can also be modeled as a network.

In general, a network can be represented as a weighted digraph $D(V,E)$ with two key points: the source and the sink. Each edge e has a non-negative integer capacity $c(e)$ representing its maximum load. The directed graph $D(V,E)$ is the underlying structure of the network N, if $e = \vec{uv}$ is an arc in N, then $c(e) = c(\vec{uv})$ is its capacity. For example in Figure 11.19, the capacity $c(e)$ of the edge $\vec{u_3u_5}$ is 4. Armed with this understanding, we may now formally define the network and allied definitions as follows:

DEFINITION 11.6.1 A *network* $N(V,A)$ is a weakly connected simple digraph consisting of vertices of V and arcs of A in which every arc $\vec{a}$ of N has been assigned a non-negative integer $c(a)$, called the capacity of $\vec{a}$.

DEFINITION 11.6.2 A vertex s of a network N is called a *source* if its indegree is zero and a vertex t of N is called a *sink* if its outdegree is zero. Any other vertex of N is called an *intermediate vertex*.

Given any vertex u of the network $N(V,A)$, we denote the set of arcs going into u by $I(u)$ and going out of u by $O(u)$.

In connection to the e-commerce problem discussed above, given a network N, the business is the source and the warehouse is the sink. The remaining vertices represent different channels of delivery, such as large delivery trucks, third party delivery services etc. The capacity on each arc is the maximum amount of goods that can be sent from the tail of the arc to its head. When flying guests to a destination wedding, we may use a network to model the situation. The problem is to determine the maximum number of guests that can be flown from city s to city t in a given day while not exceeding the capacity of any arc. If a flight cannot accommodate more than 5 wedding guests, we must find alternate flights to transport the remaining guests. A flow is a plan that aids in the creation of a structure while adhering to capacity constraints.

DEFINITION 11.6.3 Let N be a network with the source s and the sink t. Then any *flow* in N from s to t is a function f which assigns a non-negative integer to each of the arcs $\overrightarrow{a}$ in N such that

(i). $f(a) \leq c(a)$ for each arc $\overrightarrow{a}$ (capacity constraint),

(ii). The total flow into the sink t equals the total flow out of the source s, that is, $\sum_{a \in O(s)} f(a) = \sum_{a \in I(t)} f(a)$ and

(iii). For any intermediate vertex x, the total flow into x equals the total flow out of x (flow conservation), i.e., $\sum_{a \in O(x)} f(a) = \sum_{a \in I(x)} f(a)$.

DEFINITION 11.6.4 Let f be a flow in network N with the source s and the sink t. Then the value $d = \sum_{a \in O(s)} f(a) = \sum_{a \in I(t)} f(a)$ is called the *value of the flow f*.

DEFINITION 11.6.5 The *net flow out of a vertex x* is defined as

$$\sum_{a \in O(x)} f(a) - \sum_{a \in I(x)} f(a).$$

The *net flow into a vertex x* is defined as

$$\sum_{a \in I(x)} f(a) - \sum_{a \in O(x)} f(a).$$

For an arc $\overrightarrow{a}$ of N, the value of $f(a)$ is called the *flow in* along arc $\overrightarrow{a}$ and can be imagined to be the amount of material transported or the number of guests flown under the flow f along the arc $\overrightarrow{a}$. The capacity constraint deters any flow f along the arc $\overrightarrow{a}$, $f(a)$ from exceeding its capacity $c(a)$.

EXAMPLE 11.6.6 Consider the network shown in Figure 11.20. This is similar to Figure 11.19, in which the capacity $c(a)$ for each arc is shown, but in Figure 11.20, we have accommodated the flow $f(a)$ values too in each arc. In each arc, the first value is the

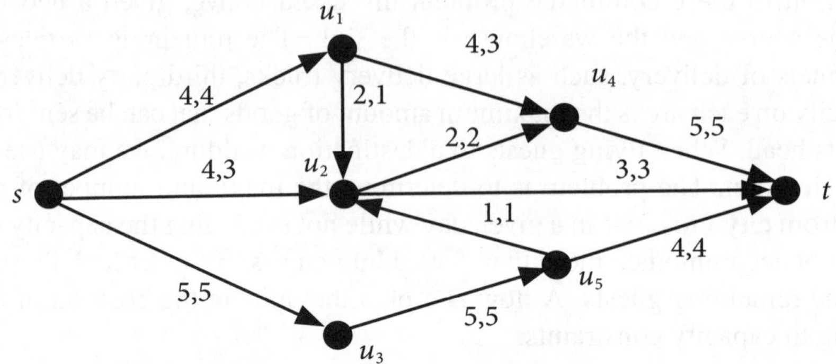

Figure 11.20 A flow f for the network N shown in Figure 11.19

capacity, while the second represents the flow of that arc. Then the function f defined for this network is actually a flow as $f(a) \le c(a)$ for all arcs a of N. Also

$$\sum_{a \in O(s)} f(a) = 4+3+5 \qquad = 12 = 4+3+5 \qquad = \sum_{a \in I(t)} f(a),$$

$$\sum_{a \in O(u_1)} f(a) = 3+1 \qquad = 4 = 4 \qquad = \sum_{a \in I(u_1)} f(a),$$

$$\sum_{a \in O(u_2)} f(a) = 2+3 \qquad = 5 = 3+1+1 \qquad = \sum_{a \in I(u_2)} f(a),$$

$$\sum_{a \in O(u_3)} f(a) = 5 \qquad = 5 = 5 \qquad = \sum_{a \in I(u_3)} f(a),$$

$$\sum_{a \in O(u_4)} f(a) = 5 \qquad = 5 = 3+2 \qquad = \sum_{a \in I(u_4)} f(a),$$

$$\sum_{a \in O(u_5)} f(a) = 1+4 \qquad = 5 = 5 \qquad = \sum_{a \in I(u_5)} f(a).$$

Thus f is a flow with value 12.

Let D be a digraph and let X and Y be non-empty subsets of $V(D)$. The set $A(X,Y)$ is a subset of $E(D)$, such that

$$A(X,Y) = \{a = \overrightarrow{xy} \in E(D) | x \in X, y \in Y\}.$$

Thus $A(X,Y)$ is the set of all arcs directed from some vertex in X to some vertex in Y. For a flow f or capacity function c in a network N, we define

$$f(X,Y) = \sum_{a \in (X,Y)} f(a).$$

$$c(X,Y) = \sum_{a \in (X,Y)} c(a).$$

Let f be a flow on the network $N = (V,A)$ and for any proper subset X of the vertex set V of N, let $\overline{X}$ denote the complement of X in V, that is $\overline{X} = V - X$.

DEFINITION 11.6.7 Let $N = (V,A)$ be a network and $X \subset V$ such that the source $s \in X$. Then the set of arcs $A(X,\overline{X})$ is called a *cut* in N.

If c is the capacity function of N for a cut $(X,\overline{X})$, the number $c(X,\overline{X})$ is called the capacity of the cut. Further, if f is a flow in N, then the flow from X to $\overline{X}$ is $f(X,\overline{X})$ and the flow from $\overline{X}$ to X is $f(\overline{X},X)$. For example if we take $X = \{s, u_1, u_2, u_3\}$ of the network shown in Figure 11.20, then $\overline{X} = \{u_4, u_5, t\}$. Here $A(X,\overline{X}) = \{u_1 u_4, u_2 u_4, u_2 t, u_3 u_5\}$ is a cut in N with each edge of $A(X,\overline{X})$ connecting a vertex of X with $\overline{X}$. Then $c(X,\overline{X}) = c(u_1 u_4) + c(u_2 u_4) + c(u_2 t) + c(u_3 u_5) = 4 + 2 + 3 + 5 = 14$. Also $f(X,\overline{X}) = 3 + 2 + 3 + 5 = 13$. If we were to consider the flow $A(\overline{X},X) = \{u_5 u_2\}$, then $f(\overline{X},X) = 1$

We may now present the first result of this section which gives the formula for flow value and also shows that it never exceeds the total capacity of that cut.

THEOREM 11.6.8 *Let f be a flow on a network $N = (V,A)$ with flow value d. If $A(X,\overline{X})$ is a cut in N, then $d = f(X,\overline{X}) - f(\overline{X},X)$ and $d \leq c(X,\overline{X})$.*

Proof. Let s be the source of flow f. Then $f(\{s\},V) = d$ and $f(V,\{s\}) = 0$. Let u be any vertex different from both source s and the sink t. Then $f(\{u\},V) = \sum_{a \in O(u)} f(a)$ and $f(V,\{u\}) = \sum_{a \in I(u)} f(a)$. Since f is a flow, $\sum_{a \in O(u)} f(a) = \sum_{a \in I(u)} f(a)$.

Therefore, $f(V,\{u\}) - f(V,\{u\}) = 0$ for all $u \neq s,t$. Then for the cut $A(X,\overline{X})$, we have

$$\sum_{x \in X} \{f(\{x\},V) - f(V,\{x\})\} = \sum \{f(\{s\},V) - f(V,\{s\})\}$$

$$+ \sum_{u \in X, u \neq s} \{f(\{u\},V) - f(V,\{u\})\}$$

$$= d - 0 + 0$$

$$= d.$$

Implies $f(X,V) - f(V,X) = \sum f(\{x\},V) - f(V,\{x\})$

$$= d.$$
$$\text{Also, } f(X,V) = f(X, X \cup (\overline{X}))$$
$$= f(X,X) + f(X,\overline{X}).$$
$$\text{Similarly, } f(V,X) = f(X,X) + f(\overline{X},X).$$

Therefore $d = f(X,V) - f(V,X)$
$$= f(X,X) + f(X,\overline{X}) - f(X,X) - f(\overline{X},X)$$
$$= f(X,\overline{X}) - f(\overline{X},X)$$
$$\leq f(X,\overline{X}).$$

Since $f(a) \leq c(a)$ for all arc $\vec{a}$ in N, we get

$$f(X,\overline{X}) = \sum_{a \in A(X,\overline{X})} f(a) \leq \sum_{a \in A(X,\overline{X})} c(a) = c(X,\overline{X}).$$

Hence $d \leq f(X,\overline{X}) \leq c(X,\overline{X})$. $\square$

COROLLARY 11.6.9 Let N be a network and let f be a flow in N. Then the value of the flow in N cannot exceed the capacity of any cut $(X,\overline{X})$ in N. In other words,

$$d \leq min\{c(X,\overline{X})\},$$

where the minimum is taken over all cuts $(X,\overline{X})$ in N.

Proof. The proof is left to the reader. $\square$

COROLLARY 11.6.10 Let N be a network and f a flow in N. Then the value of the flow in N equals the net flow into the sink t of N. That is,

$$d = \sum_{a \in I(t)} f(a) - \sum_{a \in O(t)} f(a).$$

Proof. Let $D(V,A)$ be the underlying digraph of N and let $X = V(D) - \{t\}$. It follows that $\overline{X} = \{t\}$. Then an arc $a = \vec{xy} \in (X,\overline{X})$ if and only if y is the sink t and $x \in I(t)$. Moreover, the arc $a' = \vec{yx} \in (\overline{X},X)$ if and only if $y = t$ and $x \in O(t)$. The result follows. $\square$

As an e-commerce owner, you would prefer to have a flow with value equal to the upper bound given in the above inequality as it is a loss to transport merely 50 kgs of walnuts through a channel that can transport 500 kgs of walnuts. Hence such flows where we are at capacity or close to meeting capacity constraints are called maximum flows, as we define in the following section.

11.6.2 The Max-Flow Min-Cut theorem

The Max-Flow Min-Cut theorem lays the base for an algorithm that helps to identify the maximum flow possible in a network. This theorem was first proposed in 1956 by *L.R. Ford Jr.* and *D. R. Fulkerson* in their research paper, *Maximum flow through a network*. This theorem can be verified by the Ford–Fulkerson method (which is not a fully specified algorithm). The fully specified algorithm, called the *Edmond–Karp algorithm* based on the Ford–Fulkerson method is due to *Jack Edmonds* and *Richard Karp* which was published in 1972. It is sometimes called the Ford-Fulkerson algorithm in acknowledgment of the underlying Ford–Fulkerson method. Before we study the algorithm itself, we must become familiar with the theorem. But to understand the theorem, we must introduce some concepts related to the flow of a network.

A flow f is called a *maximum flow* if d, the value of the flow f, is such that $d \geq d'$ for every flow f' in N. A cut $(P,\overline{P})$ is a minimum cut of N if $c(P,\overline{P}) \leq c(X,\overline{X})$ for every cut $(X,\overline{X})$ of N. If D is the underlying digraph of a network N with source s and sink t, then $(\{s\}, V(D) - \{s\})$ is a cut of N. Clearly this cut is the minimum cut because the capacity of the cut $(\{s\}, V(D) - \{s\})$ is the least among all cuts of N. Therefore N has a minimum cut. Likewise, we may define a flow in which $f(a) = 0$ for every arc a of N. Clearly f satisfies the conditions of a flow, as required by definition 11.6.3. Also a maximum flow for N exists.

In any graph, it is easy to see that the value of the maximum flow must equal the capacity of a minimum cut. For example, consider Figure 11.21, in which $I(t)$ consists of three edges $\overrightarrow{xt}$, $\overrightarrow{ut}$ and $\overrightarrow{vt}$, each of which has a capacity of 2, 1, 2 respectively, which, in total, equals 5. By corollary 11.6.10, any flow f defined on this network cannot exceed 5. If f_1 is flow defined on N such that $f_1(X,\overline{X}) = 5$, then the value of this flow equals the capacity of the minimum cut, making it the maximum flow possible in this network. To prove that the value of the maximum flow in N equals the capacity of a minimum cut in N, we must show that the value of some maximum flow in N is as large as the capacity of some minimum cut in N.

DEFINITION 11.6.11 A flow f in the network N is called a *maximal flow* if the value of the flow

$$d = \min\{c(X,\overline{X}) : A(X,\overline{X}) \text{ is any cut}\}.$$

Given any network N, one can ask the question, "Is it possible to guarantee the existence of the maximum flow f in any N?" The answer is yes and the reason for this answer is the theorem called the max-flow min-cut theorem. We will introduce some more definitions for clarity before we delve into this theorem.

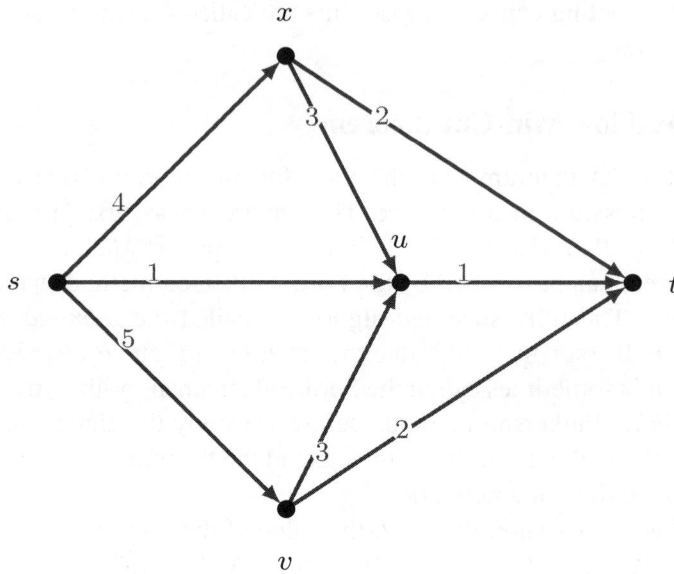

Figure 11.21   A network with flow

DEFINITION 11.6.12 A *semi-path* in a digraph D is a path in the underlying graph (undirected graph) but not a directed path in D. In other words, a path in D is a directed path in D while a semi-path in D is a path only in the underlying graph of G.

The significance of semi-path is the observation that an arc which flows in a certain way, say $\overrightarrow{uv}$ is sometimes used as v, u in the semi-path. For example, in Figure 11.21 s, x, u, t is a directed path of N while s, v, u, t is a semi-path of the underlying graph of N. It is possible to extend this definition to semi-trail, semi-walk, semi-cycles etc.

DEFINITION 11.6.13 Let $W = u_o u_1 \ldots u_k$ be a semi-path of the underlying graph of N. Then any associated arc from u_{i-1} to u_i is called a *forward arc* of W and any arc from u_j to u_{j-1} is called a *reverse arc* of W.

DEFINITION 11.6.14 Let f be a flow in a network N with an underlying digraph D and capacity function c. A semi-path $u_0, u_1, u_2 \ldots u_{n-1}, u_n$ in D is said to be *f-unsaturated* if, for each i, $1 \leq i \leq n$, either

(a). the arc $a_i = \overrightarrow{u_{i-1}, u_i}$ is a forward arc and $f(a_i) < c(a_i)$ or

(b). $a_i = \overrightarrow{u_i, u_{i-1}}$ is a reverse arc and $f(a_i) > 0$

Intuitively, if condition (a) holds, then we can increase the flow from u_{i-1} to u_i along a_i and if condition (b) holds, then we can "push back" flow from u_i to u_{i-1} along a_i.

DEFINITION 11.6.15 If Q is an f-unsaturated $s - t$ semipath, where s and t are the source and sink of N respectively, then Q is called an *f-augmenting semipath*.

DEFINITION 11.6.16 Let Q be a semi-path in the underlying graph G of the network N and f be a flow of N. Then the *increment* of Q is a non negative integer Δ, given by $\Delta = \min\{\Delta_i : a$ is an arc associated with $Q\}$, where

$$\Delta_i = \begin{cases} c(a) - f(a) & \text{if } a \text{ is a forward arc in } Q \\ f(a) & \text{if } a \text{ is a reverse arc in } Q. \end{cases}$$

That is Δ is the largest value that can be increased on the flow f subject to the capacity constraint.

To understand the max-flow min-cut theorem, one must understand how the procedure came to be. Let us understand the ideas with an example.

EXAMPLE 11.6.17 Consider the network N and a flow f in N with the first value as capacity value $c(a)$ and the second value as $f(a)$ for each arc a of N as shown in Figure 11.21. Consider the semi-path $Q = s, v, u, x, t$. Then the arcs sv, vu, xt are forward arcs while the arc ux is a reverse arc of network N. Let $u_0 = s, u_1 = v, u_2 = u, u_3 = x$ and $u_4 = t$. Let $a_1 = sv, a_2 = vu, a_3 = ux, a_4 = xt$. If a_i satisfies the condition (a), then we let $\Delta_i = c(a_i) - f(a_i)$. Otherwise, a_i satisfies the condition (b) and we let $\Delta_i = f(a_i)$. Set $\Delta = min\{\Delta_i | 1 \leq i \leq 4\}$.

$$\Delta_1 = c(a_1) - f(a_1) = 5 - 2 = 3, \quad \Delta_2 = (a_2) = c(a_2) - f(a_2) = 3 - 2 = 1.$$
$$\Delta_3 = f(a_3) = 3, \quad \Delta_4 = c(a_4) - f(a_4) = 2 - 1 = 1.$$

Hence $\Delta = 1$.

We now define a function f^* on the arc set of N as follows:

(i) If a is found in the arc set of N but not on Q, then $f^*(a) = f(a)$.

(ii) If a is in N and also on Q, then

     I If a satisfies condition (a) of definition 11.6.14, then $f^*(a) = f(a) + \Delta$.

    II If a satisfies condition (b), then $f^*(a) = f(a) - \Delta$.

It is clear that f^* is also a flow for the network in Figure 11.21.

The reason for defining a new flow f^* on the network is to create a maximal flow since an unsaturated semi-path can do better by increasing the flow. We will now formally define these concepts.

In common parlance, the value of Δ is the *bottle-neck capacity* of a semi-path. Assume that you are transporting 500 kilos of walnuts in a network channel, where it is possible to transport them half-way to a certain point, beyond which you can only transport 100 kilos of goods. In that case, it must be assumed that the network channel can transport only 100 kilos, because of its bottle-neck capacity.

DEFINITION 11.6.18 Let Q be a semi-path in the underlying graph of the network N and f be a flow of N. Then Q is called f-*saturated* if $\Delta = 0$ and f-*unsaturated* if $\Delta > 0$.

So an f-unsaturated semi-path is one which is not being fully utilized and its load can be increased to Δ.

In Example 11.6.17, since $\Delta > 0$, the semi-path Q is f-unsaturated. As Q starts from the source s to sink t, Q is an f-augmenting semi-path.

The following theorem due to Ford and Fulkerson, identifies a relationship between augmenting semi-paths and maximum flows.

THEOREM 11.6.19 *Let N be a network with an underlying digraph D. A flow f in N is a maximum flow if and only if there is no f-augmenting semi-path in D.*

Proof. Let s and t be the source and sink of N. Let us assume that D contains an f-augmenting semi-path

$$Q : s = u_0, a_1, u_1, a_2 \ldots u_{n-1}, a_n, u_n = t.$$

Then for each $i, (1 \leq i \leq n)$ either

(a) a_i is a forward arc and $f(a_i) < c(a_i)$.

(b) a_i is a reverse arc and $f(a_i) > 0$.

For the forward arcs, we compute $\Delta_i = c(a_i) - f(a_i)$ while for the reverse arcs, $\Delta_i = f(a_i)$. We then set $\Delta = min\{\Delta_i | 1 \leq i \leq n\}$, ensuring that $\Delta \geq 1$.

We now define a new function f^* on the edges of D, denoted by $E(D)$ as follows:

$$f^*(x) = \begin{cases} f(a) + \Delta & \text{if } a \text{ is a forward arc in } E(Q) \\ f(a) - \Delta & \text{if } a \text{ is a reverse arc in } E(Q) \\ f(a) & \text{if } a \notin E(Q). \end{cases}$$

We must now prove that f^* is indeed a valid flow for N and that $d^* > d$.

Since Δ is an integer and f is integer valued, f^* is also an integer. The definition of f^* ensures that $0 \leq f^*(a) \leq c(a)$ for every arc $a \in E(D)$. Let x be an intermediate edge of flow f. If $x \notin Q$, then $f^*(a) = f(a)$ for all $a \in O(x)$ and $f^*(a) = f(a)$ for all $a \in I(x)$. Since f is a flow, it follows that

$$\sum_{a \in I(x)} f(a) = \sum_{a \in O(x)} f(a).$$

Similarly,

$$\sum_{a\in I(x)} f^*(a) = \sum_{a\in O(x)} f^*(a).$$

If $x \in Q$, then $x = u_i$ for some $i(1 \leq i \leq n-1)$.

We will now examine three cases:

Case 1: Assume $a_i = \overrightarrow{u_i u_{i-1}}$ (reverse arc) and $a_{i+1} = \overrightarrow{u_i u_{i+1}}$ (forward arc). Then

$$\sum_{a\in I(x)} f^*(a) = \sum_{a\in I(x)} f(a)$$

Further

$$\sum_{a\in O(x)} f^*(a) = f^*(a_i) + f^*(a_{i+1}) + \sum_{a\in S} f(a),$$

where $S = O(x) - \{u_{i_1}, u_{i+1}\}$. Since $f^*(a_i) = f(a_i) - \Delta$ and $f^*(a_{i+1}) = f(a_{i+1}) + \Delta$, we have $f^*(a_i) + f^*(a_{i+1}) = f(a_i) + f(a_{i+1})$. Therefore

$$\sum_{a\in O(x)} f^*(a) = \sum_{a\in O(x)} f(a).$$

We can prove that

$$\sum_{a\in I(x)} f^*(a) = \sum_{a\in O(x)} f^*(a)$$

in each of the remaining two cases too.

Case 2: Suppose that $a_i = \overrightarrow{u_{i-1}, u_i}$ and $a_{i+1} = \overrightarrow{u_{i+1}, u_i}$. We leave it to the reader to solve this case.

Case 3: Suppose that either $a_i = \overrightarrow{u_{i-1} u_i}$ and $a_{i+1} = \overrightarrow{u_i u_{i+1}}$ or $a_i = \overrightarrow{u_i u_{i-1}}$ and $a_{i+1} = \overrightarrow{u_{i+1} u_i}$. Then f^* is a flow in N and

$$d^* = \sum_{a\in O(s)} f^*(a) - \sum_{a\in I(s)} f^*(a).$$

Thus f^* satisfies the conditions for it to be a valid flow of the network N. We know d is $\sum_{a\in O(s)} f(a) - \sum_{a\in I(s)} f(a)$. We will now show that $d^* > d$. If $a_1 = \overrightarrow{su_1}$, then $f^*(a_1) = f(a_1) + \Delta$. Additionally, $f^*(a) = f(a)$ for all $a \in O(s) - \{u_1\}$. Therefore, we have $d^* = d + \Delta$ in this case.

Assume that $a_1 = \overrightarrow{u_1 s}$. Then $f^*(a_1) = f(a_1) - \Delta$ and $f^*(a) = f(a)$ for all $a \in I(s) - \{u_1\}$. Once again, we find that $d^* = d + \Delta$. Thus $d^* = d + \Delta > d$ in either case, which implies that f is not the maximum flow.

Conversely, assume that there is no f-augmenting semi-path in D. We will show that there exists a cut $(P, \overline{P})$ in N such that

$$f(a) = c(a) \quad for \quad every \quad a \in (P, \overline{P})$$

and

$$f(a) = 0 \quad for \quad every \quad a \in (\overline{P}, P)$$

Let P denote the set of all vertices x in D for which there exists an f-unsaturated $s - x$ semi-path. Then $s \in P$ and by our assumption $t \notin P$. Hence $(P, \overline{P})$ is a cut in N. Suppose the arc $\overrightarrow{yw} \in (P, \overline{P})$. Since $y \in P$, there exists an f-unsaturated $s - y$ semi-path Q in D. Therefore $f(a) = c(a)$ for $a \in O(y)$; otherwise, the semi-path $Q^* : Q, \overrightarrow{yw}, w$ is an f-unsaturated $s - w$ semi-path, which contradicts the fact that $w \notin P$. Similarly if $\overrightarrow{yw} \in (\overline{P}, P)$, then $f(a) = 0$. Thus

$$d = f(P, \overline{P}) - f(\overline{P}, P) = f(a) - 0 = c(P, \overline{P})$$

If f^* is a maximum flow and $(X, \overline{X})$ is a minimum cut, then it follows from Corollary 11.6.9 that $d^* \leq c(X, \overline{X})$. Hence

$$d \leq d^* \leq c(X, \overline{X}) \leq c(P, \overline{P}) = d$$

which implies that $d = d^*$. Hence f is the maximum flow. $\square$

THEOREM 11.6.20 (The max-flow min-cut theorem) *In every network, the value of the maximum flow equals the capacity of the minimum cut.*

Proof. Let f be a maximum flow in a network N and let D be an underlying digraph of N. By Theorem 11.6.19, there is no f-augmenting semi-path in D. The proof of Theorem 11.6.19, implies the existence of a minimum cut $(P, \overline{P})$ such that the flow in each arc of $(P, \overline{P})$ equals its capacity, and the flow in each arc of $(\overline{P}, P)$ is 0. Thus $f(P, \overline{P}) = c(P, \overline{P})$ and $f(\overline{P}, P) = 0$. Hence $d = f(P, \overline{P}) - f(\overline{P}, P) = f(a) - 0 = c(P, \overline{P})$. Theorem 11.6.19 also says that if $(X, \overline{X})$ is a minimum cut in N, then

$$c(X, \overline{X}) \leq c(P, \overline{P}) = d \leq c(X, \overline{X}).$$

Hence $d = c(X, \overline{X})$. $\square$

11.6.3 The Ford–Fulkerson algorithm

We now present an algorithm, due to *Ford* and *Fulkerson*, and based on the proof of the max-flow min-cut theorem, which constructs a maximum flow for a network. It uses a labeling technique to produce a maximum flow. Starting with a known flow, for example, the flow, which has 0 assigned to each arc (that is, the zero flow), it recursively constructs a sequence of flows of increasing value, terminating with a maximal flow.

The Ford-Fulkerson algorithm finds the maximum flow of a network by repeatedly finding new augmenting semi-paths from s to t until the value of the flow d equals the capacity of the source and sink. To make the understanding of this method easier, let us redefine some concepts.

DEFINITION 11.6.21 The *residual graph* N_f is the same as the underlying digraph of the network N, where the capacity of each arc is updated in each iteration of the algorithm.

DEFINITION 11.6.22 The *residual capacity* of residual graph N_f is the updated capacities after a flow is determined. Given an arc a in an augmenting semi-path Q with capacity of the arc $c(a)$ and flow $f(a)$, the residual capacity is $c(a) - f(a)$.

For example, an arc a with capacity $c(a) = 5$, after a flow $f(a) = 3$, has residual capacity 2.

The reason for creating a residual graph with residual capacities is to show precisely the differences between the capacity of different paths and their current flow. It helps to determine future augmenting paths accurately. The Ford-Fulkerson algorithm works by following these steps.

Step 1: Construct the underlying digraph $D(V,E)$ of a network $N(V,A)$ with source s, sink t, and capacity function $c(a)$ and initial flow $f(a)$ with $f(a) = 0$ for all arcs $a \in E(D)$. This is the residual graph N_f. Residual capacity $c_f = c(a) - f(a)$.
Update the residual capacities after a new flow f is assigned. Redraw N_f after a new flow f.

Step 2: Apply a breadth-first search or depth-first search in D to find a shortest $s-t$ semi-path. If D does not contain an $s-t$ path then proceed to Step 5. Else construct the semi-path $Q : s = u_0, u_1, \ldots, u_n = t$ as the shortest $s-t$ path in D.

Step 3: Let $Q : s = u_0, a_1, u_1, a_2, u_2, \ldots, u_{n-1}, a_n, u_n = t$ be a shortest $s-t$ path in D where a_i may be a forward arc or a reverse arc of D and $f(a_i) > 0$. For $i = 1 \ldots n$, let
$$\Delta_i = \begin{cases} c(a_i) - f(a_i) & \text{if } a_i \text{ is a forward arc in } Q \\ f(a_i) & \text{if } a_i \text{ is a reverse arc in } Q \end{cases}$$
Let $\Delta = min\{\Delta_i | 1 \leq i \leq n\}$. For $i = 1, 2 \ldots n$, if a_i is a forward arc, then $f(a_i) \leftarrow f(a_i) + \Delta$; otherwise $f(a_i) \leftarrow f(a_i) - \Delta$.

Step 4: Go to Step 1.

Step 5: Output $f(a)$ for all a in $E(D)$. Output d.

We will now present the pseudocode of the Ford–Fulkerson algorithm in Algorithm 3.

Algorithm 3: Ford–Fulkerson Algorithm

Data: A digraph $D = (V(D), E(D))$, source vertex s, sink vertex t, $c(a)$ for all arcs
     a in $E(D)$
Result: $max - flow$, the maximum flow from s to t
Initialization:
$\Delta \leftarrow 0$
$max - flow \leftarrow 0$
$f(a) \leftarrow 0$ for all a in $E(D)$;   /* Initialize all flows to 0 */
while $augmenting - path(s,t)$ $exists$ in N_f **do**
    Procedure BFS(s, t);   /* Find an augmenting path from s to t */
    Find-Path $P(s,t)$
    for ($each$ arc in $P(s,t)$) {
        $\Delta = min(c(a) - f(a))$;   /* Find the increment/bottleneck
        capacity */
    }
    Procedure-Augmenting flows:
    for ($each$ $edge$ a in $P(s,t)$) {
        if a is a $forward$ arc **then**
        | $f(a) \leftarrow f(a) + \Delta$
        end
        if a is a $reverse$ arc **then**
        | $f(a) \leftarrow f(a) - \Delta$
        end
    }
    Update residual capacities in N_f:
    for ($each$ $edge$ a in $P(s,t)$) {
        if a is a $forward$ arc **then**
        | $c_f(a) \leftarrow c(a) - f(a)$
        else
        | $c_f(a) \leftarrow f(a)$
        end
    }
    Update the total maximum flow:
    $max - flow \leftarrow max - flow + \Delta$
end
return $max - flow$

The complexity of the Ford–Fulkerson algorithm is easy to calculate if one identifies the main routines. The algorithm takes $O(|E|)$ to search for augmenting paths. Updating flows and capacities also takes $O(|E|)$ time. The number of augmenting paths, depends on the maximum flow value. If each augmentation, increases the flow value by 1, then the number of iterations is atmost $O(F)$ where F is the maximum possible flow in the network. Hence the complexity of the Ford–Fulkerson algorithm is $O(F.|E|)$ where $|E|$ is the total number of edges and F is the maximum flow. We will discuss the process of the algorithm using an example.

EXAMPLE 11.6.23 Determine the maximum flow of the network in the graph in Figure 11.22.

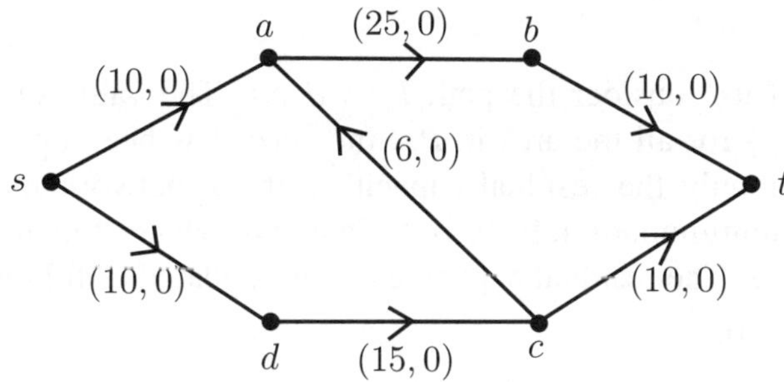

Figure 11.22   Network N

Initialization: Given a network N, we draw the underlying digraph as in Figure 11.22, with the capacities of each arc mentioned. The flow f is initialized as 0 for all the arcs. The graph in Figure 11.23 is the residual graph.

Iteration 1: We now identify an augmenting semi-path from s to t. Let us consider the path $P : s,d,c,a,b,t$. For each arc in path P, we calculate $c(a) - f(a)$ and the minimum of all these values as Δ. Hence $min\{10, 15, 6, 25, 10\} = 6$. For this path, the bottleneck capacity or the value of Δ is 6. Hence the flow is augmented from 0 to 6 for this path P. The residual capacities c_f are updated.

$c_f(\overrightarrow{sd}) = 10 - 6 = 4$

$c_f(\overrightarrow{dc}) = 15 - 6 = 9$

$c_f(\overrightarrow{ca}) = 6 - 6 = 0$

$c_f(\overrightarrow{ab}) = 25 - 6 = 19$

$c_f(\overrightarrow{bt}) = 10 - 6 = 4$

Iteration 2: We continue with another augmenting semi-path from s to t. We may use the breadth-first search or the depth-first search to identify all these shortest paths

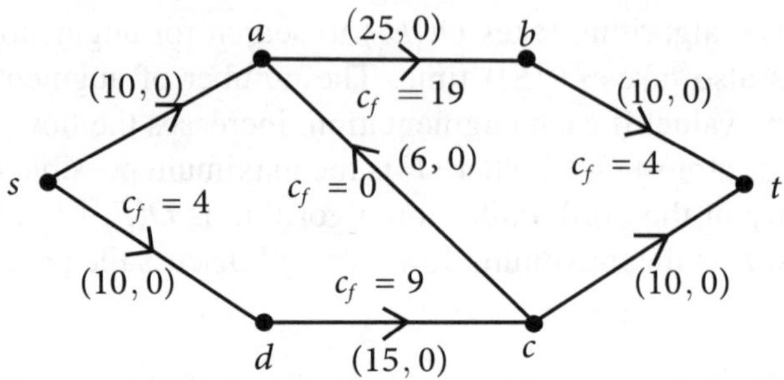

Figure 11.23   Network N after Iteration 1

from s to t. Let us consider the path $P : s,d,c,t$. The value of Δ is calculated as $min\{c(a) - f(a)\}$ for all the arcs in P. Since $c(a)$ has been updated, we must take care to consider only the residual capacities of the network instead of the initial capacities. The minimum of 4, 9, 10 is 4. Thus $\Delta=4$. Hence the flow is augmented by 4 along these arcs. The residual capacities c_f are updated as follows:

$c_f(\overrightarrow{sd}) = 4 - 4 = 0.$
$c_f(\overrightarrow{dc}) = 9 - 4 = 5.$
$c_f(\overrightarrow{ct}) = 10 - 4 = 6.$

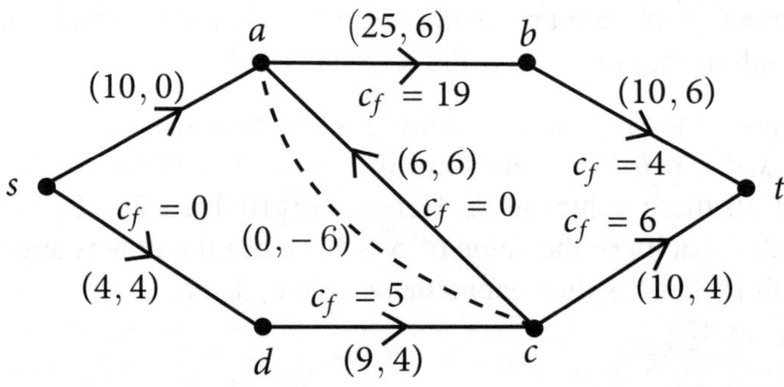

Figure 11.24   Network N after Iteration 2

Iteration 3: The next augmenting semi-path to be considered is $P : s,a,c,t$. To calculate the increment, the minimum of $10,6,6$ is considered. In this case, a reverse edge namely $\overrightarrow{ac}$ is included. Although some mathematicians and computer-scientists would like to add reverse arcs with flow $-f(a)$ alongside every arc of the semi-path, we can use it sparingly, only in situations where a reverse arc is warranted. We may think of a reverse-arc as an arc flowing in the opposite direction with capacity 0 and flow -6. The reason for this assumption is to ensure we augment flows correctly and even if we made a wrong turn somewhere, a reverse arc sets us right. The balance is also preserved since $c(a) - f(a)$ still yields a residual capacity of $f(a) > 0$. The minimum of 10, 6, 6 is 6. The flow is augmented by 6 along these arcs in P. The updated residual capacities are:

$c_f(\overrightarrow{sa}) = 10 - 6 = 4.$

$c_f(\overrightarrow{ac}) = 6 - 6 = 0.$

$c_f(\overrightarrow{ct}) = 6 - 6 = 0.$

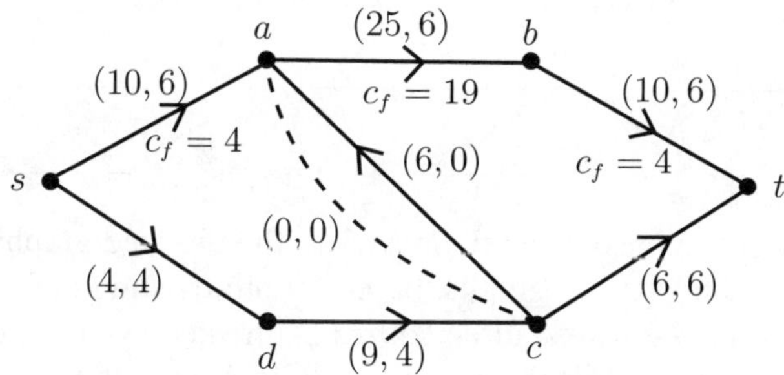

Figure 11.25   Network N after Iteration 3

Iteration 4: We will find another augmenting semi-path since the flow into the sink t still does not match the capacities of the arcs flowing into t. Consider the semi-path $P : s, a, b, t$. To calculate the increment, the minimum of the residual capacities 4, 19, 4 is found to be 4. Hence $\Delta = 4$. We now augment the flow along these arcs in P. The updated residual capacities are:

$c_f(\overrightarrow{sa}) = 4 - 4 = 0.$

$c_f(\overrightarrow{ab}) = 19 - 4 = 15.$

$c_f(\overrightarrow{bt}) = 4 - 4 = 0.$

We calculate the value of the flow as d, the sum of all $f(a)$ flowing into t, as 10+10=20. Another way to calculate the value of the flow is to find the sum of all increments Δ.

The example provides a clear description of the process behind the Ford–Fulkerson algorithm. The Edmond–Karp algorithm works just like Ford–Fulkerson, except that it determines a shortest path from s to t in each iteration.

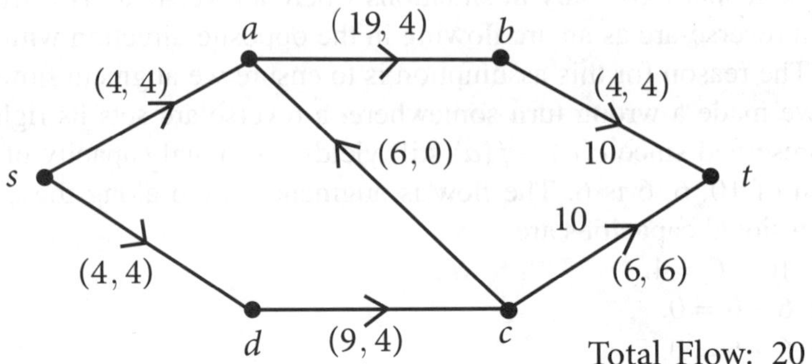

Figure 11.26   Network N after Iteration 4

Summary

In this chapter, we have introduced the reader to the directed graphs and presented new definitions for this class of graphs, based on pre-existing definitions. We have thus talked about directed walks, directed paths, directed cycles, etc. The indegree and outdegree of a directed graph have been studied through examples. A type of digraph called the tournament is introduced and its characteristics are discussed in detail. The orientable nature of a graph is explored and its existence is found to rely on the nature of the graph being strongly connected. Traversal in directed graphs has been demonstrated through two algorithms namely, the Bellman–Ford and the Floyd–Warshall algorithm. The reader is then guided through the construction of networks, the max-flow min-cut theorem and most importantly, the Ford–Fulkerson algorithm.

11.7 Exercises

Section 11.1: Definitions

1. Draw any digraph with 15 vertices and find its incidence matrix.

2. Consider the digraph D as shown in Figure 11.27.

3. Identify a directed walk of length 8. State whether it is a directed path or not.

4. Does there exist a directed trail of length 10 in D?

5. Find a directed cycle and a directed path of maximum length.

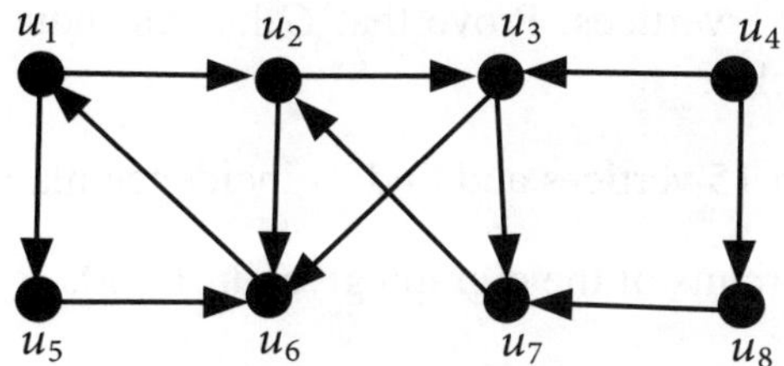

Figure 11.27   Digraph *D*

6. Is *D* strongly connected?

7. Identify the strong components of the digraph shown in Figure 11.28.

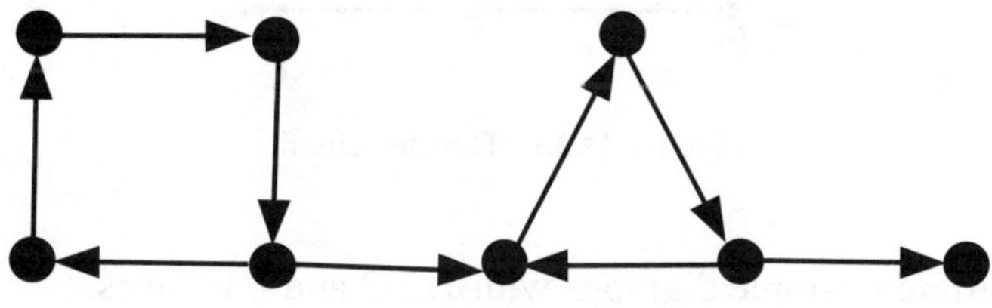

Figure 11.28   Digraph

8. Find the strong components of the digraph given in 11.29.

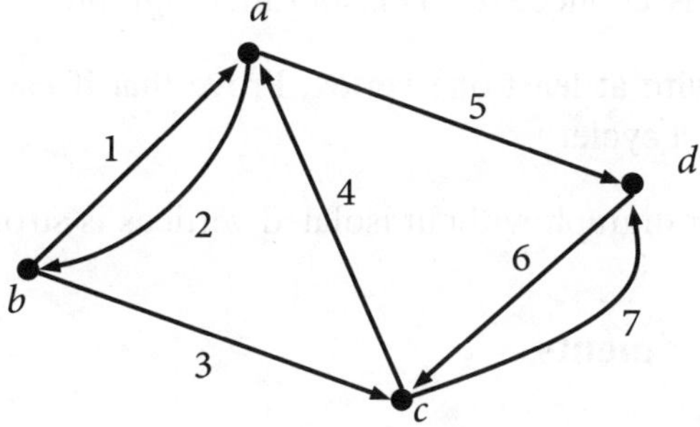

Figure 11.29   The digraph *D*

9. Let n be a positive integer. Let G be a digraph that has no cycles of length ≤ 2. Assume that G has at least 2^{n-1} vertices. Prove that G has an induced subdigraph that has n vertices and has no cycles.

10. Draw any digraph with 15 vertices and find its incidence matrix.

11. Find all the directed circuits of the digraph given in 11.30.

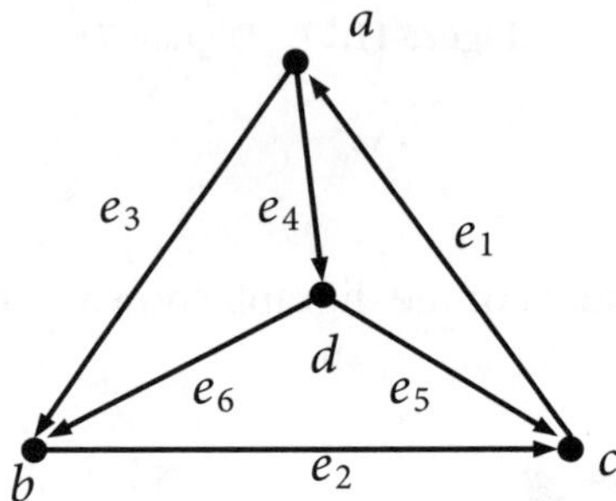

Figure 11.30 Directed circuits

12. Draw all the different simple digraphs with $1, 2, 3$ and 4 vertices.

Section 11.2: Indegree and outdegree

13. If D is a digraph having an odd number of vertices and each vertex of D has an odd outdegree, then prove that there is an odd number of vertices of D with odd indegree.

14. Prove that D is strongly connected if D is an Euler digraph.

15. Let G be a digraph with at least one vertex. Prove that if each vertex v of G satisfies $id(v) > 0$, then G has a cycle.

16. Prove that every Euler digraph without isolated vertices is strongly connected.

Section 11.3: Tournaments

17. Show that $T - U$ is a tournament where T is a tournament on atleast two vertices and U is a proper subset of vertex set of T.

18. Prove that every tournament graph contains a directed Hamiltonian path.

Section 11.4: Traffic flow

19. Show that there exist two edge disjoint paths between any two distinct vertices if G is connected with no bridges.

Section 11.5: Traversal in a directed graph

20. Determine the shortest path from s in the following figure 11.31 using Bellman–Ford algorithm.

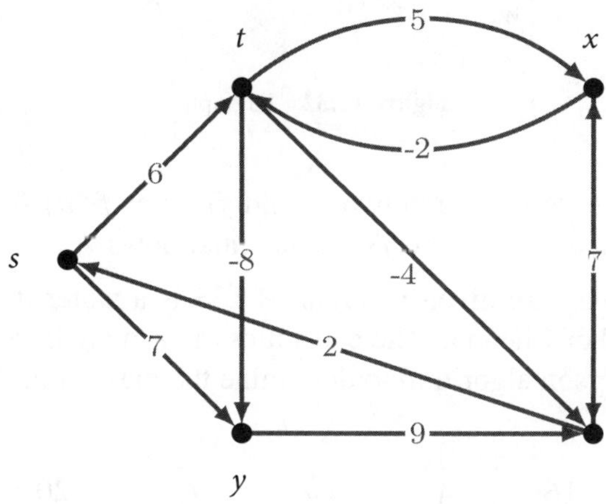

Figure 11.31 A directed weighted graph G with negative weights

Section 11.6: Networks

21. Determine all the cuts and minimum cuts for the digraph shown in Figure 11.32.

22. Let f be a flow in a network N and let $A(X,\overline{X})$ be a cut in N. Show that f is a maximum flow and $A(X,\overline{X})$ is a minimum cut in N if and only if

$$f(a) = \begin{cases} c(a) & \text{for every } a \in A(X,\overline{X}) \\ 0 & \text{for every } a \in A(\overline{X},X). \end{cases}$$

23. Let f_1 and f_2 be flows in a network N and let $A(X,\overline{X})$ be a cut in N. Prove or disprove:

   (i). If f_1 and f_2 are maximal flows, then $f_1(a) = f_2(a)$ for all $a \in A(X,\overline{X})$ and $a \in A(\overline{X},X)$.

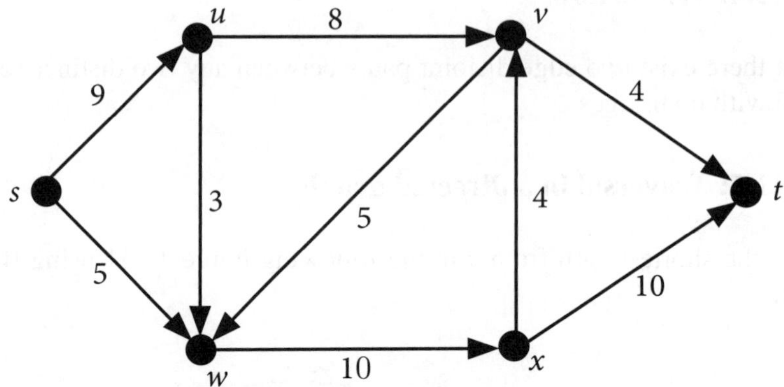

Figure 11.32   Digraph

(ii). If f_1 and f_2 are flows in a network, and $f_1(a) = f_2(a)$ for all $a \in A(X,\overline{X})$ and $a \in A(\overline{X},X)$, then both f_1 and f_2 are maximal flows.

(iii). The network diagram given in Figure 11.33 is a water distribution system in a particular ward of Chennai. The capacities of each arc have been mentioned. Use the Ford–Fulkerson algorithm to determine the maximum flow from S to T.

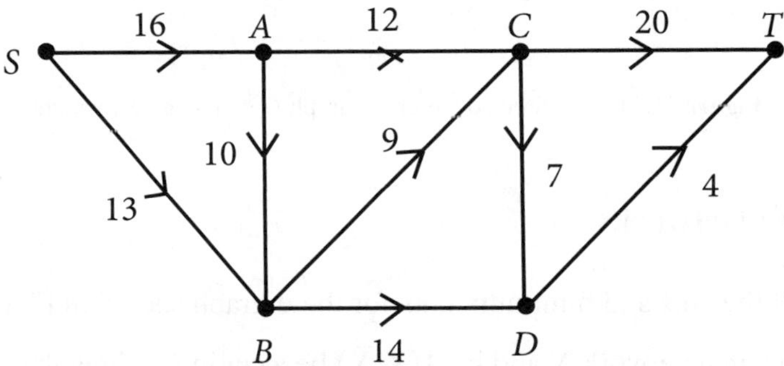

Figure 11.33   Network for water distribution system

24. A city's road network given in Figure 11.34 is represented by a flow graph, where intersections are vertices and roads are edges with capacities representing the maximum number of cars that can pass per minute. The goal is to maximize traffic flow from the northern entry point (S) to the southern exit (T).

25. A river system given in Figure 11.35 is designed with dams and reservoirs. Water can

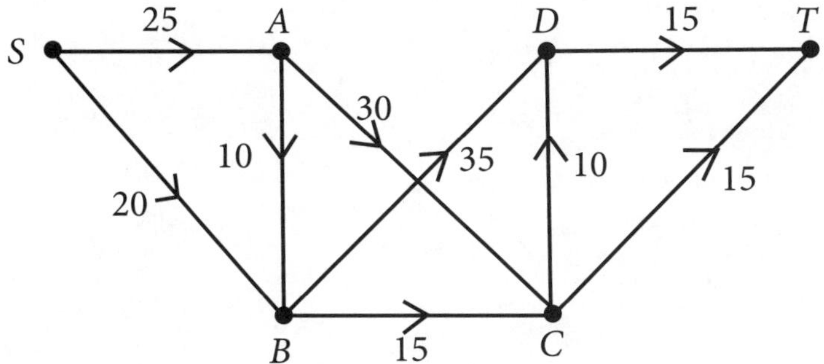

Figure 11.34   Network for city's road

flow through different paths between dams and each dam has a maximum flow capacity. The goal is to maximize the water flow from the main reservoir (S) to the final dam (T).

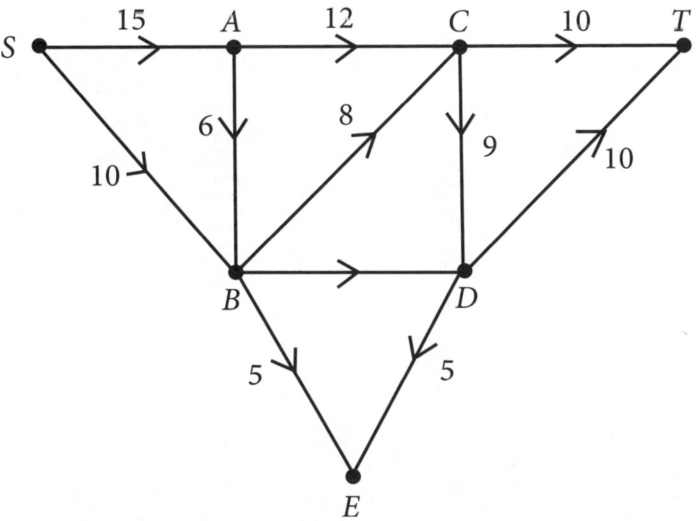

Figure 11.35   Network for dams and reservoirs

Bibliography

[1] Balakrishnan, R., and K. Ranganathan. 2012. *A Textbook of Graph Theory*. New York: Springer.

[2] Bondy, A., and U. S. R. Murty. 2010. *Graduate Texts in Mathematics: Graph Theory*. London: Springer.

[3] Bondy, J. A., and U. S. R. Murty. 1976. *Graph Theory with Applications*. London: Macmillan.

[4] Brassard, G., and P. Bratley. 1996. *Fundamentals of Algorithmics*. New Delhi: Prentice-Hall India.

[5] Chartrand, G., and O. R. Oellerman. 1993. *Applied and Algorithmic Graph Theory*. New York: Tata McGraw-Hill.

[6] Chartrand, G., and P. Zhang. 2006. *Introduction to Graph Theory*. New Delhi: Tata McGraw-Hill.

[7] Chartrand, G., and P. Zhang. 2013. *A First Course in Graph Theory*. North Chelmsford: Courier Corporation.

[8] Clark, J., and D. A. Holton. 1991. *A First Look at Graph Theory*. Singapore: World Scientific.

[9] Cormen, T. H., C. E. Leiserson, R. L. Rivest, and C. Stein. 2001. *Introduction to Algorithms*. New Delhi: Prentice-Hall India.

[10] Diestel, R. 2018. *Graph Theory*. Berlin: Springer.

[11] Gross, J. L., and J. Yellen. 2005. *Graph Theory and Its Applications*. Boca Raton: Chapman & Hall/CRC.

[12] Levitin, A. 2011. *Introduction to the Design and Analysis of Algorithms*. Boston: Pearson.

[13] West, D. B. 2001. *Introduction to Graph Theory*. New Jersey: Prentice Hall.

[14] Brubaker, B. 2024. "Scientists Establish the Best Algorithm for Traversing a Map." *Wired*. https://www.wired.com/story/scientists-establish-the-best-algorithm-for-traversinga-map/.

[15] Bari, A. 2018. *Bellman Ford Algorithm - Single Source Shortest Path - Dynamic Programming*. https://youtu.be/FtN3BYH2Zes?si=kRl9dk8M8pkhTeNO, Bellman-Ford Algorithm.

[16] Bari, A. 2018. *All Pairs Shortest Path (Floyd-Warshall) - Dynamic Programming*. https://youtu.be/oNI0rf2P9gE?si=OpKOFDa5-QznEtXI, Floyd-Warshall Algorithm - Dynamic programming.